Management and the Arts

Management and the Arts

William J. Byrnes

Focal Press
Boston Oxford Johannesburg
Melbourne New Delhi Singapore

Focal Press is an imprint of Butterworth-Heinemann.

A member of the Reed Elsevier group

Library of Congress Cataloging-in-Publication Data
Byrnes, William J.
Management and the arts/William J. Byrnes.
p. cm.
Includes bibliographical references and index.
ISBN 0-240-80131-8
1. Arts—United States—Management. I. Title.
NX765.B87 1993 92-30453
700′.68—dc20 CIP

The publisher offers special discounts on bulk orders of this book.
For information, please contact:
Manager of Special Sales
Butterworth–Heinemann
313 Washington Street
Newton, MA 02158–1626
Tel: 617-928-2500
Fax: 617-928-2620

For information on all Focal Press publications available, contact our World Wide Web page at http://www.bh.com.fp

Printed in the United States of America
10 9 8 7 6 5 4 3

Contents

Preface xiii
Acknowledgments xv

1 Management and the Arts 1

The Entertainment Business 1
Growing Businesses 2
An Uncertain Future 2
Managers and Organizations 4
The Manager 4
The Organization 5
The Process of Organizing 6
Levels of Management and Types of Managers 6
Common Elements in an Organization 8
An Effective Model? 9
The Management Process 10
Planning 11
Organizing 11
Leading 11
Controlling 11
Functional Areas 12
Key Terms and Concepts 12
Questions 13
Analysis Focus 13
References 13

2 The Evolution of Arts Organizations and Arts Management 15

The Artist-Manager 15
The Arts as Institutions 15
A Historical Overview 16
Ancient Times 16
The Middle Ages 17
The Renaissance 17
The Seventeenth Through Nineteenth Centuries 18
The Twentieth Century 19
Profile of the Arts Manager 20
Jobs for Managers Today 21
The National Endowment for the Arts 23
Government Support 24
Budget Battles and Censorship 25
The NEA and the Arts Manager 25
Arts Councils 26

The Education Revolution 26
The Slowdown 27
Conclusion 27
Summary 27
Key Terms and Concepts 28
Questions 28
Case Study: Wanted: Art Scholar, M.B.A. Required 29
Questions 32
References 32
Additional Resources 33

3 Evolution of Management 34

Management as an Art and a Science 34
On-the-Job Management Theory 34
Evolution of Management Thought 35
Preindustrialization 35
A Change in Philosophies 35
The Industrial Revolution 36
Changes in America 37
Management Trends to the Present 37
Scientific Management Today 38
Organizational Management (1916 to Present) 39
Human Relations Management (1927 to Present) 39
The Behavioral Approach 39
The Hawthorne Effect 39
Maslow's Hierarchy of Needs 40
McGregor's Theory X and Theory Y 41
Modern Management 41
Systems and Contingency Approaches 42
Rational Management 42
Conclusion 43
Summary 43
Key Terms and Concepts 44
Questions 45
References 45
Additional Resources 46

4 Arts Organizations and Multiple Environments 47

Changing Environments 49
Managing Change 49
Growing into Change 49
Where to Start? 51
Environments 53
Economic 53
Political and Legal 54
Cultural and Social 55
Demographic 56
Technological 57
Educational 58
Information Sources 59
Audiences 59
Other Arts Groups 60
Board and Staff Members 60

The Media 61
Professional Meetings and Associations 61
Consultants 61
Other Sources 62
The Impact of Future Trends on the Arts 62
Censorship 62
The AIDS Epidemic 62
Economic Pressure 63
Summary 63
Key Terms and Concepts 64
Questions 65
Case Study: The Future Looks Grim for a Theater in Los Angeles 65
Questions 67
References 68

5 Strategic Planning and Decision Making 69

Planning Basics 69
Planning, Goals, and Objectives Defined 69
Types of Plans 70
The Planning Process 71
Five Steps in Formal Planning 71
Other Planning Approaches 72
Strategic Planning 73
The Mission Statement 73
Mission Analysis 74
Formulating an Organization's Strategy 76
Limits of Planning 77
Relationship of Planning to the Arts 77
The Necessity of Planning 78
The Organization's Map and Leadership 78
Decision Making 78
Choices, Decisions, and Problem Solving 78
Steps in Problem Solving 79
Problem-Solving Techniques 79
Decision Theory 80
Conclusion 80
Summary 82
Key Terms and Concepts 83
Questions 83
Case Study: Karamu Renaissance 83
Questions 85
References 85
Additional Resources 86

6 Fundamentals of Organizing and Organizational Design 87

The Management Function of Organizing 87
Four Benefits of Organizing 87
Organizing for the Arts 88
Organizational Design Approaches 88
Mechanistic and Organic Organizational Design 89
Bureaucracy 89
Organizational Structure 90
Organizational Charts and Formal Structure 90
Informal Structure 93

Structure from an Arts Manager's Perspective 94
General Considerations 94
Departmentalization 96
Coordination 98
Vertical Coordination 98
Horizontal Coordination 101
Organizational Growth 101
Corporate Culture and the Arts 102
Theory Versus Reality in Organizations 102
Application to the Arts 102
Five Elements of Corporate Culture 103
Corporate Cultures and the Real World 106
Summary 107
Key Terms and Concepts 107
Questions 107
Case Study: An Arts Institution's Management Overhaul 108
Questions 110
References 111

7 Staffing the Organization 112

The Staffing Process 112
Planning 114
Job Description 115
The Overall Matrix of Jobs 116
Constraints on Staffing 117
Recruitment 118
Selection Process 119
Training and Development 122
Replacement and Firing 123
Unions and the Arts 123
Definition and Purpose 123
Disputes 124
Maintaining and Developing the Staff 125
Career Management 125
The "Right Staff" 126
Summary 126
Key Terms and Concepts 127
Questions 127
Case Study: Should Equal Opportunity Apply on the Stage? 127
Questions 130
References 130
Additional Resources 130

8 Fundamentals of Leadership and Group Dynamics 132

Leadership Fundamentals 132
Formal and Informal Leadership Modes 133
Theory X and Theory Y Approaches to People 133
Power: A Leadership Resource 133
Sources of Power 134
Limits of Power 135
Guidelines for Using Power 135
APPROACHES TO THE STUDY OF LEADERSHIP 136
Trait Approaches to Leadership 136

Behavioral Approaches to Leadership 136
Contingency Approaches to Leadership 136
Applications to the Arts 138
Future Leadership? 138
Motivation and the Arts Work Setting 138
Theories of Motivation 139
Content Theories 139
Process Theories 140
Theory Integration 143
Group Dynamics 143
Group Management Terms 143
Strategies for Making Groups More Effective 146
Distributed Leadership 148
Communication Basics and Effective Leadership 148
The Communication Process 148
Perception 149
Formal and Informal Communication 149
Conclusion 150
Summary 151
Key Terms and Concepts 152
Questions 152
References 153
Additional Resources 154
Additional Titles 154

9 Control: Management Information Systems and Budgeting 155

Control as a Management Function 155
Elements of the Control Process 156
Management by Exception 158
Management by Objectives 160
Performance Appraisal Systems 160
Summary of Control Systems 161
Management Information Systems 162
Data and Information 162
Management Information Systems in the Arts 163
Computers and the Management Information System 163
An Effective Management Information System 165
Common Mistakes 167
Management Information System Summary 167
The Future 168
Budgets and the Control System 168
Budgetary Centers 168
Budgets as Preliminary Controls 168
Types of Budgets 170
The Budgetary Process 170
Budget Reality 171
Budget Controls 171
Summary 171
Key Terms and Concepts 172
Questions 172
Case Study: Radio Station Officials Got Free Cruises 173
Questions 174
References 175

10 Economics and the Arts 176

Introduction to Economics 176
The Economic Problem 177
Three Problems for an Economy 178
Macroeconomics 179
The Role of Government 179
Microeconomics 180
Law of Demand 180
Demand Determinants 180
Market Demand and the Arts 183
Law of Supply 183
Supply Determinants 183
Market Equilibrium: Supply and Demand Curves 186
General Applications of Economic Theories 188
Economic Impact 188
Organizational Impact 189
The Economic Problems Facing the Arts 190
The Cultural Boom 192
The Arts Audience 192
The Productivity Issue 193
Productivity 194
Other Findings 194
Conclusion 194
Summary 195
Key Terms and Concepts 196
Questions 196
Case Study: Broadway Adopts Plan to Cut Costs and Ticket Prices 197
Questions 199
References 199
Additional Resources 200

11 Financial Management 201

Overview of Financial Management 201
For-Profit Organizations 202
Nonprofit Financial Management 202
Financial Management Information System 203
Legal Status and Financial Statements 203
Incorporation 203
Tax Exemption 205
Budgeting and Financial Planning 206
The Budget 206
From the Budget to Cash Flow 210
Developing a Financial Management Information System 213
Accounting 215
Financial Statements 218
Investment 224
Conclusion 224
Managing Finances and the Economic Dilemma 225
Looking Ahead 225
Summary 225
Key Terms and Concepts 226
Questions 227
Case Study: Dance Company Financial Report 227
Questions 228
References 228

12 Marketing and the Arts 229

An Event in Search of an Audience 230
A Means to an End 230
Marketing Principles and Terms 230
Needs and Wants 231
Exchange Process and Utilities 231
Evolution of Modern Marketing 233
Modern Marketing 233
Marketing and Entertainment 234
A Marketing Approach 234
Product Orientation 234
Sales Orientation 235
Customer Orientation 235
Marketing Management 237
The Four P's 237
Market Segments 237
Market Research 238
Marketing Ethics 239
Strategic Marketing Plans 240
Planning Process 240
Marketing Audit 242
Consultants 242
Strategies 242
The Competitive Marketplace and Core Strategies 242
Project Planning and Implementation 246
Evaluation 246
Marketing Data System 246
Conclusion 247
Summary 248
Key Terms and Concepts 249
Questions 249
For Further Discussion 250
Case Study: Outreach Gimmicks Furrow a Skeptical Highbrow 250
Questions 252
References 252
Additional Resources 253

13 Fund Raising 254

Fund Raising and the Arts 254
Fund Raising Plans 255
Preparing Fund Raising Plans 256
Strategic Planning 256
Funding Pyramid 258
Fund Raising Audit 259
Marketing and Fund Raising 259
Fund Raising Management 259
Background Work 261
What Does the Organization Do? 261
Staff and Board Participation 263
Data Management 263
Fund Raising Costs and Control 264
Fund Raising Techniques and Tools 265
Individual Donors 265

Corporate Giving 270
Foundations 272
Government Funding 274
Conclusion 275
Summary 276
Key Terms and Concepts 276
Questions 277
Case Study: Studio Arena to Seek Funding Sources 277
Questions 278
References 279
Additional Resources 279

14 Integrating Management Styles and Theories 280

Management Styles 280
The Irrational Arts Manager: A Model in Overextension 280
The Rational Manager: Changing the Culture 282
The Institutional Manager 284
The Organic Manager 285
Management Theories 285
Scientific Management 285
Human Relations 287
The Open System 287
The Contingency System: An Integrating Approach 288
The Management Functions 288
Planning and Development 288
Marketing and Public Relations 289
Personnel Management: Staff, Labor, and Board Relations 290
Fiscal Management 291
Government Relations 292
Conclusion 293
Questions 293
Case Study: Spoleto Feud Ends; Menotti Wins 294
Questions 296
Reference 297

15 Final Thoughts 298

Current Trends and Future Directions 298
An Uncertain Future 298
Modest to No Growth 299
Arts Facilities and Social Engineering 299
Conclusion 300
The Partnership: Artist and Manager 300
Questions 300

Index 301

Preface

When I first began teaching arts management, I had to use several textbooks to build the kind of interdisciplinary approach to the field I wanted. I set about writing this text with the goal of blending management theory and practice, economics, personnel management, marketing, and fund raising with the performing and visual arts. The focus of the book is on the process of managing an arts organization through integrating many different disciplines. After covering a brief historical perspective, we will examine all of the functional and operational areas involved in operating an arts organization. Our study will focus on performing arts organizations in theater, dance, music, and opera, and on museums.

This is an introductory text and is intended for use in an undergraduate course on the arts. I have assumed that the student has had some course work in the arts, even if only at the introductory level. Although every topic may not receive all of the attention it deserves, it is hoped that the reader's interest in a specific topic will lead to an exploration of the other resources suggested at the end of most chapters. I have tried to find case studies that offer thoughtful application of the material in each chapter.

Finally, this text was written with the underlying belief that it is important to develop managers in the arts who have sensitivity, use common sense, and apply skills from such disciplines such as business, finance, economics, and psychology. The central premise of this text is that an arts manager's specific purpose is to help an organization and its artists attain their goals and objectives. This lofty-sounding aim is grounded in the assumption that an effective arts manager helps bring to others the unique benefits of the arts experience. There are different ways to describe this experience. For example, when a note is sung or played perfectly, or a movement defies gravity or triggers an emotion or creates a realization, we experience something unique. Sometimes a painting, sculpture, or photograph provides an indescribable pleasure as we stand there viewing it. When we go to the theater and witness a scene that is acted with such power and conviction that it gives us chills, we are enriched.

Working to bring these experiences to others is a worthwhile endeavor to pursue. Although this book makes no pretense of having all the answers about how best to go about maximizing the arts experience or operating the perfect organization, it is hoped that it will provide information and guidance about how an arts manager can be as effective as possible given the resources available.

Organization of the Text

Chapter 1 provides an overview of types and levels of management found in arts organizations. The management process is also discussed.

Chapter 2 examines the historical origins of arts organization as well as profiling the evolution of arts management.

Chapter 3 introduces the reader to the evolution of management theory from ancient times to the present. The basic concepts of systems and contingency management are introduced.

Chapter 4 discusses the relationship of the arts organization to the many external forces that shape how our society functions today.

Chapter 5 begins the examination of the process of management by explaining strategic planning and the decision-making process.

Chapter 6 analyzes the principles of organizing and how organizations are designed. The concept of organizations as complex cultures is discussed.

Chapter 7 integrates strategic planning with organizational design to show various methods for designing jobs, recruiting employees, selecting staff, and providing job enrichment.

Chapter 8 outlines the major concepts of leadership theory. Trait, behavior, and contingency leadership approaches, group dynamics and committee behavior are discussed.

Chapter 9 integrates leadership, planning, and organizing with the management information systems required to effectively operate an arts organization. Concepts of control and resource allocation are introduced.

Chapter 10 examines basic economic theory as it applies to the arts. Concepts in the areas of supply and demand are related to arts organizations. Baumol and Bowen's classic study of performing arts economics is also highlighted.

Chapter 11 relates the economic environment directly to how an organization must organize itself to manage its finances. Such concepts as the balance sheet, the income statement, budgeting, cash flow, and financial planning are reviewed.

Chapter 12 reviews the basic principles of marketing. Marketing is related to the financial planning system and the overall strategic planning process of the organization. The concepts of a marketing audit, marketing management, segmentation, and audience development are discussed.

Chapter 13 focuses on ways that an organization can increase its revenues to meet its mission. The fund raising audit, strategic planning, working with different categories of funders and the techniques of fund raising are discussed.

Chapter 14 develops approaches to integrating management styles, theories, and operations. Dysfunctional, rational, humanistic, and scientific management techniques are discussed.

Chapter 15 asks the arts manager to look to the future and consider where the arts will be in 20 years.

Most chapters conclude with a list of terms and concepts, questions, a case study, and a list of references for further reading in related topics.

A detailed course syllabus with additional project assignments is available by writing to me at Florida State University, School of Theatre 2008, Tallahassee, FL, 32306.

Acknowledgments

I would like to thank some of the people who assisted with this project. First, I owe much to my wife Christine for the many long hours she spent proofing my drafts and for her suggestions. My student assistant Stephanie Goss also was tremendously helpful in pulling together many of the sources used in this text and in handling the permissions. The ideas and suggestions of Steve Roth and Claudia Chouinard were a big help in the early stages of planning this book. I would also like to thank Professor James Zinser of the Oberlin College Department of Economics for his insightful advice; my daughter Alison for her help at the copying machine; Oberlin College for its help and resources; William Patterson and James Schempp for their many helpful comments on the early drafts; and last but not least, Sharon Falter, Kris Smead, and Pat McLaughlin and the staff at LeGwin Associates for their excellent work in producing this book.

1

Management and the Arts

The Entertainment Business

By a unique combination of historical circumstances and the existence of what is often referred to as a market-driven or consumer-driven economy, the United States has created a multibillion-dollar entertainment industry that is a mix of profit and nonprofit businesses. Unlike many other nations, the arts and entertainment industry in the United States is minimally supported by the federal government. Some museums and many performing arts centers are owned by cities or states, but the vast majority of performing arts organizations, media companies, and sports teams are privately owned businesses, public companies with stockholders, or tax-exempt nonprofit corporations.

Both popular entertainment and nonprofit businesses depend on admission sales and other investments for income and tax benefits. For-profit arts organizations are able to take advantage of numerous laws that allow them legally to minimize their tax liability. Nonprofit organizations enjoy the additional benefits of being exempted from paying taxes and being permitted to raise money through the solicitation of tax-deductible contributions.

The roots of the current system of profit and nonprofit arts businesses were established around the beginning of the twentieth century as advances in technology began to change the way people experienced entertainment. The new technologies created what would later be dubbed the *mass media audience*. People tuned in to the radio, went to the movies, and eventually stayed home to watch television or videotapes. The profits attained by being able to package and distribute entertainment to millions of people led to the creation of an industry based on appealing to the broadest possible audience. Meanwhile, the live performing arts groups continued to face the inherent limitation of seating capacity and the rising costs of delivering the product. Fortunately, the rising levels of education, population, and income fed by unprecedented growth after World War II, along with contributions by major foundations in the 1950s and 1960s, helped support the art forms abandoned by audiences for the mass media.

On paper, the future looks bright. For example, recent surveys by the National Endowment for the Arts (NEA), which has been responsible for much of the growth in the arts since its inception in 1965, found more than eight thousand profit and nonprofit arts organizations operating in America (excluding motion picture companies).[1] As recently as 1990, consumers spent more than $4.9 billion on admissions to performing arts events, $5 billion on tickets to motion pictures, and $3.5 billion on spectator sports in America.[2] More than 1.5 million people were employed in the arts and entertainment industry, according to the NEA.[3]

Management in Practice

The typical production process for a performing arts event provides a good example of management in practice. A director working to prepare a production draws on many of the same techniques and principles applied every day in the highly competitive world of business. Such concepts as teamwork, project management, and performance appraisal are fundamental ingredients in a show. The leadership skills of a director determine how well the entire production will go. For example, preparing a play is a group management effort and therefore requires careful attention to the changing, complex dynamics of the cast, designers, and production staff. Motivation levels must be maintained, conflicts must be resolved, and effective time-management skills are required if the show is to benefit from the best work of all participants.

Growing Businesses

With the new ways of experiencing live and prerecorded entertainment and the increase in wealth among the general population came the proliferation of both profit and nonprofit businesses designed to meet the rising demand for entertainment. New jobs for managers were created by the thousands as companies expanded their operations. Each of these enterprises needed people with special skills and knowledge to ensure that the product was created and distributed in a way that realized the organization's goals, as stated by the owners or boards of directors.

For-profit theater, film, television, videos, nightclubs, popular music, radio, and spectator sports are big businesses employing highly visible "stars" and hundreds of thousands of support people. In *Jobs in Arts and Media Management*, Stephen Langley and James Abruzzo estimate that there are 77,655 businesses in the entertainment industry.[4] Their total includes theaters, opera companies, musical theater groups, music groups of all types, dance companies, performing arts-presenting organizations, arts councils, broadcasting and cable companies, film companies, museums, unions, and recording companies. See Figure 1–1.

Nonprofit arts organizations in theater, music, dance, and opera and nonprofit museums make up a great many of these organizations and provide year-round employment at all levels of management. As was noted, these two sectors of the entertainment market account for more than 1.5 million workers. These people in turn contribute to the national economic system with their purchases of goods and services. The arts help to foster economic growth in communities across America. Chapter 10 will elaborate on the economic impact of the arts.

An Uncertain Future

Despite all of the growth and development, many people in the arts are anxious about the future. Some of these concerns stem from the changing demographics in America, which lead to questions about where the future audiences will come from. Others see the increasing debt at all levels of government as a sign that the arts will experience more funding cuts. Issues relating to censorship, the empowerment of minorities, more conservative views about funding (see "Other Viewpoints"), and the AIDS epidemic all have an impact on art groups. The commercial entertainment industry shares many of these concerns. Rising production costs are driving ticket prices up to levels that exceed all expectations.

New technology has permitted entertainment to become more personalized and miniaturized. The change from mass media to individual entertainment systems, coupled with the decreasing resources for arts in the schools, is creating an audience with different attitudes about what they see and hear.

The often predicted dramatic increase in leisure time has failed to materialize. With less leisure time available, consumers are making careful choices about how they spend their entertainment dollar. Many organizations fear that there will be too many arts groups chasing after too few patrons.

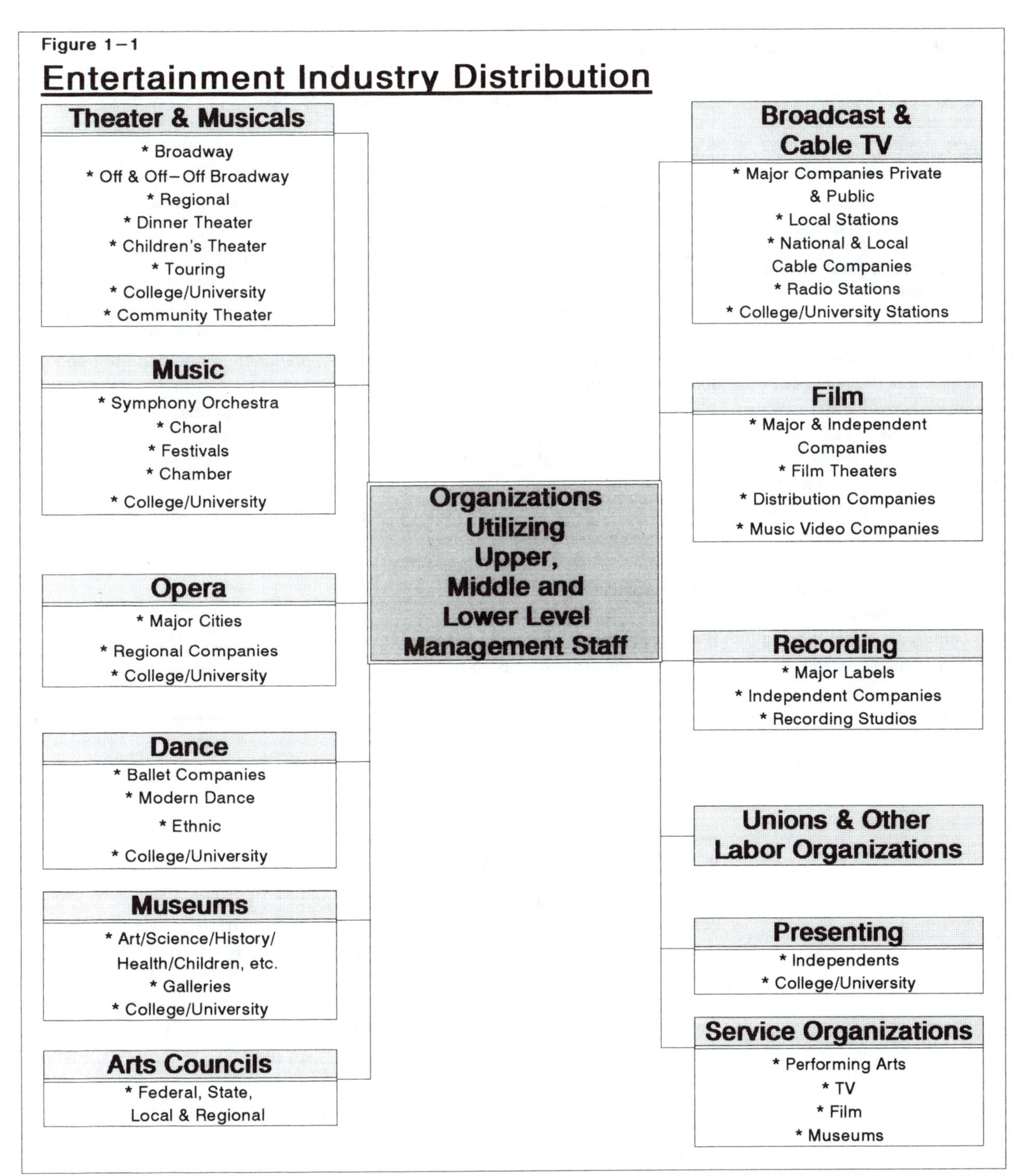

Source: Stephen Langley and James Abruzzo, Jobs in Arts and Media Management, (New York: American Council for the Arts, 1990).

Other Viewpoints: The Arts Subsidy Challenged

As the following excerpt demonstrates, arts managers should not assume that everyone agrees that state, local, or federal governments should subsidize the arts. Dr. van den Haag has been a longtime critic of any subsidy for the arts.

Involuntary Patrons: Taxpayers' Rights Versus Government Art

Dr. Ernest van den Haag

European governments have traditionally subsidized art and religion because both glorified God, king and country and helped governance: Art and religion were the principle means of indoctrination in social values before TV, radio and print. They were quite indispensable in forging the social bond that makes a nation. Subsidies have continued in Europe, although art and religion have become marginalized. It would be hard to imagine Italy without its cathedrals or its operatic performances. They require subsidies; but they also bring in tourist dollars. In the United States art, mostly imported, never has become a central part of our social bond. Baseball is more like it—and does well without a federal subsidy.

Consider opera. (Exhibitions of paintings would do as well as an example.) Only a small proportion of Americans enjoy it. I do. Yet all taxpayers are forced to subsidize performances. I feel guilty every time I attend, thinking of the people whose taxes are used for my enjoyment, although they would rather use their money for their own enjoyment—perhaps to attend a Madonna concert (unsubsidized). Why is money taken from (low income) taxpayers to benefit (middle class) opera lovers? Congress forces taxpayers to subsidize my aesthetic preferences, because Congress thinks opera is good for taxpayers (or paintings are) even if they don't attend (luckily they are not forced to).

Taxes are often spent on things individual taxpayers do not want. But defense, or police forces, unlike art, cannot be bought individually. They protect people whether they pay or not. To avoid "free riders" everyone must pay through taxes. However opera, concerts, paintings or poetry do not benefit those who do not attend, view, or listen. Why, then, should the voluntary non-beneficiaries be compelled to pay for those who benefit from art (or at least enjoy it)? Why should the non-beneficiaries, or non-enjoyers, not be allowed to spend their money on what they prefer?

SOURCE: Ernest van den Haag, "Involuntary Patrons," *Vantage Point* (the magazine of the American Council for the Arts) Spring 1990: p 6. Excerpted with permission.

Managers and Organizations

This book will examine how the manager of the arts can use the processes of planning, organizing, leading, and controlling to solve an organization's problems and fulfill its mission in these uncertain times. These *four functions of management* are the basis for the working relationship between the artist and the manager. Because most of the activity associated with the performing arts and with museums occurs through some type of organization, this text concentrates on management in a group environment.

Let's look now at a brief overview of the manager, the organization, and the process of organizing.

The Manager

In any organization, a manager is "a person who is responsible for the work performance of one or more people."[5] The manager's basic job is to organize human and material resources to help the organization achieve its stated goals and objectives. With this definition, a director, a stage manager, a lighting designer, and a curator, are all managers. The details of their job descriptions may differ, but the responsibility of getting others to do something is the same. Leadership skills are needed to effectively direct others to accomplish the work that must be done.

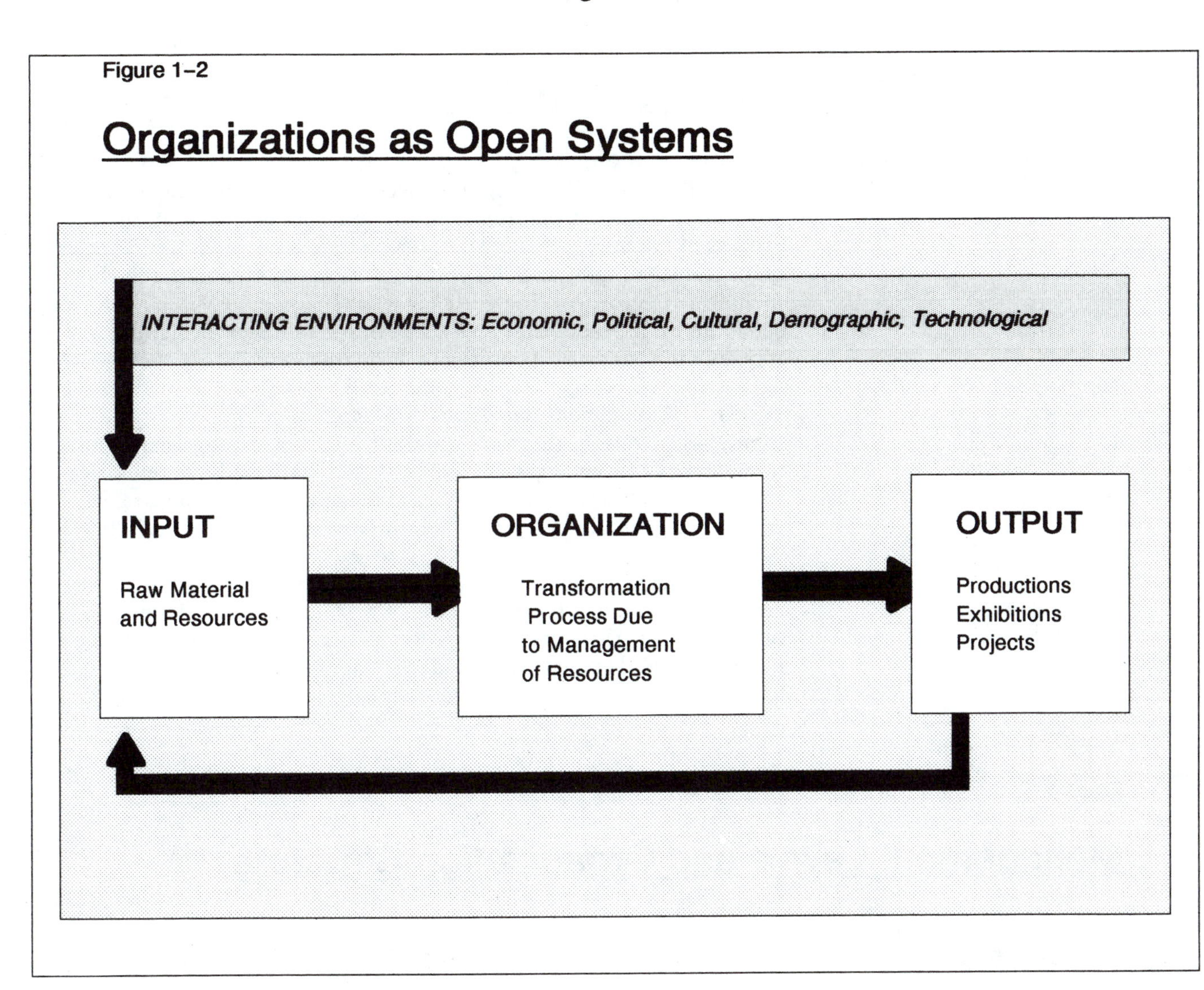

The Organization

Managers function within an *organization*, which has been defined as "a collection of people working together in a division of labor to achieve a common purpose."[6] For example, the Guthrie Theatre, a world-renowned regional theater in Minneapolis, has defined its purpose in this way: "to celebrate the shared act of imagining between audience and actors."[7] Other arts groups share equally lofty aims. However, these groups are only as effective as the managers and artists they hire to carry out the mission.

Figure 1–2 shows how organizations interact with many external environments in a process of transforming their resources (inputs) to products or services. The output of an arts enterprise may be a performance or an exhibition. The primary environments that affect all organizations are economic, political, cultural, demographic, and technological. Chapter 4 examines the impact of each of these environments on organizations. As we will see, the survival and growth of an organization depends on its being able to adapt as these environments change. Managers of organizations must use all the skills and knowledge at their disposal because these environments are always presenting new opportunities and threats.

The Process of Organizing

As we will see in Chapter 5, the process of achieving the organization's goals and objectives requires that the manager actively engage in the process of *organizing*, which has been defined as "dividing work into manageable components."[8] Typical examples of organizing in the arts include a director working with a stage manager to develop a rehearsal schedule for a production and a box office manager designing a staff schedule to cover the upcoming performances.

Levels of Management and Types of Managers

In any organization, there are different levels of management and different types of managers. Typically, organizations have operational, managerial, and strategic levels of management[9] and line, staff, functional, and general managers or administrators.[10] See Figure 1–3.

Levels of management The *operational level* of management is concerned with the day-to-day process of getting the work done. The sets must be built, the museum guards must assume their posts, the rehearsal schedule must be posted, the membership renewals must be

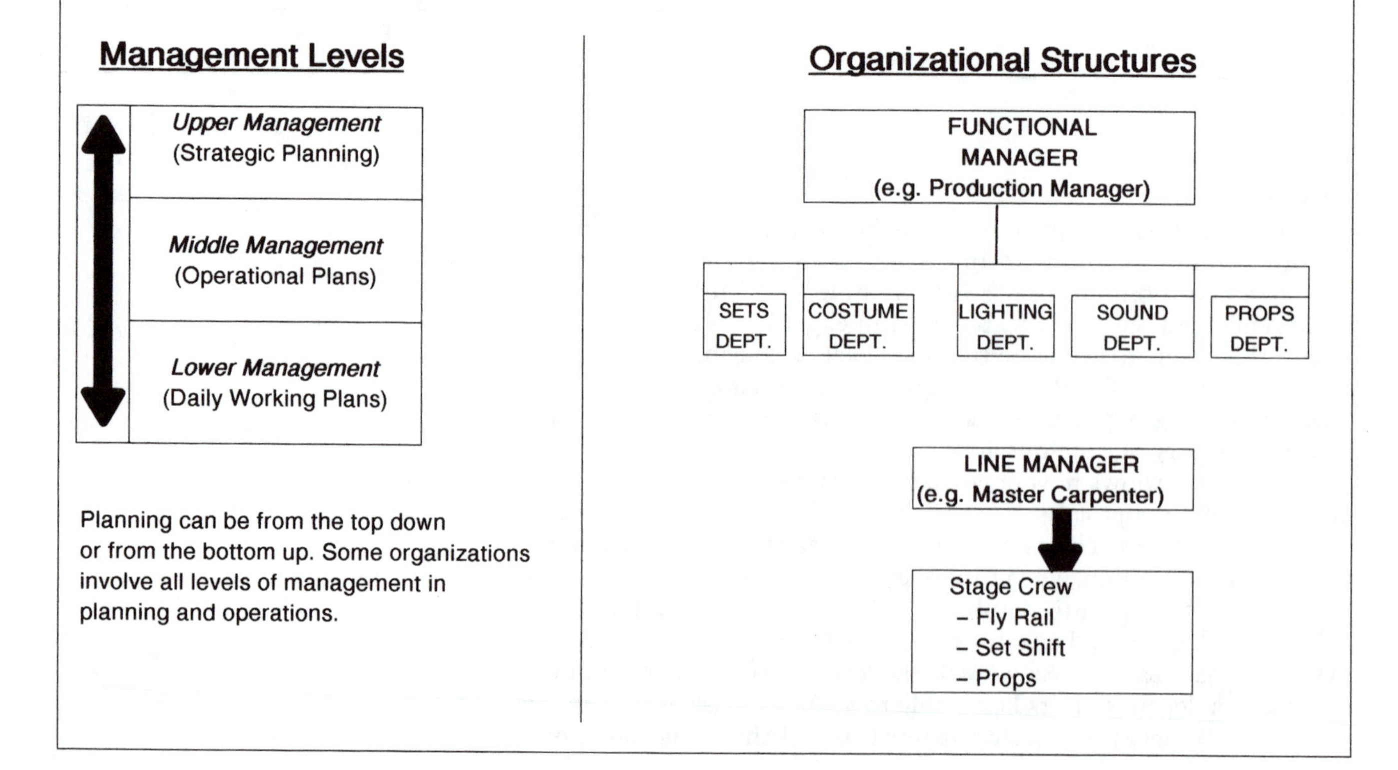

mailed, and the box office must sell tickets. The operations level is central to the realization of the organization's goals and objectives. Without the efficient and productive management of its operations, the organization faces extinction.

The *managerial level* is often called *middle-management*, because it coordinates the operations and acts as a bridge between the operational and strategic levels of management. For example, the board of directors and the artistic director of a theater company ask the production manager to evaluate the impact of adding a touring season to the company's schedule. If the plan is feasible, the production manager will have the task of coordinating the schedules, materials, and people required to initiate this program of activity. The managerial level usually functions in a one- to two-year planning cycle in the organization.

The *strategic level* of management, on the other hand, watches the overall operation of the organization with an eye toward constantly adjusting and adapting to the changing environments that affect the future of the organization. The mission, goals, and objectives are typically assessed in a three- to five-year time frame. In addition, strategic managers are responsible for looking as far ahead as 10 to 15 years in an effort to chart a path for the organization. The artistic director, general manager, general director, managing director, marketing director, or other similar senior level personnel are associated with this role. In addition, strategic managers typically present these long-range plans to a board of directors. The board ultimately oversees the organization's mission and purpose.

Types of managers The arts have evolved unique types of managers to make the organizations work. The types of managers listed in this section are found in different combinations in arts organizations depending on the purpose and design of the organization. Each art form has specialized job titles and responsibilities.

The first type of manager is the *line manager*. This person is directly responsible for getting the product or service completed. The head carpenter, who supervises a stage crew, is a good example of such a manager. The head carpenter's job is to get the set up on stage and ready for the performance.

Staff managers "use their special technical skill to support the efforts of the line personnel."[11] For example, the technical director in a performing arts group is usually given this responsibility. He or she coordinates the work of the line managers responsible for such areas as sets, props, rigging, sound, and lighting. Another example of a staff manager is the production manager. This person is given the responsibility of overseeing all of the production departments in an arts organization, such as scenery, lighting, costumes, props, and sound.

The *functional manager* has responsibility over a single area in the organization. For example, the production manager in a theater company oversees the production departments and therefore is both a functional manager and a staff manager. A company manager, who is responsible for the performers, is another example of a functional manager.

General managers are found in more complex organizations with many functional areas. For example, the general manager of an opera company oversees production, marketing, fund raising, and administration for the organization.

Another managerial title often found in organizations is the *ad-*

ministrator. Although the administrator is really a manager, based on the definition of being responsible for the work efforts of one or more people, the title is often used in nonprofit organizations to refer to someone empowered only to carry out functional tasks defined by others. Like the general manager, the administrator does not make plans or policies but is responsible for their implementation.

Common Elements in an Organization

Chapter 3 examines the history and early management theories that influenced much of today's thinking about how to accomplish the objectives of an organization. Some of these theories apply to fundamental issues of organizational design and structure and are applicable to art organizations.

A division of labor and some type of hierarchy exist in most organizations. The *division of labor* usually takes in a form that matches the organization's function. A dance company has a different division of labor than an opera company for the simple reason that the processes and techniques used in preparing a performance are different. For example, many opera companies have a small permanent administrative and fund raising staff. The singers, orchestra, director, and designers are hired for short periods of time to do a single show. Ballet companies, on the other hand, often have 30 or 40 dancers under a 40-week contract each year. They therefore require a different division of labor.

The *hierarchy of authority* in an organization is designed to ensure that the work efforts of the different members of the organization come together as a whole.[12] The typical hierarchy involves a vertical reporting, communication, and supervision system. Chapter 6 details various methods for organizing management systems.

In most arts organizations, which are small to medium-size businesses, the levels of management and the formality of the hierarchy are usually limited. However, as an organization grows in size and more staff are added, the levels of management increase, and the hierarchy tends to become more formal. A good arts manager is watchful of this development, especially if overly complex divisions of labor or a burdensome hierarchy begins to impede the accomplishment of the organization's goals and objectives.

An *informal structure* also exists in all organizations. No organizational chart or detailed plan of staff responsibilities is able to take into account all of the ways people find to work with each other. Employees often find new combinations of people to accomplish tasks that do not fit into the existing hierarchy or organizational design. Some organizations thrive on this sort of internal innovation; others become chaotic. Arts organizations—which tend to cluster toward the rigid, rather than the innovative, end of the structure continuum—often develop organizational designs aligned with functional areas. For example, the production staff, office staff, performers, and upper management develop structures to operate their own areas. The result is four organizations instead of one. At the same time, organizations, like people, can lapse into habitual behavior patterns. Tradition becomes the norm, and innovation is resisted. Again, the arts manager must keep an eye on the organization's formal and informal structure. Careful intercession can correct unproductive structures that develop.

Organizations are not neutral entities. They are microcosms of the society at large. Organizations are collections of individuals with beliefs, biases, and values. Unique myths and rituals are part of what is called the organization's *corporate culture* (see Chapter 6). Simply described, the corporate culture is the way things are done in the organization. For example, the culture of the organization usually establishes values on such things as the quality and quantity of work expected. Some organizations have a positive culture that is communicated to employees. For example, managers might say, "Our stage crew is here to make things work, and their contribution is valued and recognized." In this situation, the overall culture of the organization values the labor of its employees. Other organizations have weak or destructive cultures. Phrases such as "The crew around here is always looking for a way to get out of work, and they are not to be trusted" signal a culture based on distrust and conflict.

The founder-director organization, a model quite prevalent in the arts, can help establish a strong culture imbued with the beliefs and values of one individual. Unfortunately, the departure of this person often leaves the organization adrift.

Any arts organization, no matter how small, is ultimately a complex mixture of behaviors, attitudes, and beliefs. Because people are the major resource used in creating its products, an organization will continue to be influenced and changed in ways that no one can predict. Interaction with external environments also affect the way people inside the organization think, feel, and behave. For example, changes in laws and the social system have led to the addition of multicultural programming and the hiring of more minorities in many arts organizations. See Figure 1–4.

An Effective Model?

Arts organizations are learning to effectively integrate long-term strategic thinking while developing a sensitivity to the changing environments that shape the beliefs and values of the entire culture. Because the performing and visual arts are dependent on the creative explorations of the individual for the new material they present, the design

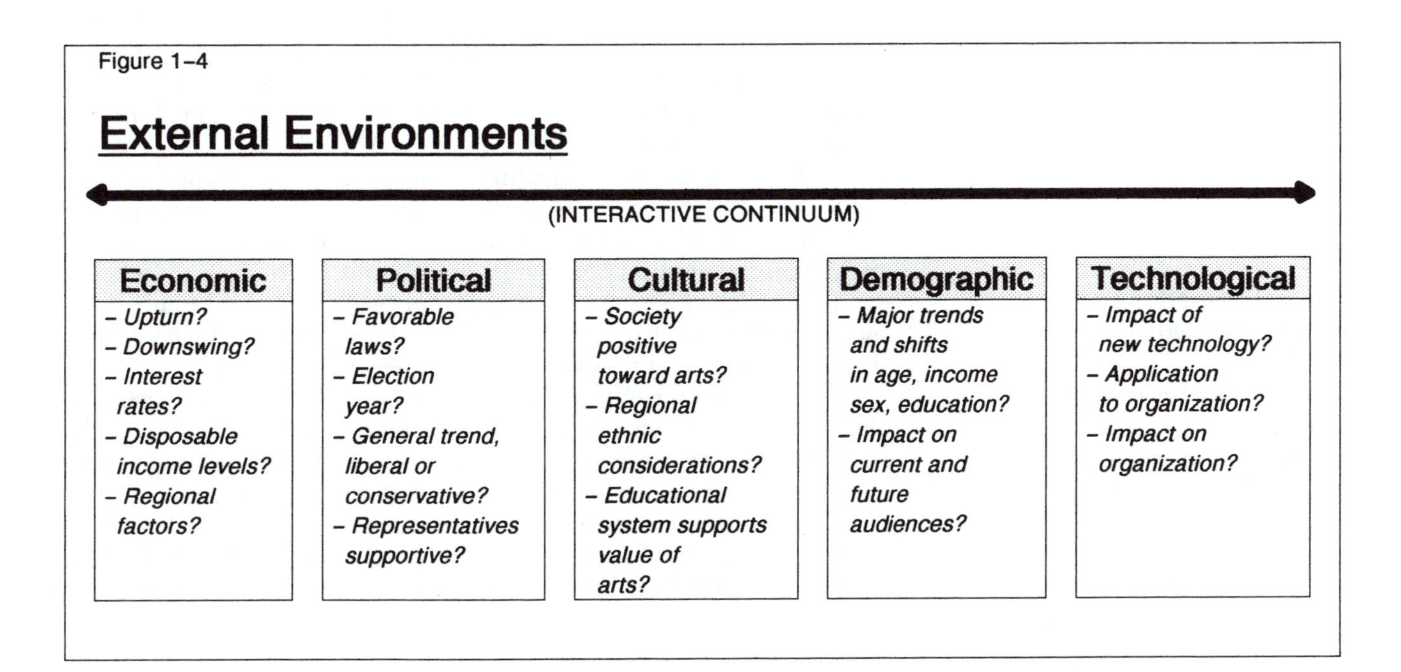

and function of these institutions should be focused on looking to what will be and not at what was. However, many artists perceive of arts organizations as institutions that are more comfortable with the past. The creation of organizations in the performing and visual arts that look like imitations of corporations with executive directors, vice-presidents, and associate directors is not universally seen as a good sign. Many artists are asking organizations to examine such fundamental questions as "What is our mission?" "Just what is it we are doing?" "What things are essential to our mission?" "Whom do we serve?" "What do people think we do?" and "What are we really contributing to the community?"

In other cases, artists are seeking a more entrepreneurial environment in which to work. They are finding that rigid structures and corporate models are not the most effective way to bring the audience and artist together.

The Management Process

The organization and systems described thus far are predicated on the assumption that there is an artistic product to manage. How does this product come into being? In many cases, an individual or a small group of people have the drive and energy to create something from nothing. For example, a playwright and director may team up to interest other people in a script. If people with money can be found to back the show, they hire designers and performers to bring the work to life. Sometimes, much less often than anyone cares to consider, the show is a hit. A long-standing love for opera may drive someone to start an opera company. Two dancers may decide that it is time to start their own company. They are tired of dancing someone else's choreography, and they have some ideas of their own that they would like to see performed. A group of visual artists may start a cooperative exhibition gallery and operate it themselves.

Whatever the circumstances, the success or failure of these artistic ambitions will be directly related in part to how well the four functions of management are fulfilled. Without proper planning, good organization, creative leadership, and some control over the enterprise, the chance of success is greatly diminished. Obviously, projects and programs succeed in this world that do not master these four functions. Poorly planned, badly organized, and uncontrolled events happen all the time. However, because of the incredible leadership skills and talents of one person, the project or the show somehow makes it to completion or to its opening. The events that suffer from various forms of dysfunctional management make for great stories, but the human toll taken by such examples of bad management is precisely why good managers are needed in the arts. There is no benefit to the art form or the community if the very people who love the arts are destroyed by it.

It is important to remember, however, that a bad play, opera, musical, ballet, symphony, or exhibition cannot be made good by excellent management. If people do not respond to a work after all the rewrites and extra rehearsals, it does not matter how well it was managed.

We will take more time to examine the evolution of the arts and how arts managers fit into the entire process in Chapter 2. For now, let's consider the four functions and relate each of them to an arts application.

The Four Functions of Management

- *Planning* is deciding what is to be done.
- *Organizing* is deciding how it is to be done and who is to do it.
- *Leading* is deciding how other people are to get it done.
- *Controlling* is deciding if it is or isn't getting done and what to do if it isn't.

Planning

This first function of management is the hardest. Deciding exactly what it is we want to do, setting realistic *goals* (what the organization wants to accomplish), and then determining the *objectives* (the specific steps to take and the time table for completing the tasks) to be used in meeting the goals is hard work.

There are various sorts of plans. Some are short-range plans: What am I going to do tomorrow? These plans usually don't present too much of a challenge. On the other hand, planning five years ahead can be an intimidating task.

Organizations and people must plan because the world is constantly changing. Audience tastes and values change over time. The arts manager's job is to recognize the elements in the world around the organization that may pose new opportunities or may be a threat. Then the manager must work with the board and the artistic leadership to chart a course of action designed to guide the organization into the future.

For example, the artistic director of the ABC Theater Company reads in the newspaper that state funds for theater companies that visit schools will soon be available. A goal is established to implement a touring program in the next year because it relates directly to the organization's mission statement. The staff researches costs and benefits. The plan and the goals are drawn up and reviewed with the board. The company agrees to set up a pilot program.

Organizing

Organizing is the process of converting the plans into a course of action. Getting the people and things together, defining the details, creating a schedule and budget, estimating the number of people needed, and assigning them their jobs is all part of organizing.

Continuing the example, the ABC Theater Company sets up a special touring department. The company hires a director of touring and puts into place the details of the plan. For the first year, the company will have a small group of six actors tour 20 schools to perform scenes and hold theater workshops.

Leading

The third function of management requires getting everyone in the organization to share a vision of what can be accomplished if everyone works together. Leadership skill and effectiveness are highly prized attributes in any situation. For the arts manager, working with the self-motivated, independent-minded people often found in the arts offers a unique leadership opportunity.

After the ABC Theater Company touring staff is hired, the artistic director meets with everyone to clarify the project's purposes and goals. The director provides an overall time table and explains where this new operation fits into the organization. The company's mission is recalled, and a challenge is issued to make this a quality touring program.

Controlling

The fourth function of management is concerned with monitoring how the work is proceeding, checking the results against the objectives, and taking corrective action when required.

After six months, the artistic director reviews the activities of the touring company and finds that bookings are down, cast turnover is

Sample Mission Statement and Objectives

The Seattle Symphony's strategic long-range plan provides a sample of two key elements of an organization's planning process: the mission statement and objectives.

Mission Statement

The mission of the Seattle Symphony is to present concerts of the highest artistic quality for the enjoyment, enrichment and education of the people of the Pacific Northwest.

1. Situation Analysis (Artistic)
   [The consultants summarize the current factors helping and hindering the symphony from fulfilling its mission.]
2. Opportunities (Artistic)
   [The consultants outline the conditions that present opportunities for the symphony in the future.]
3. Objectives (Artistic), Fiscal Years 1991–1995
   a) Maintain the highest possible standards of artistic excellence.
   b) Enhance the scheduling capabilities, performing and rehearsal spaces, and the overall excellence of its artistic programming by obtaining a performance hall.
   c) Maintain an active recording program.
   d) Expand the Symphony's artistic exposure through radio and TV broadcasts.
   e) Present innovative, stimulating and entertaining programs.
   f) Increase the number of permanent violinists to enhance the quality of the string sound.
   g) Continue to tour when economically viable.

SOURCE: Seattle Symphony Orchestra, *Strategic Long-Range Plan*, Fiscal Years 1991–1995, Draft #1 (December 1990).

high, and the budget for the year is almost gone. Meetings are held to pinpoint problems and consider solutions. Staffing changes are made, and the project is monitored on a weekly basis. After a year, many of the problems have lessened.

Functional Areas

When engaged in planning, organizing, leading, and controlling there are seven basic functions an arts manager fulfills:[13]

1. planning and development
2. marketing and public relations
3. personnel management
4. fiscal management
5. board relations
6. labor relations
7. government relations

Planning and development are linked because arts organizations are always seeking ways to increase revenue to fund new programs and to pay for the inevitable increases in operating costs. Marketing and public relations provide the organization's most visible link to the community. Good personnel management and labor relations are essential if the organization is to be productive. Neglect or abuse of the human resources available to a manager can disrupt the entire enterprise. Good fiscal management is critical if the organization's planning, marketing, and fund raising efforts are to succeed. In addition, donors prefer to make contributions to organizations that show they know how to manage their financial resources. As with personnel relations, an arts manager must effectively work with and report to a board of directors. The board and the management may sometimes have a different set of priorities. Until the differences are resolved, the organization will find it difficult to meet its goals and objectives. Finally, government relations, which includes the local, state, and national levels, grows more complex each year. New laws are passed or court rulings are enforced that change the way an organization does business. These types of changes typically add to the expenses of the organization.

Throughout this text, we will examine how external environments and internal organizational dynamics make the task of being a manager in the arts a challenging and demanding job. The almost endless variety and changing circumstances in the world around the arts organization keep the manager's job from ever getting too dull or routine.

Key Terms and Concepts

Manager
Organization
Organizing
Open system model
Levels of management: operational, managerial, strategic
Types of managers: line, staff, functional, general, administrative
Division of labor
Hierarchy of authority
Formal and informal structures
Corporate culture
Functions of management: planning, organizing, leading, controlling

Functional areas of work for an arts manager: planning and development, marketing and public relations, personnel management, fiscal management, board relations, labor relations, government relations

Questions

1. Are you aware of any arts organizations that have been particularly successful or have faced difficulty in your community? Outline the situation, and explain why you think the organization did well or faltered.

2. Can you recall a particular work situation you have been in that was either positive or negative as a direct result of the manager in charge? What type of manager was this person (line, staff, functional)? What made this manager effective or ineffective?

3. What are some recent changes in the five environments cited in the open system model (economic, political, cultural, demographic, technological) that may have an impact on arts organizations?

4. Do you agree or disagree with the author of the "Involuntary Patrons" article? Why? Do you agree that art "never has become a central part of our social bond" in America? The author argues that because the arts are not available to everyone, subsidies are inappropriate. What is your opinion? Can you think of situations where professional sports teams have received government "subsidies"?

5. List some examples of how you "manage" your life. Have you used any combinations of the management functions of planning, organizing, leading, or controlling to achieve objectives you have set for yourself?

Analysis Focus

Select a local arts organization, and request a copy of its mission statement, bylaws, and other relevant planning documents (for example, a five-year plan) for a discussion by the class. Based on some of the topics covered in this chapter, answer the following questions:

1. Is the organization fulfilling its stated mission. If yes, how? If no, why not?

2. Is the organization facing financial problems? Did it have a deficit or surplus in the last budget year?

3. Based on the information gathered, is it possible to ascertain if this is a well-managed organization? If yes, what evidence supports this position? If no, what are the management areas that need improvement (planning and development, marketing and public relations, personnel management, fiscal management, board relations, labor relations, government relations)?

References

1. Research Division Report #23, NEA (Washington, DC 1987).
2. Research Division Report #34, NEA (Washington, DC 1990).
3. Research Division Report #33, NEA (Washington, DC 1990).
4. Stephen Langley and James Abruzzo, *Jobs in Arts and Media Management* (New York: American Council for the Arts, 1990), 10–11.

5. John R. Schermerhorn, Jr., *Management for Productivity*, 2d ed. (New York: John Wiley and Sons, 1986), 7.
6. Ibid., 8.
7. Guthrie Theatre, *Long Range Plan* (1988), II.2.
8. Schermerhorn, *Management for Productivity*, 161.
9. James H. Donnelly, Jr., James L. Gibson and John M. Ivancevich, *Fundamentals of Management*, 7th ed. (Homewood, IL: BPI-Irwin, 1990), 28–29.
10. Schermerhorn, *Management for Productivity*,13–15.
11. Ibid., 13.
12. Ibid., 12.
13. Paul DiMaggio, *Managers of the Arts*, Research Division Report #20, NEA (Washington, DC: Seven Locks Press, 1987).

The Evolution of Arts Organizations and Arts Management

2

❑❑❑❑❑

In this chapter, we will review the evolution of the job of arts manager. We will explore how the responsibilities have changed to meet the increasingly complex demands placed on arts organizations and artists. We will also touch on the impact the National Endowment for the Arts has had on the arts scene in the United States. The chapter concludes with an examination of the national network of performing arts centers in communities and on university campuses in the United States.

The Artist-Manager

For more than two thousands of years, the artist-manager has been the person who created and arranged the meeting of artist and public. Creative drive, leadership, and the ability to organize a group of people around a common goal remain the foundation on which all arts management is built. The traditional role of the artist-manager has been split into separate jobs to cope better with the increasingly complex demands placed on managers. However, this split does not mean that a division or barrier must be erected between these two roles. Instead, the separation should be viewed in much the same way as the human brain functions: the two hemispheres are linked and communicate with each other while each side continues to do what it does best.

The Arts as Institutions

One result of the political and social upheaval of the last four hundred years has been the establishment of institutions designed to provide continuing support and recognition for the artist and the arts. In much of the world, the performing arts are part of a state-supported system and are operated by resident managers with extensive administrative staffs. Performing and visual arts centers for opera, dance, theater, and music as well as museums reserved exclusively for art, history, and science are integral parts of many communities in the world.

In the United States, token governmental backing for the arts is a recent phenomenon. Fund-matching grants, special project support, and a taxation system designed to promote deductible donations by individuals and corporations continue to be the extent of government involvement in the arts. More recently, in the last one hundred years, the U.S. government opted for an alternative system that encouraged the creation of tax-exempt, nonprofit corporations to supply and distribute the arts and culture in society.

The increasing complexity of an industrially based society hastened the shift from the artist-manager as the dominant approach to organizing and presenting the arts. As many communities began to establish arts institutions late in the nineteenth century (museums, symphony orchestras), year-round management experts began to emerge. Many arts institutions now appear to be organized along patterns similar to large business corporations.

Today, the role of the artist and the manager and the degree of control each has over their respective domains vary from art form to art form. Many small arts organizations are still created and managed by artist-managers. However, the norm is a corporate structure with a board of directors and multiple levels of staff arranged in a hierarchy.

A Historical Overview

Let's examine some selected points in Western history to trace the development of the management function in the arts. As has been noted, the artist-manager is a well-established pattern in the arts. Although this pattern of management has not changed much in the last two thousand years, the demands placed on this individual have increased to the point where the artist-manager format is now only one of many ways to organize the presentation of arts events.

Ancient Times

As the centers of civilization grew, so did those functions we associate with the arts. The first examples of performance management were the public assemblies associated with religious rites in early societies. These "performances" were "managed" by the priest and were enmeshed in the fabric of a society. The theatrical trappings of costumes, dramatic settings, music, movement, and so on, all supported and heightened the event. Ultimately though, these events were not an expression of the creative drive of a people, but rather a way of controlling and molding a culture. However, these staged events did provide a model for organizing large-scale public gatherings.

The beginnings of a system of state-sponsored play festivals can be traced to the Greeks around 534 B.C. These festivals required the management skills of planning, organizing, leading and controlling, much as they do today. Typically, a principal magistrate, the *archon eponymous*, supervised the production of the play festivals sponsored in Athens. Financial support came from the richer citizens (*choregoi*), and the cities provided the facilities. The playwright functioned as the director and had something akin to total artistic control over the show.[1]

The Romans also produced state-sponsored arts festivals as part of an overall cycle of public events throughout the year. City magistrates were responsible for screening and coordinating the entertainment for their communities. The managers (*domini*) acted as producers, bringing the play and the performers to the festivals. These early managers arranged all the elements needed for the production with the financial support of the local magistrate. According to research in theater history, as many as one hundred days a year were committed to the various theater festivals of ancient Rome. If this schedule is indeed true, a great deal of managerial skill must have been required to coordinate and produce these events.[2]

With the decline of Rome came the dissolution of the state-sponsored festivals. The breakup of the Empire did not mean that all artistic activity came to a halt. However, the transition into the Dark Ages left society without a developing dramatic literature generating works to perform. The disappearance of organized financing and facilities also made it impossible to sustain an ongoing arts community. Performance groups therefore resorted to touring as a means of survival. The management of the troupe was done by a member of the performing group. Smaller-scale community festivals helped provide opportunities for the itinerant artists to eke out a living. On the whole, Western history has not provided much evidence of significant artistic activity in Europe during this time.

Other cultures were of course developing indigenous forms of music, dance, and theater. The arts were very much a part of Byzantium, India, and China. While Europe was struggling, other cultures were establishing forms of dance, theater, music, and visual arts that are with us today. Varying degrees of state and private sponsorship were involved. The role of the manager did not radically differ in these cultures because the functions required to organize and coordinate arts events were the same.

The Middle Ages

The Church was the producer of many sanctioned performances during the Middle Ages. The performance of liturgical drama, which served as a type of religious instruction, originally resided within the management structure of the Church. As communities developed and the overall economic environment improved, this drama moved outdoors and became part of public pageants (using stages mounted on portable wagons) and festivals. Nonliturgical drama and various forms of popular entertainments, such as jugglers and mimes, were part of a rebirth of performance.

By the fourteenth century, the Church had little control over the proliferating performances. A system of patronage and sponsorship by the trade guilds led to an expanding role for the manager-director. Historian Oscar Brockett notes that during the fifteenth and sixteenth centuries:

> complex productions required careful organization, for the handling of casts that sometimes included as many as 300 actors, of complex special effects, and large sums of money could not be left to chance. Consequently, the director (or stage manager, or pageant master) was of considerable importance. Often this position was given to a member of the guild, but in some instances a "pageant master" was put under contract for a number of years at an annual salary. The pageant master secured actors, arranged rehearsals, and took charge of every phase of production.[3]

The Renaissance

The continuing surge of the arts was dramatic throughout the Renaissance. The social, political, economic, and cultural environments were undergoing changes that fundamentally altered people's perceptions of the world. The rediscovery of the Greeks opened up the creative spirit of the times. During the fourteenth to sixteenth centuries, neoclassical theater began to flourish, opera and ballet were born, and the role of the arts manager burgeoned.

In theater, the expansion of literature was accompanied by the construction of performance spaces that took advantage of the new stage technology of the time. This in turn led to the rise of stage crew specialists in such areas as rigging, lighting, special effects, and costumes. The coordination required of the increasingly complex productions helped solidify many of the traditional roles in backstage management.

In the late sixteenth century, opera was born in Italy out of the *intermezzi*, which was a form of entertainment that occurred between the five-act dramas of the time. In 1594, the first opera, *Dafne*, premiered and laid the foundation for an entire art form.[4]

The court dance of the thirteenth and fourteenth centuries helped forge a path for the creation of ballet. The court "dance masters of the [fourteenth and fifteenth centuries] began to develop a theory of dance instruction that systematized its various movements and styles."[5] The first ballet, *Ballet Comique de la Reine*, was performed in 1581 in the court of Henry III of France.[6] As with opera, specialized production and management techniques evolved over the centuries to support the art form.

Like today, finding financial support was an ongoing activity of the early artist-managers. Church support, royal patronage, and shareholder arrangements were the chief means of financing work. The shares sold to people helped provide the resources needed to pay for salaries and production support. Management functions were expanded to include overseeing the distribution of any profits to the shareholders.

The other major problem managers and artists grappled with was censorship. Throughout history, the performing and visual arts have had to contend with varying degrees of control from both the church and the state. The selection of plays, the access to performance spaces, and sometimes even the selection of performers have been subject to very severe constraints. The arts manager is often placed in the middle of the battle between an artist seeking an avenue of expression and a state or religious group attempting to suppress the work. We see the legacy of the sometimes uneasy relationship between the arts and society in the recent controversy over the reauthorization of the National Endowment for the Arts.

The Seventeenth Through Nineteenth Centuries

In many European countries during this time, the arts continued to grow and flourish. Playwrights, directors, composers, musicians, dancers, and singers found work in newly created companies and institutions. In France, the theater, opera, and ballet companies were being organized in state-run facilities, and the performers received salaries and pensions. Germany established a state theater by 1767. It became the foundation for a national network of subsidized arts institutions. England also had a thriving performing arts community. The Education Act of 1870 and the Local Governments Act of 1888 helped promote the growth of museums and performing arts facilities throughout Great Britain.[7] British support for museums was well rooted in the nineteenth century. However, the first Arts Council in England was not created until 1945.[8] Throughout the seventeenth to nineteenth centuries, especially on the Continent, the formalization of management structures and systems to operate the state theaters solidified the role of the arts manager.

In the United States, theatrical presentations were made up of touring groups performing varied programs in cities across the nation. The local theater venue often contained stock sets that were used by the performers, who brought their own costumes. The spreading rail system of the mid-nineteenth century helped support an extensive touring network of performing groups. Companies were formed and disbanded almost constantly, and no permanent theater companies were established. The management structure was dominated by the producers and booking agents who arranged the tours. The control of most theaters eventually fell into the hands of these booking agents. A monopoly known as *The Syndicate* controlled what was available for viewing around the country. This monopoly was supplanted by another group of theater owners, the Shuberts. The Shubert brothers created a management dynasty that lasts to this day.[9]

Unlike the impermanent theater, symphony orchestras and opera companies began to secure a more stable place in the larger metropolitan areas in the United States. For example, the support of wealthy patrons made it possible to establish symphony orchestras in New York City (1842) and Boston (1881). Opera, which had been performed in the United States since early in the eighteenth century, found its first home in the Metropolitan Opera in 1883.[10] Dance was often included in touring theatrical productions in the eighteenth and nineteenth centuries. European dance stars also regularly toured the country. However, permanent resident dance companies were not a regular part of the arts scene until the twentieth century.[11]

The Twentieth Century

The role of management increased as the continued growth of the arts accelerated. Despite two world wars, European arts institutions expanded into smaller communities, developing national networks of performing spaces and providing jobs for managers and artists. Seasons expanded, repertories grew, and new facilities were constructed—especially after World War II—in an overall environment of support from the government. As noted, England eventually established a state-supported system for the arts after the war.

In Europe and the United States, the new technologies of radio and film significantly changed attendance patterns at live performance events. The theater in the United States, for example, saw a rapid decline in attendance by the 1920s.[12] Because there were no resident theater companies, it was difficult to keep a loyal audience base such as existed for the few opera and symphony groups in the country.

The rise of the Off Broadway and regional theater system helped renew the theater and, at the same time, helped build a base for what were to become established organizations. The more experimental, but still profit-driven Off Broadway system was born in the early 50s. The nonprofit regional theater network was built from the Barter Theater in Virginia (1932), the Alley Theatre in Houston (1947), The Arena Stage in Washington, DC (1950) and the Actor's Workshop in San Francisco (1952). These theaters formed the nucleus of the new distribution system for theater in America.[13] The need for good managers escalated in the professional world, and because of the unprecedented baby boom after the war, the educational system—especially colleges and universities—expanded offerings in the arts. Community and campus performing arts centers helped establish a new network for touring and provided local groups with venues to use. Managers were needed to operate the new multimillion dollar complexes and to book events throughout the year.

Opera first spread beyond New York into the major metropolitan areas of Chicago, San Francisco, Philadelphia, St. Louis, and New Orleans. However, after the Great Depression, only New York and San Francisco were able to hold onto their companies.[14] The support in the 50s by the Ford Foundation helped bring opera to the American arts scene. By the early 70s, 27 opera companies were in operation.[15] Today, the Central Opera Service reports that there are 113 major opera companies in existence. Part of this growth was due to the NEA matching grant programs, which enabled many companies to professionalize their management.

Until the early 60s, dance companies were in limited supply in the United States. The American Ballet Theatre, the New York City Ballet, and the San Francisco Ballet topped the list of professional companies. Ballet West in Utah and Ruth Page's dancers, who were associated with the Chicago Lyric Opera, offered regular programs with their semiprofessional companies.[16] At the same time, modern dance companies were being operated on very tight budgets by such pioneers as Martha Graham, Alvin Ailey, Merce Cunningham, José Limón, and Paul Taylor. Their staff resources and their seasons were very limited.

The Ford Foundation in the 60s and the NEA in the 70s helped create a new national support system for ballet and later for modern dance. Although these groups still struggle, there are now more than five hundred dance companies that operate 20 weeks or more a year, according to Dance/USA.[17]

Symphony orchestras have also grown in number over the last 30 years. According to the American Symphony Orchestra League, there are 88 orchestras with yearly expenses of more than $1.5 million in the United States and Canada.[18] It is estimated that the number of United States art museums currently stands at 938. There are also at least 3350 history and 1200 science museums in the United States.[19]

The expansion period in the arts seems to be slowing now that most communities have established visual and performing arts institutions. The long-term struggle for operating funds has been accelerated in recent years as competition for support has increased. Additional funding from the state and federal government appears to be an unrealistic expectation. Demand is increasing for resources to assist with social programs, medical research, and education. Foundation, corporate, and individual support is being tapped by increasingly sophisticated fund raisers from hospitals to day-care centers. Meanwhile in Europe, the government subsidy is being reevaluated. Ironically, the model being adopted is the U.S. approach of private and public support for the arts. Performing and visual arts organizations are scrambling to develop the expertise to become successful fund raisers to maintain their current levels of operation. In England, for example, bitter battles have been fought with the conservative government over the level of support for the arts.

Profile of the Arts Manager

The growth in the arts over the last 30 years has created a tremendous demand for managers at all levels and in all disciplines. However, arts managers are not clearly identified as a work group when counting the 1.5 million people employed in the arts and entertainment industry. The Census Bureau counts performers, architects, composers, printmakers, and instructors in the arts but does not include people in arts management, sales, consulting, or promotion or public television employees.[20] It is not clear whether the people who do not directly make art were counted in the census data, but they are obviously a central part of the culture industry in the United States.

One source that provides substantial information on the arts manager is Paul DiMaggio's 1987 book, *Managers of the Arts*. Originally created for the NEA under the official title *Research Division Report #20*, the book outlines the background, training, salaries, and attitudes of arts managers in theater, orchestra, and museum management and community arts associations.

Unfortunately, DiMaggio does not include data about opera or dance managers. In addition, the survey was conducted in 1981, which may make the data somewhat irrelevant to today's market. DiMaggio's book only samples a limited number of people. With these limitations in mind, let's take a look at some of the highlights of this report.

DiMaggio's book revealed the following profile of arts managers: upper-middle class, highly educated individuals who either majored in the subject they were managing or were humanities majors in English, history, or foreign languages. DiMaggio found that a limited number of managers had management or arts management degrees. The upper management jobs tended to be held by men in museums (85%), theater companies (66%), and orchestras (66%), but women held the majority of positions in community arts associations (55%).[21] The data also indicated that there were a wide variety of ways to enter the career path in arts management, thus making it a fairly open system.

The section of DiMaggio's report on training offers some interesting insights into the opinions of those surveyed regarding their preparation for their jobs. Figure 2–1 shows the results of a survey that asked how well prepared participants felt to handle various aspects of the job, including fiscal and personnel management, planning, and board, labor, and government relations. The data indicates that "few managers felt they were well prepared to assume many of [the] functions" required for their jobs.[22] Labor relations consistently stands out as an area for which respondents felt poorly prepared. The survey results show that in many areas, less than 40 percent felt they had "good preparation" for budgeting and finance, planning and development, personnel management, and government relations.

DiMaggio also asked arts managers how they learned to do their jobs. An overwhelming number of the respondents indicated that they learned how to manage while on the job. These managers included 95 percent in theater and orchestra management, 90 percent in museum management, and 86 percent in community arts agency (CAA) management.[23] Typically, around 20 percent said they had learned through university arts administration courses.

In the 80s, the diversification of arts institutions increased the opportunities for women and minorities in the field of arts management. As a result, today's arts manager profile would probably be more representative of our society. However, no recent studies have been issued to support this supposition.

Jobs for Managers Today

When scanning a publication like *ArtSEARCH*, an employment service bulletin that is issued 23 times a year by the Theatre Communications Group (TCG),[24] it is possible to gain an overview of job titles. The job listings also reveal organizations' expectations about staff qualifications for an arts manager in today's work place. For example, a recent issue listed six openings for executive directors ranging from an Afri-

Figure 2–1

Self–Evalution of Preparedness at the Time of First Managership by Function
(in percent)

	Fiscal Management	Personnel Management	Board Relations	Planning and Development	Marketing and Public Relations	Labor Relations	Government Relations
THEATERS							
Good preparation	27.45	42.57	30.69	37.62	39.60	20.00	NA
Poor preparation	25.49	13.86	29.70	23.76	16.83	16.83	
(Respondents)	(102)	(101)	(101)	(101)	(101)	(95)	
ART MUSEUMS							
Good preparation	25.60	30.40	45.83	32.52	29.27	15.25	21.95
Poor preparation	40.80	24.00	14.17	23.58	30.89	55.00	43.09
(Respondents)	(125)	(125)	(120)	(123)	(123)	(118)	(123)
ORCHESTRAS							
Good preparation	26.42	36.89	43.14	33.33	47.06	22.00	NA
Poor preparation	23.58	15.53	23.53	19.61	20.59	49.00	
(Respondents)	(106)	(103)	(102)	(102)	(102)	(100)	
ARTS ASSOCIATIONS							
Good preparation	29.46	39.84	42.64	52.71	53.13	11.02	37.01
Poor preparation	20.16	13.28	17.83	14.73	11.72	50.85	25.20
(Respondents)	(129)	(128)	(129)	(129)	(128)	(118)	(127)

NOTE: NA = not asked/not applicable

Source: Paul DiMaggio, Managers of the Arts, 1987, Research Division Report #20, National Endowment for for the Arts (Washington, DC, Seven Locks Press).

can American Dance Ensemble to the Toledo Repertoire Theatre. The dance company was seeking someone who would be "responsible for long-range planning, marketing/public relations, and fund raising."[25] In addition, the person would supervise a company manager and an administrative assistant and would work with an artistic director, a music director, and a technical staff. The theater company emphasized that it was seeking someone with "the ability to relate to the business and artistic communities."[26]

Other upper-level job titles advertised included managing director, administrative director, development director (fund raising), and general manager. Some of the titles for middle- and lower-level management positions included director of marketing, director of special products, assistant to the artistic director, annual fund manager, director of group sales, box office manager, and box office staff. The range of responsibilities listed for many positions was quite long, and the emphasis was on communications skills for middle- and upper-level jobs and computer skills for middle- and lower-level positions.

In the News—The Arts Manager in Training

This excerpt from the *New York Times* illustrates a contemporary version of the artist who becomes manager.

A Utility Player Runs the Show Off Broadway
Ellen Pall

A radiance surrounds Don Scardino when he talks about his new job: the sweet, heady radiance of early love. [In] January [1992], Mr. Scardino will become artistic director of Playwrights Horizons, one of Off Broadway's most prestigious developmental theaters.

A theater dedicated to nurturing contemporary American playwrights in the 20 years since its founding, Playwrights has evolved from a scruffy near-insolvency to mature, prosperous eminence. The driving force behind much of this transformation has been Andre Bishop, who became the theater's literary manager in 1975 and for the last decade has been its artistic director.

In both experience and achievement, the 43-year-old Mr. Scardino is a very different man from his low-profile, Harvard-educated predecessor. The youngest child of two jazz musicians, he grew up in Astoria, Queens, and describes his education as "school of hard knocks all the way down the line." While Mr. Bishop has devoted almost the whole of his professional life to Playwrights, there is scarcely an area of the performing arts Mr. Scardino hasn't ventured into: he has acted, sung, danced, directed, composed and recorded music and produced for television—practically everything except what he has just been hired to do.

He was not alone in being surprised by his new circumstances.

"When I heard the name, it wasn't one I expected," said Rocco Landesman, president of Jujamcyn Theaters, which owns five Broadway houses. "This is not 'round up the usual suspects.' It's a bold choice. They're obviously taking a risk, since he's had no experience as an artistic director. He hasn't been at an established institution for 10 years, he hasn't produced five plays a season, there's no body of work at a resident theater. He's had no history dealing with a board, with fundraising, with planning a season, with union negotiations."

"But nothing ventured, nothing gained," he added. "The safe choice is always the less interesting one."

Though he admits to "a bit of nervousness" about meeting the standards set by Mr. Bishop, Mr. Scardino has no intention of merely following in his footsteps. On the contrary, he expects to "kick Playwrights into the 90's." For example, while Playwrights has come to be associated with satiric comedies under Mr. Bishop, Mr. Scardino will be looking for "darker" plays, more serious "quote unquote dramas."

"I enjoy the 'Playwrights Horizons' play. My own tastes might run to something that is a little more political," he said. "I like a theater that challenges your assumptions, that wakes you up a little bit. Works that truly hold the mirror up to life and shake you up socially."

He also believes that, as an institution, "a good theater should have a large dose of social responsibility, social conscience." He hopes to establish several outreach programs, sending a company made up of students from Playwrights Theater School into hospitals, schools and neighborhoods to perform works for the Young Playwrights Festival. The public needs to be "re-educated," he says, to come to the theater expecting to be challenged.

SOURCE: Ellen Pall, "A Utility Player Runs the Show Off Broadway," *New York Times*, August 25, 1991.

The overall work environment in the arts has improved in the last 25 years as organizations have grown and matured. Full-time positions with full benefits are now the norm. Salaries may range from as little as $7000 a year for an administrative assistant to $100,000 or more for an artistic director. Typically, the salary ranges are in the low teens to the mid-50s. There are also numerous internship programs designed to provide administrative help for an organization at practically no cost.

We will examine in more detail issues relating to staffing in Chapter 7.

The National Endowment for the Arts

The role of the arts manager in the United States was further defined by the passage of legislation establishing the National Endowment for the Arts and Humanities on September 16, 1965.[27] The struggle to create a modest system for promoting growth and excellence in the arts took several years, numerous congressional hearings, and incredible dedication by a few people. Since its establishment, the NEA has helped shape the arts scene in the United States by organizing an identifiable arts constituency, stimulating donations through matching grants, and providing guidance to arts groups on ways to manage their limited resources effectively. Although the NEA budget was only $176 million in 1992, or roughly 70¢ per person in the United States,[28] the endowment regularly generates millions more through various matching grants.

The NEA's mission statement is worth noting because it helps shape the numerous grant categories created to support the arts. The following is from the NEA's own guidebook:

> The mission of the National Endowment for the Arts is:
>
> - to foster the excellence, diversity and vitality of the arts in the United States; and
> - to help broaden the availability and appreciation of such excellence, diversity, and vitality.
>
> In implementing its mission, the Endowment must exercise care to preserve and improve the environment in which the arts have flourished. It must not, under any circumstances, impose a single aesthetic standard or attempt to direct artistic content.[29]

The creation of the NEA led to the development of a support system for performers, performing arts organizations, museums, and film, design, and humanities projects over the last 25 years. The NEA has grant programs in more than 88 areas. Education, opera, theater, museums, music, and other disciplines have several subcategories each. In addition, the NEA supports grants for international projects and arts administration internships.

The typical application process moves through a system of staff screening, review by a committee of peers in the discipline, review of the peer group recommendation by the National Council on the Arts, and final decision by the chair of the endowment. Applications can take from six months to a year to work their way through the system. The chance of receiving funding is dependent to a large degree on how well the proposed project matches the criteria the NEA has set for the funding area. Chapter 13 provides more information about the overall process.

Government Support

The pros and cons of government support for the arts have not changed significantly since the inception of the NEA. The supporters of the legislation that led to the creation of the NEA saw it as an opportunity to make the arts more available to people throughout the United States and to enrich the nation's cultural life. Programs were designed to promote a type of cultural democracy through very modest grants to a wide range of projects and institutions. It was deemed important to support the creative spirit and at the same time promote new work. The preservation of a cultural heritage was a high priority, and the support of work that might not otherwise exist in a market-driven economic system was thought to benefit everyone.

The critics of the legislation believed that the establishment of a government subsidy system would eventually result in a general mediocrity creeping into the arts. There was fear that centralizing the power of the subsidy in the hands of a few would lead to less, not more, creative work in the country. Others believed that it was wrong to give the taxpayers' money to projects and programs with no appeal beyond a limited number of people. Some people argued that a type of cultural dictatorship would result from the peer review system. Others argued that if the government started subsidizing the arts, private and corporate philanthropy would dry up.

Budget Battles and Censorship

In the end, the astute shepherding of the legislation through the House and Senate by Livingston Biddle and others helped neutralize critics in the early days of the endowment. The NEA flourished and survived the annual congressional budget hearing process until 1981. Under the budget planning guidance of David Stockman, the new Reagan administration, proposed a 50 percent cut in the NEA budget for 1982 and additional cuts in 1983–1986.[30] The new administration saw in the NEA an example of the government creating a disincentive for private support for the arts. When confronted with the increase in private giving that had been generated by the endowment, the Reagan administration backed away from massive budget cutting, and reductions of 6 percent were adopted by Congress.

The budget battles were minor in comparison with the fire storm that erupted with the reauthorization legislation in 1989 and 1990. The reauthorization of the NEA became the focal point for a political struggle over censorship and the whole concept of funding for the arts. In the fall of 1990, arts lobby groups pleaded with arts groups across the country to support the NEA's reauthorization. Telegrams and letters were sent to Washington to show the members of Congress that there were constituents who supported the arts.

The compromise legislation eventually enacted required grant recipients to return the grant if the work they produced was found to be obscene by the courts. This compromise did not sit well with the artistic community. Controversy continued to follow the NEA as artists and organizations sued over the obscenity pledge. Several organizations, among them the Public Theater in New York City, turned down substantial grants rather than agree to the terms that the NEA established.

Censorship charges continued to be leveled at the NEA when grant recommendations by the National Council on the Arts were overturned by the acting director of the endowment in the spring of 1992. The resignation of peer review panels and key staff have disrupted the operations of the endowment.

The NEA and the Arts Manager

The granting process implemented by the NEA in the late 1960s helped to stimulate the growth of many careers in arts management and to hasten the professionalization of the field. The specialized skills required to seek out grants were in great demand. Because all organizational grants were at least a one-to-one match, meaning that for every federal dollar of a grant a matching dollar of other money must be found, fund raising staffs and development experts began to be hired. Typically, grants to large organizations required three dollars of private money for every dollar of federal money over a three-year period. This further necessitated establishing a staff support system to run the initial campaign and to continue bringing the money in after the grant expired.

The development of the now-common board of directors and

For Further Consideration

Livingston Biddle's comprehensive personal history of the NEA, entitled *Our Government and the Arts: A Perspective for the Inside*, is filled with hundreds of anecdotes about the struggle to establish and maintain what may be one of the most cost-effective organizations in government. In addition to telling interesting stories, the author takes the reader inside the legislative system as well as the management structure of the endowment. The book was published in 1988 by the American Council for the Arts, 1285 Avenue of the Americas, Third Floor, New York, NY 10019; (212) 245-4510 or 1-800-321-4510.

management staff structure was a product of the new accountability that arts organizations faced. Organizations had to prove that they could responsibly manage the funds they were given. Annual reports, financial statements, and five-year plans became standard operating procedures for organizations that wanted to be considered by the federal, corporate, and foundation funders. The net result was an increase in staff openings, which provided jobs for the baby boomers graduating from the colleges and universities across America in the 70s and 80s.

Arts Councils

The NEA legislation provided funds for the creation of state agencies to distribute 20 percent of the endowment's overall budget. The state and local arts agencies created another network of funding opportunities for artists and arts groups and created staff positions for arts managers. Today, there are more than 50 state arts councils and 670 local arts agencies in operation in the United States. The budgets range from a few thousand dollars to over $20 million.[31]

The Education Revolution

Arts managers became an intregral part of the unprecedented growth in the higher education industry. The baby boom and the resulting expansion programs undertaken by many schools to accommodate the incoming students led to the creation of numerous fine arts programs integrated into the overall mission of the institution.

There are more than 1800 administrative heads of university or college theater programs in the United States today.[32] Langley and Abruzzo list 961 music programs, 409 opera programs or workshops, 1622 visual arts programs, and 266 dance programs graduating future artists each year.[33]

In many larger university graduate programs, the performing arts evolved into preprofessional training programs in the late 60s and 70s. The number of management jobs grew as fine arts centers were built to house the visual and performing arts programs across the country. In many cases, at least one large facility was constructed to serve the dual purposes of supporting the academic programs and providing a venue for community events. Subscriptions to various arts series were offered to the community and to students and provided the performance experience that arts students needed. Even small college and university programs adopted more professional approaches to producing shows due, in part, to the influx of faculty trained in the graduate programs of the large universities. The late 70s and early 80s were a period of tremendous growth in the smaller performing arts programs.

Educators within the arts community began to recognize the need to develop trained managers as well as performers, designers, and technicians. Several schools, such as Yale, Northwestern, University of Wisconsin, and University of California at Los Angeles, created training programs that became models for developing art managers. More than 40 U.S. schools currently offer master's degrees in arts or nonprofit management.

By the early 70s, the network of facilities managers had developed into an organization now known as the Association of Performing Arts Presenters (APAP). The APAP sponsors an annual conference that is an important booking opportunity for theater groups, dance companies, music ensembles, and soloists. Student associations also hold annual meetings to book acts on campuses around the country.

The Slowdown

The educational boom began to slow at the end of the 1980s as the number of college-age students dropped. The consolidation of arts programs and the outright elimination of arts departments began to be discussed. By the early 90s, the talk had turned to action. Budget cutting, little or no facility maintenance, and staff reductions began to sweep across the U.S. campuses. Many programs experienced difficulty filling their graduate classes even when they offered full scholarship support.

Conclusion

The evolution of the role of the arts manager continues as thousands of arts organizations undergo the arduous process of adapting to the changing cultural environment. As we will see in Chapter 4, arts groups must constantly assess the opportunities and threats that present themselves in the world around them. In theory at least, an arts manager should be trained to serve the needs of his or her particular discipline by effectively solving the problems of today and anticipating the significant changes of tomorrow. Unfortunately, the day-to-day struggle for financial survival that goes on in most organizations leaves little time for planning for the future.

Whatever changes take place in the next 20 years, arts managers working with artists, boards, and staffs will play a central role in the future of the arts in the United States. Dynamic vision and articulate leadership will be required if the arts are to build on the growth of the last 50 years.

Summary

Over the last two thousand years, the basic functions of the artist-manager have remained the same. Bringing the art and the public together is the continuing objective.

In ancient times, simple religious ceremonies evolved into full-scale state-sponsored arts events that lasted from several days to a few weeks. The functions of management (planning, organizing, leading, and controlling) were distributed between artist-managers and the public officials who acted as arts managers. The rise of the Church and the decline of Rome created a shift away from state-sponsored events.

The late Middle Ages produced an economic growth that allowed the expansion of population centers. The rise of guilds and community-sponsored celebrations helped fuel changes in the overall arts climate. Complex pageants often needed people with management expertise to organize the large casts and the various sets associated with the productions.

Continued changes in society and the birth of more democratic forms of government eventually led to changes that became the foundation of many modern organizations. The Renaissance fostered the rebirth of drama and contributed to the development of the first operas and ballets. Problems with financing, patronage, and censorship also accompanied the growth in the arts. The additional art forms created additional jobs for arts managers.

In the seventeenth and eighteenth centuries, some countries began to establish national dance, opera, music, and theater companies. Permanent staff members and performers received salaries and pension benefits. By the nineteenth century in Europe and the United States, the arts had expanded into smaller population centers. However, there were no state-recognized arts institutions in the United States comparable to those of Europe. As communities became cities, orchestras, opera companies, and museums became permanent institutions. Most were supported by a small group of philanthropists. The role of the arts manager in the United States expanded with the continued development of touring which was made possible by an extensive rail system. Monopolistic enterprises took control of many of the theaters at the end of the nineteenth century. The invention of movies and radio contributed to a decline in attendance at arts events by the 1920s.

The last 90 years have been shaped by major wars, slowly improving economic conditions, the new technologies of television and videotape, and a population boom. As a profession and a recognized field of work, arts management is a product of changes in United States national policy since the 50s. Ford Foundation funding and, beginning in the late 60s, the National Endowment for the Arts helped make private and public support for the arts a priority. The expanding arts market resulting from the population increase and the education boom has also contributed to the creation of thousands of new jobs in the arts.

The typical arts manager is a highly educated, upper-middle-class person with a background in the humanities. A limited number have done course work in management while in school. Survey results show that many arts managers had to learn the functions of their positions on the job.

The NEA was created in 1965 to promote excellence, broaden the availability of the arts, and preserve work identified as part of the United States national heritage. It offers more than 88 different grant programs in all disciplines in the performing, visual, and media arts. The NEA has assisted many groups in organizing and professionalizing their staffs. In addition to promoting the growth of the arts at a national level, the NEA was also instrumental in establishing 55 state arts agencies and 670 local arts agencies.

The field of arts management has also grown as a result of the boom in arts programs at colleges and universities across the United States. Degree programs, new facilities, and often extensive staffs have created a national network of arts centers, but recent shifts in demographics have depressed the education market and created shortages of students in some areas.

Key Terms and Concepts

artist-manager
Archon eponymous
choregoi
domini
the Syndicate
the NEA, its mission

Questions

1. Summarize the major arts management activity associated with the following time periods:

a. Ancient Greece
b. Ancient Rome
c. Middle Ages
d. Renaissance
e. Seventeenth through nineteenth centuries
f. Twentieth century

2. What changes have taken place in the job market that might alter DiMaggio's profile of the arts manager?

3. The news excerpt about Playwrights Horizons describes the newly appointed artistic director. Do you think Playwrights Horizons will benefit or suffer from having an inexperienced artistic director? Explain.

4. What can the NEA do to head off continuing congressional battles over the obscenity issue? Should limits be placed on the types of projects that the NEA funds?

5. How much control should management have over the artistic product of an organization? For example, how much input should management have when it comes time to select the season's titles? Can you think of a situation in which too much or too little control was exercised by the management of an arts organization? What were the results?

Case Study

The following article presents a model for change in the world of museum management. As you will see, not everyone agrees that this change is good. A related article about museum leadership can be found in Chapter 7, "Whitney Museum Cans Director."

Wanted: Art Scholar, M.B.A. Required

Alexander Stille

When Jack Lane took over as director of the San Francisco Museum of Modern Art three years ago, one of the first things he did was shut his office door. "The previous director's door was open to everyone who wanted to talk to him," Mr. Lane explains. "That just seemed unworkable. I wanted to clarify the chain of command, so we divided the museum into four major areas, and the heads of those areas report directly to me."

It was that kind of tough-minded decision the museum's trustees had hired Mr. Lane, who holds a master's degree in business from the University of Chicago as well as a Ph.D. from Harvard, to make. They wanted a director with a degree in business as well as art to lead the museum through a period of major expansion.

Their choice is representative of a growing perception that leadership at museums requires business acumen as well as connoisseurship. Although business management techniques have been used in museums for at least two decades, in the last few years those skills have been more heavily emphasized—to the dismay of some in the art world.

While many cultural institutions have long had business departments to take care of marketing, public relations and accounting functions, the emphasis at the top had been on scholarship. But now, at some major museums, even directors are being chosen for their business training as well as for their fine arts backgrounds.

Perhaps the most conspicuous and controversial appointment has been that of Thomas Krens at the Guggenheim Museum in New York.

The 43-year-old director, who has a master's degree from the Yale School of Management, has come under criticism for his plans to create a multinational museum with branches in this country and abroad, and for selling paintings from the museum's permanent collection to buy newer works.

Museums run by art historians are also making increasing use of modern management techniques. "You have to know how to negotiate with unions, hire contractors, run a restaurant and a bookstore," says Patterson Sims, associate director of the Seattle Art Museum, who does not hold an M.B.A. "Those are not issues that connoisseurship prepares you for."

To help meet the demand, several of the country's best business schools—Harvard, Yale, Stanford, and the University of California at Los Angeles—offer courses in arts management. And the American Federation of Arts, a New York–based service organization that offers both exhibitions and professional training to 600 member museums, runs a summer program to train museum curators in business administration.

Perhaps not surprisingly, this mingling of art and business tends to polarize the museum world. "Administrative imperatives are taking precedence over artistic judgments," says Hilton Kramer, art critic and editor of *The New Criterion*, a magazine of cultural criticism. Others disagree. "A well-run operation should free the curator," counters Graham W.J. Beal, director of the Joslyn Art Museum in Omaha, who made the jump from curator to administrator after participating in the American Federation of Arts' training program.

Whether they see it as a blessing or a curse, there is widespread agreement that a business orientation in a museum is here to stay. "It's not a question of whether it will happen," says Roger M. Berkowitz, deputy director of the Toledo Museum of Art. "It is happening."

The Changing Profile of the Curator

According to museum professionals, the trend is linked to the increased size and complexity of many arts institutions.

"Museums are demonstrably more complex," says J. Carter Brown, director of the National Gallery in Washington and perhaps the first museum head to combine a joint M.B.A.-art history track as a calculated career strategy. He decided to take a business degree from Harvard while pursuing a master's degree in art history at the Institute of Fine Arts in New York. "When I became director in 1970," says Mr. Brown, "we had 350 employees. Now we have 1,000. Back then the museum was all in one building; we have doubled the space. Our budget was $5.9 million in 1970; today it's $53.9 million.

"My predecessor's fund-raising activities were simple," he continued. "He had lunch with Ailsa Mellon, and if he needed a Leonardo she would write a check." During Mr. Brown's tenure, the National Gallery has raised $40 million from 92 corporations.

As a result of such changes, the profile of the typical museum director has changed since the mid-70s. Directors of the old school tended to be socially prominent connoisseurs, as knowledgeable in wooing wealthy collectors as in evaluating art. "It was a hobby," says Thomas Hoving, director of the Metropolitan Museum from 1966 to 1977 and now editor in chief of *Connoisseur* magazine. "When I first got out of graduate school, people told me that the biggest prerequisite for becoming a curator was an outside income."

Although Mr. Hoving has no degree in business (nor does his successor, Philippe de Montebello), many credit—or blame—him for starting the focus on business in museums. "It was during Hoving's tenure at the Met that museums became broad, popular institutions, with blockbuster shows, expanded education programs, gift shops, bookstores and restaurants maximizing income," says Mr. Lane, who has raised more than $65 million for the San Francisco museum and will double its exhibition space with a new building.

Last year the Minneapolis Institute of Art hired Daniel E. O'Leary as deputy director, in part because of his experience turning around a financially troubled museum. After earning a joint Ph.D. and M.B.A. at the University of Michigan, Mr. O'Leary became director of Artrain, a Michigan-based regional museum that presents traveling exhibitions in a railroad car. "I applied classical business school analysis," he says. "We booked exhibitions of greater appeal, extended the touring schedule to create a greater economy of scale. We managed to more than double the revenues and cut expenses by a third, and audiences went from 40,000 annually to 130,000."

One of the first things he did at Minneapolis was to conduct a market survey. "We live in a competitive world. We have to justify ourselves to government, to foundations, to corporations."

Some critics of the trend toward a business mentality fear that museums' preoccupation with attendance is inimical to their artistic purpose. "Museums are market-driven," says Hilton Kramer. "Museum directors have become salesmen."

Others insist that ignoring the bottom line can jeopardize a museum's artistic mission.

"A number of museums have become more or less insolvent," says Myrna Smoot, director of the American Federation of Arts. "They reduce their hours, they don't change their collections, they reduce maintenance and security, stop paying to conserve the collection, in order to keep the doors open."

Does 'Thank You' Mean Forever?

The problem of continual growth is one that preoccupies Mr. Krens at the Guggenheim, the most outspoken advocate of the so-called M.B.A. revolution, who has sought to challenge the assumptions on which museums have traditionally operated. "It's increasingly expensive to store and preserve works of art," he says. "When you accept an object into your collection, do you take on an obligation to preserve the work forever?" Older museums, he says, are at a critical juncture; they cannot continue indefinitely as both custodians of the past and purveyors of the new. "Do you go on collecting forever? And if not, where do you stop?"

Mr. Krens clearly answered his own question with his decision to sell a Kandinski, a Modigliani and a Klee last year to pay for a major collection of Minimalist art from the Italian Count Giuseppe Panza.

Most M.B.A. curators warn against taking this trend too far. "If M.B.A.'s are running museums without an art history background, we're in for a grim future," says Mr. Nash.

For the moment, anyway, the low wage scale of the not-for-profit world ensures to some degree that the M.B.A.'s entering the museum world will have to be passionate about more than money.

SOURCE: Alexander Stille, "Wanted: Art Scholar, M.B.A. Required," *New York Times*, January 19, 1991.

Questions

1. List at least two reasons why a museum curator needs business management skills.

2. Do you agree with Hilton Kramer's judgment that museums have become "market-driven" and museum directors have become "salesmen?" Explain.

3. What alternatives can museums pursue to increase membership other than high-visibility exhibitions?

4. Do you agree with the notion of selling off museum holdings to raise money for other works? What should museums do about preserving their holdings?

References

1. Oscar G. Brockett, *History of the Theatre,* 3d ed. (Boston: Allyn and Bacon, 1977), 15–47.
2. Ibid., 51–73.
3. Ibid., 105.
4. Ibid., 136.
5. Richard Kraus and Sarah Alberti Chapman, *History of the Dance in Art and Education,* 2d ed. (Englewood Cliffs, NJ: Prentice-Hall, 1981), 62.
6. Ibid., 67.
7. John Pick, *Managing the Arts? The British Experience* (London: Rhinegold, 1986), 23.
8. Ibid., 45.
9. William J. Baumol and William G. Bowen, *Performing Arts: The Economic Dilemma* (Cambridge, MA: MIT Press, 1966), 20.
10. Ibid., 29.
11. Kraus and Chapman, *History of the Dance,* 93–95.
12. Baumol and Bowen, *Performing Arts* 29.
13. Ibid., 27–28.
14. Martin Mayer, "The Opera," in *The Performing Arts and American Society,* ed. W. McNeil Lowry (Englewood Cliffs, NJ: Prentice-Hall, Spectrum Books, 1977), 45.
15. W. McNeil Lowry, ed. *The Performing Arts and American Society* (Englewood Cliffs, NJ: Prentice-Hall, Spectrum Books, 1977) 14.
16. Ibid., 11.
17. Dance/USA provided this information based on its detailed surveys of the field. The five hundred dance companies had to have legal nonprofit status, pay their dancers, sell tickets, and maintain dance activity for at least 20 weeks a year.
18. American Symphony Orchestra League, Washington, DC.
19. Stephen Langley and James Abruzzo, *Jobs in Arts and Media Management* (New York: American Council for the Arts, 1990),10-11.
20. John Naisbitt and Patricia Aburdene, *Megatrends 2000* (New York: Avon, 1990), 67.
21. Paul DiMaggio, *Managers of the Arts,* Research Division Report #20, NEA (Washington, DC.: Seven Locks Press, 1987), 12.
22. Ibid., 42.
23. Ibid., 46.
24. *ArtSEARCH* is published by the Theatre Communications Group, Inc., 355 Lexington Ave., New York, NY 10017.
25. *ArtSEARCH* II (August 15, 1991):1.

26. Ibid.
27. Livingston Biddle, *Our Government and the Arts* (New York: American Council for the Arts, 1988), 180.
28. The per person cost was calculated by dividing the current budget for the NEA (fiscal year 1992) by the total population according to the 1990 census.
29. National Endowment for the Arts, *Guide to the National Endowment for the Arts* (Information book published by government printing office, Washington, DC:1990), 2.
30. Biddle, *Our Government and the Arts*, 492.
31. Based on information provided over the phone by the National Assembly of State Arts Agencies (NASAA) and the National Assembly of Local Arts Agencies (NALAA).
32. Association of Theatre in Higher Education (ATHE).
33. Langley and Abruzzo, *Jobs in Arts and Media Management*, 10–11.

Additional Resources

The *Journal of Arts Management and Law*, volume 13, number 1, Spring 1983, is devoted to the arts and public policy. The topics covered in this issue give the reader a good sense of today's important topics in arts management.

3 Evolution of Management

In this chapter, we will scan the evolution of management thought. We will review early management practices and then examine the management concepts that grew out of the shift to mass production during the Industrial Revolution. Finally, we will look at the impact of scientific management and the application of psychological theories to the work place.

Little mention will be made of arts organizations in this chapter because the primary objective is to provide the reader with a general historical background on management. Many of the terms and concepts noted in Chapters 1 and 2 have developed from classic and contemporary management theory and practice. Before moving into the specific areas of the external environments, planning, organizational design, and human resource management, it seems appropriate to explore the source of the current management systems used to operate all organizations.

Management as an Art and a Science

A basic assumption of this text is that management is an art. In this case, art is used to mean an ability or special skill that someone develops and applies. Studying the theories of management, synthesizing the application of these theories to a practical work environment, and then creating a workable system for a specific organization requires a tremendous amount of thought and effort. It is often a lifetime job.

Management can also be considered to be a science. As we will see, the concept of *scientific management* is not universally welcomed in the work place. The term describes a particular approach to maximizing productivity by applying research and quantitative analysis to the work process. The creation of general and specific management theories to explain and predict how organizations and people behave is also integral to thinking of management as a science.

At the center of any theory is the ability to predict an outcome if given a specific set of circumstances. A scientist develops a theory, conducts experiments, establishes an outcome that can be repeated by others, and provides proof of the theory. Management theory tries to achieve the same goal: predictable outcomes given specific inputs. Unfortunately, the science of management, as with any social science, is sometimes subject to unanticipated outcomes. In management science, the unpredictability of human beings can quickly render a theory inoperative.

On-the-Job Management Theory

When studying management theory and practice, which are often examined by using case studies, it becomes apparent that managers often enter into the practice of managing with virtually no theoretical background. For example, Katherine Graham, who once owned the *Washington Post,* had no formal training in business management. The sudden death of her husband thrust her into the role of chief executive officer. Nonetheless, she was able to successfully operate a major newspaper using her personal ability and adaptability. She was able to learn on the job and to further develop her own operating theories and practices to maintain a successful business.

In Paul DiMaggio's study for the National Endowment for the Arts, more than 85 percent of the arts managers in theaters, art museums, orchestras, and arts associations said that they learned from on-the-job training.[1] The university-trained arts managers surveyed claimed that their schooling did not adequately prepare them for many of the demands of running an organization. Therefore, it seems safe to say that the experience of the work place is required to complete the education of any arts manager.

Regardless of how an individual learns the techniques of management, an effective manager must eventually be able to analyze variables and predict outcomes based on experience. In other words, the manager must find a set of operating principles that can be used. For example, an arts manager might have to say to the board, "If we raise prices, orders will drop. If we change our subscription plans, fewer people will order because any change creates confusion. If we perform a concert of avant-garde music, subscribers will stay home." Obviously, actual outcomes depend on the particular community and circumstances, but arts managers should be able to articulate their expectations based on their understanding of the variables.

The rest of this chapter focuses on some of the major theories and principles that shape management today.

Evolution of Management Thought

Preindustrialization

For the last several thousand years, organized social systems have managed the resources needed to feed, house, and protect people. The evolution of management is intertwined with the development of the social, religious, and economic systems needed to support cities, states, and countries. The church and state provided the first systems for planning, organizing, leading, and controlling. These management systems were predicated on philosophies that placed people within complex hierarchies.

History provides many examples of management systems established by the Egyptians, Romans, and Chinese. Many basic principles of supervision and control evolved from the projects undertaken by these societies. Building temples, pyramids, and other massive structures required extensive management and organizational skill. Modern management concepts expanded on the skill needed to implement public works projects as the world shifted from an agrarian to an industrial base.

A Change in Philosophies

The decline of the Catholic super state in the fourteenth and fifteenth centuries and the subsequent religious struggles created by the rise of Protestantism slowly changed the fundamental relationship of people to their government and religious systems. The seeds of the Protestant work ethic were planted in the new order. The expansion of trade and the creation of a permanent middle-class grew out of the changes brought about by the national and international economic systems. The effects of the Renaissance and the Reformation extended far beyond rediscovering the thoughts and philosophies of antiquity. The development of new political and social theories of government and management by such theorists as Niccolo Machiavelli, Thomas Hobbes, John Locke, and Adam Smith led to crucial changes in thinking about the individual and the society at large. For example, Adam Smith's *Wealth of Nations*, published in 1776, moved economic theory beyond the mercantile system with Smith's now-famous economic principles. The "invisible hand" of the marketplace is the core concept of the system of economic self-regulation that survives today.

The Industrial Revolution

Four principal changes in the management of the work place are often attributed to the Industrial Revolution:

1. mechanization of work
2. centralization of production
3. creation of the labor class
4. creation of the job of manager

The elements of science and technology, changes in government policies, population growth, improved health conditions, and the more productive use of farmland were all part of the changes that occurred during the seventeenth, eighteenth, and nineteenth centuries. The early entrepreneurs who established manufacturing businesses using the new technologies of the time (for example, the steam engine) needed others to supervise the laborers hired to operate the equipment. Essentially, the industrial manager was created to watch over the laborers. The problem of treating people as nothing more than extensions of machines and the subsequent abuses of labor—long hours, low pay, no job security, health and safety hazards, child labor, and so on—has left a legacy we still grapple with today. For example, the concept of "the carrot and the stick," which was used as a motivational management method in the factories, survives in the minds of many managers today. The positive inducement (the carrot) to earn more by working harder and faster was set off against a punishment (the stick), which included such things as a cut in wages or a more dangerous task, as a method of motivating people. However, not all owners and managers approached labor and production with the same attitude.

One of the early pioneers of a more enlightened approach to management was Robert Owen (1771–1858). At age 18, Owen operated and supervised a cotton mill, where he observed problems occurring in the manufacturing process. He tried to improve overall working conditions and changed the equipment to reduce the hazards to workers. However, due to a shortage of labor, he hired children to work 13 hours a day.[2]

Charles Babbage (1792–1871), the inventor of the world's first computer in 1822, is also credited with creating the first research techniques to study work.[3] His early work was the forerunner of what is now called *scientific management*. In *The Evolution of Management Thought*, Daniel Wren notes that Babbage attempted to establish salary systems that showed the mutual interest labor and management shared in the process of production. "Babbage's profit-sharing scheme had two facets: one, that a portion of wages would depend on factory profits; and two, that the worker 'should derive more advantage from applying any improvement he might discover,' that is, a bonus for suggestions."[4]

Changes in America

The early stages of the Industrial Revolution in the United States depended on borrowing management and organizational techniques from England and Scotland. However, by the mid-nineteenth century, U.S. manufacturers began to show the world how to mass-produce interchangeable parts for a variety of equipment.[5] Late-nineteenth century America's management system development was, in large part, due to the engineer. The mechanical, industrial, and civil engineers were the primary force behind the development of "systems" for doing work.

The railroads and the new technology of the telegraph created a climate for rapid business expansion in America. Daniel Craig McCallum (1815–1878), a manager for the Erie railroad, is credited with such things as creating a formal organization chart (it was shaped like a tree), matching authority with responsibility, and using the telegraph system to provide feedback about the location of trains.[6]

Henry Varnum Poor (1812–1905), the editor of the *American Railroad Journal*, wrote extensively about management organization and systems. Wren describes Poor's three-part philosophy as follows:

> [First,] organization was basic to all management; there must be a clear division of labor from the president down to the common laborer, each with specific duties and responsibilities. Second, communication was devising a method of reporting throughout the organization to give top management a continuous and accurate accounting of operations. Finally, information was "recorded communication"; Poor saw the need for a set of operating reports to be compiled for costs, revenues, and rate making.[7]

As noted in Chapter 2, the railroads played an important part in changing how entertainment was distributed in the United States. As we saw, the railroad brought to the arts the need for a specialist to manage the logistics of moving the company from city to city. The complexity of railroad schedules (time zones were not in place until the late 1880s) also demanded a large portion of a manager's time.

Although management concepts may have been growing in sophistication and depth during this period, the treatment of employees lagged behind. The safety and well-being of workers were not high priorities. Child labor, extremely low wages, and the lack of job security helped to create powerful labor unions later in the nineteenth and early twentieth centuries.

Management Trends to the Present

One of the founders of modern management is Frederick W. Taylor (1856–1915). Taylor is credited as the founder of scientific management. His efforts to change the work place often faced bitter opposition. In 1912, Taylor stated his principles before a special congressional committee created to investigate the effects of scientific management on the worker. His words speak clearly of a management theory that is far different from the highly efficient assembly line many people imagine as the realization of his principles. Taylor's ultimate goal was to use his methods to achieve a "great mental revolution."[8] His testimony makes a convincing case:

> Scientific Management is not any efficiency device, not a device of any kind for securing efficiency; nor is any bunch or group of efficiency devices. It is not a new system of figuring costs; . . . it is not holding a stop watch on a man and writing things down about him; it is not time study; it is not motion study nor an analysis of the movement of men. . . .
>
> Scientific management involves a complete mental revolution on the part of the working man engaged in any particular establishment or industry. And it involves the equally complete mental revolution on the part of those on the management's side—a complete mental revolution on their part as to the duties toward their fellow workers in the management, toward their workmen, and toward all of their daily problems.
>
> Frequently, when the management have found the selling price going down they have turned toward a cut in wages . . . as a way of . . . preserving their profits intact. Thus it is over the division of the surplus [or profits] that most of the troubles have arisen; in the extreme cases this has been the cause of serious disagreements and strikes.[9]

The drive toward making the work place and the work process as efficient as possible by careful analysis of all phases of manufacturing continues to the present. Time and motion studies, for example, which were first done in the early twentieth century, are now regular fixtures in examining how an organization is accomplishing its tasks, from building cars to making hamburgers.

Some of the other pioneers of the scientific management field were Carl G. Barth, Henry L. Gantt, Frank and Lillian Gilbreth, Harington Emerson, and Morris L. Cooke.[10]

Scientific Management Today

The rise of research universities and graduate schools of business and management and the increased application of scientific management to the work place have come together in the last 50 years to form a strong theoretical base for the study of management. Wharton was the first undergraduate school to offer a degree in business (1881), and Dartmouth (1900) and Harvard (1909) were the first universities to offer graduate programs in management.

Scientific management techniques have undergone further refinement with the assistance of computer models to help design the most efficient and productive work place. The worldwide application of these techniques is well documented. Terms such as *operations research* (OR), the application of quantitative analysis to all parts of a business operation, are now common.[12] The critical path method (CPM), for scheduling and controlling work on projects is part of standard operating procedures in many businesses. Scientific management techniques have been applied by the Japanese in much of their manufacturing, and the resulting gains in productivity have propelled them ahead of the United States in many areas. Such concepts as "just-in-time inventory" (or *Kanban*), computer-aided design (CAD), computer-assisted manufacturing (CAM), and computer-integrated manufacturing (CIM) are natural extensions of the work started by Taylor one hundred years ago.

Arts groups may see limited application of sophisticated scientific computer models in their day-to-day operations, but the fact is that whatever limited gains in productivity are to be achieved will result from integrating specific quantitative techniques in the organization. For example, inventory and accounting (scenery, costumes, props) can easily be computerized and linked to networked office computer systems. The process of assembling sets can be streamlined if time is spent analyzing how the work is being done. Almost any routine procedure is worth examining. There is often a faster way to do almost any work, whether it is counting ticket stubs, sorting color media, or hanging lights.

Organizational Management (1916 to Present)

Henri Fayol (1841–1925), a mine engineer, was a pioneer in the field of modern organizational management. His *Fourteen Principles* (Figure 3–1) helped to form the first comprehensive approach to management theory. Although many of Fayol's Fourteen Principles seem straightforward today, they broke new ground in 1917 by helping to establish a basis for organizational management.

Fayol also postulated that an individual with more skill in management than in technical expertise would not necessarily be bad for a company. In fact, he believed that an engineer with no aptitude for management would do more harm than good in an organization.[13] He also saw that management could be studied separately from engineering, and he noted that every organization required management: "Be it a case of commerce, industry, politics, religion, war, or philanthropy, in every concern there is a management function to be performed."[14]

Human Relations Management (1927 to Present)

The Behavioral Approach

The major failure of the classic approaches to management mentioned thus far was their lack of understanding of the human factor in work. The most efficient way of accomplishing a task was often thwarted by what the scientific management theorists thought was stubborn resistance to change. The researchers began to apply principles and concepts from the new field of psychology in an effort to understand workers better and to make organizations and people more productive. The basic assumptions behind much of this research were that (1) people desire satisfying social relationships and derive satisfaction from accomplishing specific tasks; (2) they respond to group and peer pressure in their work output; and (3) they search for individual fulfillment in their work. One of the classic studies in the early research on work behavior was conducted at the Hawthorne Wire Works in Illinois.

The Hawthorne Effect

In 1924, Vannevar Bush of MIT undertook a study of worker productivity at the Hawthorne Wire Works. The employees wound wires on motor coils or inspected small parts. Bush and his colleagues experimented with different lighting conditions on the assumption that different intensities of light would affect worker output. They found that the lighting level had no effect. Worker output increased despite wide variations in brightness.

Elton Mayo and Fritz Roethlisberger, professors at Harvard, began

Figure 3–1

Fayol's Fourteen Principles of Management

1. DIVISION OF LABOR. Labor should be divided to permit specialization.
2. AUTHORITY. Authority and responsibility should be equal.
3. DISCIPLINE. Discipline is necessary to develop obedience, diligence, energy, and respect.
4. UNITY OF COMMAND. No subordinate should report to more than one superior.
5. UNITY OF DIRECTION. All operations with the same objective should have one manager and one plan.
6. SUBORDINATION OF INDIVIDUAL INTEREST TO GENERAL INTEREST. The interest of one individual or group should not take precedence over the interest of the enterprise as a whole.
7. REMUNERATION. Reward for work should be fair.
8. CENTRALIZATION. The proper degree of centralization–decentralization for each undertaking is appropriate.
9. SCALAR CHAIN. A clear line of authority should extend from the highest to the lowest level of an enterprise.
10. ORDER. "A place for everything and everything in its place."
11. EQUITY. Employees should be treated with kindness and justice.
12. STABILITY OF TENURE OF PERSONNEL. Turnover should be minimized to ensure successful goal accomplishment.
13. INITIATIVE. Subordinates should be allowed the freedom to conceive and execute plans in order to develop their capacity to the fullest.
14. ESPRIT DE CORPS. Harmony and union build enterprise strength.

Source: Arthur G. Bedeian, Mangement, (New York, Dryden Press, 1986), Copyright (c) 1986 by Dryden Press. Used with permission.

the second phase of the study in 1927. A group of workers were carefully monitored for five years using a special test facility built for the experiment. The researchers gave the workers physicals every six weeks, monitored their blood pressure, recorded weather conditions, noted their eating and sleeping habits, and so forth. No matter what changes were instituted, worker productivity kept increasing. It became clear to the researchers that other factors were influencing the employees. Mayo and Roethlisberger surmised that the extra attention being paid to the experimental group combined with such things as changes in the supervision system, the creation of a small social system in the work groups, and the creation of a type of esprit de corps among the workers contributed to the increased output.[15] The *Hawthorne Effect* stresses the importance of human interaction in the work place.

Maslow's Hierarchy of Needs

Another theory that helped shape the human relations approach to management was Abraham Maslow's *hierarchy of needs.* Maslow's 1943 paper, "A Theory of Human Motivation," was quickly incorporated into management theory. Chapter 8 discusses ways to apply his approach in the work setting. In summary, the theory suggests that part of the manager's job is to provide avenues leading to employee satisfaction and that managers must work to remove obstacles that prevent employees from accomplishing their jobs.

McGregor's Theory X and Theory Y

Douglas McGregor gave a speech in 1957 at the Sloan School of Management called "The Human Side of Enterprise." His presentation included an idea about work that changed the relationship of manager to employee. McGregor's theory is based on the concept that managers develop "self-fulfilling prophecies" about people that affect all of their interactions with employees.[16] He identified two major perspectives held by managers: Theory X and Theory Y. Theory X assumes that (1) people generally dislike work and avoid it when possible; (2) they must be coerced, controlled, and threatened with punishment to get them to work; and (3) they want to be directed and avoid taking responsibility. On the other hand, Theory Y assumes that (1) people are generally willing to work; (2) they are willing to accept responsibility; (3) they are capable of self-direction; and (4) they have creative and imaginative resources that are not effectively utilized in the work environment.[17] The Theory Y approach to management has become a part of current trends toward what is called *participative management.* Companies are now beginning to ask employees what they think, rather than treating them simply as labor. McGregor believed that any enterprise can flourish if there is a partnership between the workers and the managers.[18]

Modern Management

Harold Koontz coined the phrase *management theory jungle* in 1961 when summarizing the six major approaches to the study of management.[19] He identified six main schools of management thought, which are summarized here by Wren:

1. Management Process—This school perceives management as a process of getting things done through and with people operating in organized groups. [It is] often called the traditional or universalist approach.
2. Empirical School—The use of the case studies for comparative purposes and to draw conclusions are the techniques common to this school of thinking.
3. Human Behavior School—The human relations or behavioral approach to work and managing utilizes various principles of psychology to concentrate on people in the organization.
4. Social System School—In this approach the workplace is seen as an extension of the cultural interrelationships people bring to their work. The manager's objective is to then achieve a level of cooperation and interaction among employees.
5. Decision Theory—With its sources in scientific management, this school concentrates on how decisions are made and processes used in reaching decisions. By refining the decision-making process the managers should be able to more effectively meet the stated objectives.
6. Mathematical School—This school takes the science of management and integrates it into a series of mathematical models of the organization. Operations research and analysis and statistical reporting are integrated into the decision-making processes of the organization. Work performance and output are measured against quantifiable standards set

by management. This technique is particularly applicable and is widely used in manufacturing.[20]

Systems and Contingency Approaches

Since Koontz's article, the management theorists have been busy integrating the different schools of thought into two major approaches to operating organizations. The first, the *systems approach*, was noted in Chapter 1. It assumes that an organization functions in a complex world influenced by multiple environments as it goes about gathering inputs and transforming them into outputs in the form of goods or services. The output, or performance of the organization, is not the sum of its parts, but rather the result of the interaction of the parts. Ideally, an organizational synergy results, and the whole becomes greater than the sum of its parts.

The *contingency approach* to managing an organization works on the assumption that there is no one way that works best in all circumstances facing an organization. The management must therefore be adaptable and capable of understanding the different mixes of management techniques that may be required at different times. This approach also recognizes that the people who make up the organization have differing styles of work and management. The top management must therefore expect that different work groups will have alternative ways of achieving the stated objectives. Rather than seeing this as a threat, diversity must be perceived as a strength. Synergy can once again be achieved if the management is capable of effectively coordinating the different work groups.

Rational Management

The upper management in any organization must therefore help shape the perception of the organization's goals and objectives in relation to a combination of an open and a contingency system of operation. To put it simply, people want to know the big picture and to feel as if their efforts toward helping the organization reach its goals are valued and recognized. Both the employees and the management should believe that the organization *is* its people. Success for the organization would therefore translate into success for the individuals working there. This may seem to be the only rational way for organizations to operate, but the tradition of excluding labor from management hampers the effectiveness of the open and contingency systems.

What does the future hold for management theories? One place to look is the business section of any bookstore. There you will find the latest trends and fads in management thinking. Authors like Tom Peters (*In Search of Excellence, Thriving on Chaos*), James O'Toole (*Vanguard Management*) or John Naisbitt and Patricia Aburdene (*Megatrends, Megatrends 2000*) offer a variety of perspectives about future trends and ways to effectively manage organizations in the current business community. Peter Drucker, the author of more than 20 books on management and economics, politics, and society, is excellent at synthesizing current practices and offering guidance for the manager of the future.

University MBA programs are also a source for ideas about future directions in management. Many of the journals found in college and university libraries provide an academic view of all of the major fields of management. Specialty journals are published regularly on such topics as operations, systems analysis, human resources, organizational psychology, and marketing.

Additional Readings to Consider

The works listed below are an excellent resource for further study in the field of management. The articles sited are a sample of a wide range of topics.

Matteson, Michael T., and Ivancevich, John M., eds, 4th ed., 1989. *Management and Organizational Behavior Classics.* Homewood: BPI/IRWIN, 1989.

Suggested readings:

Argyris, Chris, "Interpersonal Barriers to Decision Making," 1966, p. 461.

Follett, Mary Parker, "Management as a Profession," 1927, p. 7.

Herzberg, Frederick, "New Approaches in Management Organization and Job Design," 1962, p. 229.

McClelland, David C., "That Urge to Achieve," 1966. p. 392.

Mintzberg, Henry, "The Manager's Job: Folklore and Fact," 1975, p. 299.

Also consider:

Pierce, Jon L. and Newstrom, John W., eds, 2nd ed., 1990. "The Manager's Bookshelf: A Mosaic of Contemporary Views." New York: Harper Collins, 1990.

Conclusion

Several thousand years of the evolution of management theory have led to the open system and contingency approaches to organizational management. During this time, societies have created organizations capable of accomplishing an incredible range of activities. Cities, roads, dams, hospitals, schools, and churches have been built by organized groups of individuals using the techniques and theories of management. At the same time, it is important to remember management techniques have been and continue to be used to organize and implement unimaginable amounts of destruction and suffering.

Organizations and systems of management are still evolving today. The nineteenth century organizational model, with its rigid hierarchy and complex chains of command, has been proved to be incapable of responding quickly enough to change. Newer information-based organizational models with fewer levels of management are forming. As we head into the twenty-first century, political and economic upheaval will continue in ways we cannot foresee. Change seems to be the only constant on which organizations and individuals can count. If change is managed wisely as part of the planning process, the resources needed to provide for the future of the organization will be available. However, it is also possible to envision a world overwhelmed by the problems of population, pollution, and hunger. The images of an unmanageable world that come to us from both science and fiction writers may provide the incentive people need to solve the problems around us. Ultimately, it will be the people who make up the organizations that will determine the type of future we all share. Cooperation and collective action among these people and organizations hold the key to the future.

In the next chapter, we will examine how all organizations are affected by the social and political systems within which they must function. These and other external environments shape how the organization defines its mission and what the people in the organization believe.

Summary

Management is an integral part of all social systems, from a family to a multinational corporation. Whether the objective is gathering food or taking over another corporation, managers are required to coordinate the interactions of people carrying out designated tasks. Although many people have learned to manage while on the job, a body of knowledge accumulated over the last two thousand years constitutes management theory and practice.

Preindustrial societies developed laws, rules, myths, and rituals to control and direct people. The Renaissance and the Reformation created many new dynamics in the Western world. The opening of trade, the expansion of city centers, the rise of the middle class, and the major changes in political and social philosophy led to the formation of more sophisticated concepts of managing.

The Industrial Revolution produced fundamental changes in the nature of work and production, and it transformed Western societies. The mechanization of work in factories created the need for managers to supervise the activities of the factory workers.

The railroads, telegraph communication, the ability to manufacture precise interchangeable parts, and other new inventions and advances in technology radically altered the work place in the nineteenth century. As new production methods were devised, techniques for managing employees and organizing work began to be documented. The early systems of organizational design, production supervision, and data recording that were used in the railroads and factories became the basis of modern systems of scientific management.

Frederick Taylor was the first to document techniques for improving work output and streamlining antiquated manufacturing techniques. Henry Gantt, Frank and Lillian Gilbreth, and others expanded applications to the work place. Scientific research was quickly adopted by the business world. Computer models and simulations are now used regularly to improve productivity and output in factories.

Other major management practices focused on organizational design and optimal ways to structure the operation. The basic principles expressed by Henri Fayol and others about such things as chain of command, lines of authority, and rules and policies in business were thought to be applicable to any organization.

Another branch of management theory falls under the heading of human relations management. The premise underlying this research is that people want socially satisfying work situations. Studies by Elton Mayo and others verified that work output increases if employees are given more control over their jobs. Abraham Maslow's hierarchy of needs and Douglas McGregor's Theory X and Theory Y articulated many of the complex needs and interrelationships people bring to the work place. Harold Koontz identified six different major management theories that had evolved by the early 1960s in what he called the *management theory jungle*.

Contemporary management practices are based on integration models. Two systems make use of the theories identified by Koontz and others. One model assumes that organizations are open systems affected by external environments in the process of transforming inputs into outputs. The other model, the contingency approach, assumes that there is no one best way to operate an organization; managers must therefore be flexible and find the best match between the resources available and the problems to be solved.

Key Terms and Concepts

Robert Owen
Charles Babbage
Daniel Craig McCallum
Henry Varnum Poor
Frederick W. Taylor
Scientific management
Operations research
Critical path method
Computer-aided design, computer-assisted manufacturing, and computer-implemented manufacturing
Henri Fayol's Fourteen Principles
Human relations management
The Hawthorne Effect
Elton Mayo
Fritz Roethlisberger
Abraham Maslow's hierarchy of needs
Douglas McGregor's Theory X and Theory Y

Management schools of thought
Systems management
Contingency management
Synergy

Questions

1. List examples from antiquity that demonstrate the use of the basic management functions of planning, organizing, leading, and controlling.
2. Describe some of the legacies of the Industrial Revolution in manufacturing today.
3. Which of Fayol's Fourteen Principles can be most easily applied in an arts organization? Which principles seem inappropriate?
4. Have you ever worked for a Theory X or Theory Y manager? To which theory do you subscribe?
5. Summarize the six major schools of thought in management.
6. How does a college or university fit into the open system model? What are the inputs? What happens in the transformation process? What are the typical outputs?
7. Do you think government, business, and social service organizations in the United States are capable of solving the problems facing society? If not, what changes must be made in these organizations to meet the demands?

References

1. Paul DiMaggio, *Managers of the Arts*, Research Division Report #20, NEA (Washington, DC: Seven Locks Press, 1987), 46.
2. Daniel Wren, *The Evolution of Management Thought*, 3d ed. (New York: John Wiley and Sons, 1987), 56.
3. Ibid., 58–62.
4. Ibid., 61.
5. Ibid., 68–72.
6. Ibid., 74–76.
7. Ibid., 78.
8. Michael T. Matteson and John M. Ivancevich, eds., *Management and Organizational Behavior Classics*, 4th ed. (Homewood, IL: Richard D. Irwin, 1989), 4.
9. Ibid., 3–5.
10. Wren, *Evolution of Management Thought*, 132.
11. Ibid., 199.
12. Ibid., 397.
13. Ibid., 180.
14. Henri Fayol, *General and Industrial Management*, trans. Constance Storrs (London: Sir Isaac Pitman and Sons, 1949), 15.
15. Wren, *Evolution of Management Thought*, 195.
16. Arthur G. Bedeian, *Management* (New York: Dryden Press, 1986), 51–55.
17. Warren Bemis, Foreword to *The Human Side of Management*, by Douglas McGregor (New York: McGraw-Hill, 1960), iv.
18. Douglas McGregor, *The Human Side of Management* (New York: McGraw-Hill, 1960), 33–57.
19. Wren, *Evolution of Management Thought*, 355.
20. Ibid.

Additional Resources

Loden, Marilyn. *Feminine Leadership.* New York: NY Times Books, 1985.

Morrison, Ann M., Randall P. White, Ellen Van Velsor, and the Center for Creative Leadership. *Breaking the Glass Ceiling: Can Women Reach the Top of America's Largest Corporations?* Reading, MA: Addison-Wesley, 1987.

Taylor, Frederick W., *The Principles of Scientific Management.* New York: W.W. Norton, 1967. Originally published by Harper and Row in 1911.

Arts Organizations and Multiple Environments

4

The opening night of the Metropolitan Opera's 1991-92 season, a gala performance with Luciano Pavarotti, Placido Domingo and Mirella Freni that celebrates the company's 25th anniversary at Lincoln Center, will be broadcast live on cable television on a pay-per-view basis, it was announced Tuesday.

New York Times
August 1, 1991

The news release above is a good example of an organization experimenting with new ways of reaching an audience through changes made possible by technology. For $34.95, which is no doubt substantially less than the cost of a single ticket to the gala concert, millions could enjoy what formerly only 4000 would have been able to see. The Met's experiment illustrates the ways that an organization might exploit an opportunity made available by the changing technological environment in which it functions. As we will see, arts organizations must adapt to the changes in many other areas of society if they are to survive in an increasingly competitive marketplace.

An arts organization, like any business, must work within what are called *changing environments*. This term is used throughout this text to denote external forces that interact with organizations. We will examine six environments: economic, political and legal, cultural and social, demographic, technological, and educational. We will assess the impact of each of these environments on arts organizations, and later in the chapter, we will examine some of the trends that may reshape the arts in this decade. In addition, we will study how arts organizations interact with these environments based on the information received from six major sources: audiences, other arts groups, board and staff members, the media, professional meetings and associations, and consultants.

Figure 4–1 provides a graphic representation of the organization, the information sources (inputs), and the environments. The organization's relationship to the information sources and to the different environments may vary from organization to organization. Some organizations are more responsive to audiences and patrons, and others are more responsive to their boards or staffs. The process used in evaluating input from various sources and external environments will therefore vary with the predominant operating approach of the organization.

The information provided by the external environments and obtained from the major source areas helps the organization in the vital

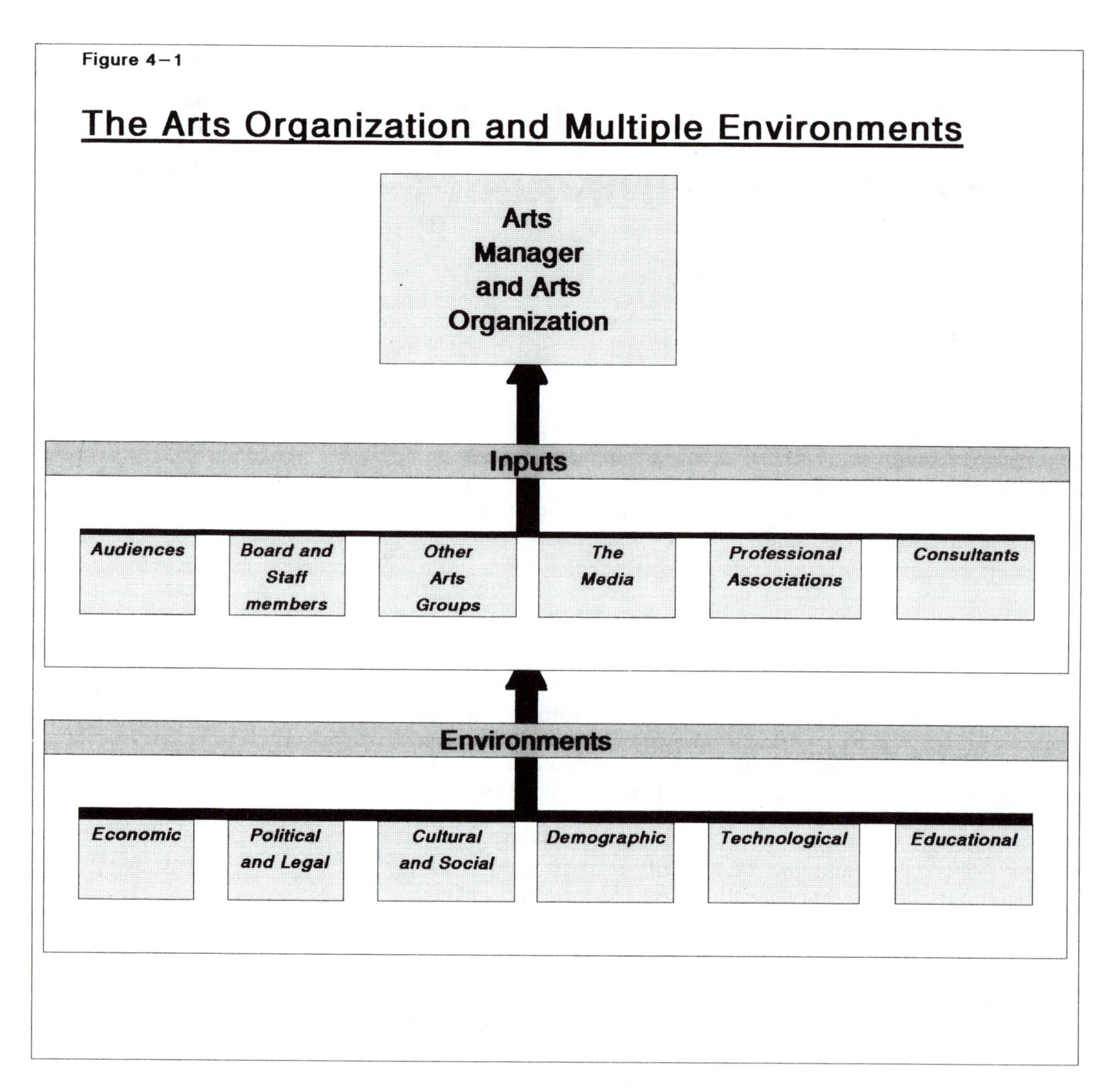

process of strategic planning, which is covered in Chapter 5. In addition, how an organization is designed relies on the input received from these sources. The areas of hiring, leadership, economics and financial management, marketing, and fund raising use input from the environments and information sources to help shape decisions made about programming, advertising campaigns, and fund raising strategies. Nearly all of the activities that an arts organization undertakes can be related to the interaction between external environments and the functions of management: planning, organizing, leading, and controlling. Based on these relationships, the manager's goal is to help fulfill the stated mission of an organization by dynamically balancing the external environments and the internal input with the changing opportunities and threats posed by each.

Changing Environments

As depicted in Figure 1–2, organizations are open systems that receive inputs from various sources: the inputs are then processed and transformed to outputs. This process applies whether the organization is an arts organization or a company that manufactures automobiles. The mission of the organization and its interpretation of the data may differ, but this fact remains: If the enterprise does not adapt to changes in its environments, it is likely to suffer. If it is too rigid or too slow to change, it will cease to exist. Ultimately, it is not the abstract "organization" that will suffer if the business is shut down. It is the people who work within the organization who will pay the price for the lack of adaptability and for poor management decisions. For example, the thousands of unemployed auto workers are the victims of businesses that were slow to adapt to new processes and new thinking. Effective management can help keep an organization thriving in the changing environments. As should be evident, the process of managing change is an integral part of the manager's job.

Managing Change

A manager of an arts organization is responsible for more than helping to get the show or exhibition opened. He or she must manage change. One of the most challenging aspects of management is keeping a sense of perspective on the growth and development inside the organization while watching the changing world around the arts organization for opportunities and threats. Unfortunately, change is difficult to monitor accurately because, like the hands of a clock, you only notice the movement after the fact.

In addition to navigating the organization through uncertain external circumstances, the manager must also attend to its internal needs. For example, in most organizations, people feel comfortable with a certain degree of routine or predictability. For this reason, the enterprise establishes specific rules of operation and detailed procedures for getting the work done. For example, payroll procedures and basic work conditions are established as a routine part of the organization. People cannot function anywhere near their potential if they are worrying about whether they will be paid or about other conditions of their employment. Once conditions are stabilized, these people will tend to be bored if their jobs begin to take on a routine that seldom varies. Subtle changes in work patterns can be just as important to the long-term health of an organization as major shifts to new programs. An important part of the manager's job is to remain aware of the overall direction and mood of the organization while helping people do their day-to-day jobs.

Growing into Change

As we saw in Chapter 2, arts organizations are often the result of a person or a small group of people who are willing to commit themselves to the task of bringing an idea to life. The same could be said for many of the major corporations in existence today. It is possible to trace the founding of many companies to one or two people with the drive and the ambition to make it happen. Many arts organizations are a product of the post–World War II boom in births coupled with increased access to education and financial support from foundations and the government. An unplanned mix of circumstances from the 50s to the 80s supported the unprecedented growth in performing and visual arts organizations across the nation. New theater, dance, and opera companies were founded, symphony orchestras spread throughout the nation, and museums opened in large and small cities. Once established, the arduous task of maintaining arts institutions began to take center stage.

Let's examine a hypothetical example. A small founder-director opera company with an annual budget of $35,000 in the early 70s grows to become an arts institution in its community with a budget of $3.5 million in the late 80s. As the organization evolves, a board of directors is added, and new staff members are hired to do what the original founder and one or two people were doing. The tiny storefront office becomes a large storefront office, and when that space is suddenly too small, a suite of offices are found in a high-rise. Slowly, but inevitably, the small opera company begins to function at a scale that requires longer-term planning and careful analysis of its future. In other words, the company begins to move from a year-to-year planning cycle to a three- to five-year planning cycle. The original founder-director no doubt had a vision of what the opera company could become, but now that it has grown into a full-fledged business and nearly 20 years have passed, the world around the organization has changed.

As an organization matures, it begins to take on characteristics that make it less responsive to change. After all, when a company finds something that works, it continues making more and more of the product. The opera company finds that a particular pattern of performances (two grand operas, two operettas, one musical) sells well, and so it repeats that cycle with different titles every year. Suppose now that the same community also becomes home to a theater company, a ballet company, and a symphony orchestra. The creation of each new organization will have an impact on the opera company, which may find itself competing for arts revenues because audiences have more choices. The theater company, for example, may decide to do two or three musicals each season. The change in the cultural environment of the community requires that the opera company's decisions about programming be made in the context of three other groups struggling for the same entertainment dollar.

To better adapt to its new circumstances, a process of continual evaluation should become the company's operating norm. Asking questions about where the opera company stands with respect to the six environments and the other arts groups and sifting through the feedback from its information sources becomes as critical as mounting a season of high-quality productions. Does adopting this process offer any guarantee that the opera company will fare better than a business that doesn't? Unfortunately no, but it can give the opera company an edge if the management and artistic teams cooperate.

It is important to remember that the process of continual evaluation is nothing more than a tool for survival and that it will only be as effective as the managers and artists who use it. If the managers are not skilled at using the process, they may chart a course for the organization that leads to ruin. On the other hand, successful assessment and planning should lead the opera company to develop at a pace that fits well within the parameters of the community and the six environments.

Organizations discover that a process of ongoing assessment requires the development of techniques for gathering and analyzing information. Technique, as any performer will tell you, is acquired through long hours of rehearsal. Just as dancers, singers, and actors learn to master their art through developing techniques for approaching a role, analyzing a musical score, or examining a play text, so too can an arts manager master the art of organizational evaluation. Once mastered, this technique can then be applied to the four functions of management. For example, consultation with the theater company on upcoming titles and dates would become a routine part of the opera company's planning procedures to avoid duplication or conflicts. This may seem obvious, but many organizations don't think of their work as part of a larger cultural context in their community.

Before discussing techniques for exploring individual environments, let's examine a general approach to organizational evaluation and assessment. As we will see, the biggest problem in assessing the opportunities and threats to an organization is the conflicting input facing a manager.

Where to Start?

An arts manager can start gathering the vital information from books, newspapers, magazines, and the broadcast media. The basic methodology, which is called *content analysis*, simply involves identifying sources for clues about current practices and possible future trends. Gathering input from sources external to the organization is complicated by the cyclical patterns of the print and electronic media. Topics come and go from the front page and evening news with incredible speed. The key component in an arts manager's quest for information through content analysis is therefore sufficient variety. The manager must also differentiate between trends and fads. For example, shifts in population growth establish trends that ripple through a society for years: more people, more services, more houses, more apartments, and so forth. Fads, on the other hand, tend to die out more quickly. Arts organizations that react to fads sometimes find themselves scheduling programming that is out of step with current topical stories. In the time between the decision to produce a particular program and its actual production, a new hot issue may arise to take its place.

The arts manager must therefore use caution when trying to sort through what the future may bring. For example, it is not unusual to find contradictory opinions expressed about a particular topic. Let's compare two distinctly different views of how the arts will fare in the 90s. While reading the extracts below, consider the different directions that an organization might follow in its planning if it subscribed to one or the other speculation.

> In the final years before the millennium there will be a fundamental and revolutionary shift in leisure time and spending priorities. During the 1990's the arts will gradually replace sports as society's primary leisure activity. This extraordinary megatrend is already visible in an explosion in the visual and performing arts that is already well under way:
>
> - Since 1965 American museum attendance has increased from 200 million to 500 million annually.
> - The 1988–89 season on Broadway broke every record in history.
> - Membership in the leading chamber music associations grew from 20 ensembles in 1979 to 578 in 1989.
> - Since 1970 U.S. opera audiences nearly tripled.

> From the United States and Europe to the Pacific Rim, wherever the affluent information economy has spread, the need to reexamine the meaning of life through the arts has followed.[1]

When John Naisbitt and Patricia Aburdene offered this and other equally optimistic assessments about the future of the arts in the United States in *Megatrends 2000*, people struggling in the trenches of many arts organizations were stunned. Were the authors talking about the same country where many arts groups were grappling with insolvency on a daily basis? Because the economy lingered in recession in the early 90s, state and local governments were slashing their support for the arts, and corporations were putting thousands of white and blue collar workers out of jobs. The glowing picture painted by Naisbitt and Aburdene seemed a cruel hoax to many in the arts industry. Yet, the authors reached this prediction by using many of the same sources that any inquisitive arts manager would use, including *Variety*, Theatre Communications Group publications, the *Wall Street Journal*, Dance/USA, NEA Research Reports, *Business Week*. In fact, Naisbitt and Aburdene's research found enough positive signs to lead them to title their chapter "Renaissance in the Arts." They discuss one art form after another and conclude that things are only going to get better. The art market, museums, opera companies, and so on, are all headed for a future filled with patrons resulting from the strong support of corporate America.

Meanwhile, in other quarters, a darker picture of the future was painted.

> Whatever the balance of this decade holds, our assumptions about the role of the arts in the complex and contradictory scheme of American life have begun to unravel. And, our assumptions about the strength, endurance and resilience of our arts institutions have been challenged by the hard economic realities of the Reagan/Bush era. While we will ultimately view this period in the light of many economic, social, environmental and political events, the situation for the arts has been, and still is, defined by two crises.
>
> The crisis that has most captured our attention has been the attack on the National Endowment for the Arts led by Jesse Helms and Donald Wildman. It began with seemingly isolated cases of criticism of certain NEA-supported arts projects by certain conservative members of Congress. Today, it has become a major debate over government funding in the arts and over freedom of expression.
>
> The second crisis afflicting the arts is much harder to see or understand, but it is at least as threatening as the first. The quiet crisis is about mounting debt and organizational dysfunction.
>
> A confluence of factors, financial and otherwise, that have developed during the last 10 to 15 years has changed the world by happenstance and by careful, calculated design. The three most dramatic factors are (1) financial shifts and the debt culture; (2) the shrinking human capital pool, and (3) over-regulation and stagnation in the arts structure and community.[2]

Nello McDaniel and George Thorn's book, *The Quiet Crisis in the Arts*, depicts a very different world from *Megatrends 2000*. McDaniel and Thorn predict that unless organizations face the fact that debt financing, shrinking labor pools, and unrealistic fund raising expectations have become the distorted norm, the trickle of bankrupt arts groups will increase to a flood. They point out that arts organizations acquired debt by expanding beyond a reasonable funding base. They stress that organizations are extended beyond the means of their supporters and their staff. They counsel a complete reevaluation of revenue and expense patterns and urge organizations to seek a balance between what they can realistically do with the resources they have and the programs and projects they undertake.

Who is right? Is the future full of promise or problems—or both? In the remainder of this chapter, we will try to answer this question by exploring the environments that affect arts groups. Our objective will be to piece together a composite of the forces that shape the arts organization.

Environments

To effectively manage change, the arts manager must identify the environments that will have the most direct impact on the organization. Let's review some of the environments that interact with the organization and try to establish some basic guidelines about what constitutes significant input.

Economic

Arts organizations, as part of a national economic system, experience the effects of expansions and contractions in the national economy. In addition, regions, states, and cities exhibit different reactions to changes in the national economic environment. (See "Recession Begins to Pinch Art Museums, Study Finds" for some real-world examples of the effects of a recession.) Some of the factors that may have an impact on an arts organization include the federal banking system (raising and lowering interest rates), new tax increases or cuts, revisions in existing tax legislation (which may promote or hinder donations), and inflation (price increases). This last factor can be the most destructive to an arts organization in the long run. When the cost of doing business continues to escalate, the organization faces tremendous pressure to increase revenue from either more sales or more donations. Chapter 10 reviews some basic principles of economics and explores the unique economic dilemma that increasing costs and limits on productivity impose on arts groups.

The process of evaluating the economic environment is often subject to contradictory reports by experts in the field. One expert may issue a news release announcing that the recession is over while another says it will continue for six more months. If a manager is trying to plan a budget based on projections of income and expenses in uncertain economic times, the most practical approach is to plan with contingency budgets. In other words, the organization's budget would be subject to constant revision depending on whether the economy is growing, is stable, or is slowing down.

One of the myths of the entertainment industry is that when times are tough, people seek escape by spending what they have on entertainment. The facts indicate that when the economy goes into a recession, people in the middle income levels reduce their spending, people in the upper income levels do not radically change their spending, and people in the lower income levels curtail what little spending they do on the arts.[3] Donation frequency seems to follow similar patterns. If a recession extends beyond a year, arts organizations will generally see a slowdown in ticket purchases and donations from upper-income patrons. In more severe economic conditions, such as a depression, all spending and donation activity will slow dramatically. Knowing this, an arts organization can plan for reduced revenues, plan to increase fund raising activity, or both. The key to working with the economic environment is to always have alternative budgets ready to implement should conditions change.

Political and Legal

The arts in the United States today are very much a part of the political scene. However, with the heightened visibility of the arts has come the added responsibility of artists and arts organizations to lobby continually to protect the support gained over the last 25 years. Up until the creation of the NEA in the mid-1960s, the arts were only minimally involved with the political system. The only other time that the government and the arts had joined forces was during the 1930s when work projects that employed actors, singers, dancers, painters, and so on, became part of the U.S. economic recovery plan. The Federal Theatre Project, for example, burst on the scene in 1935; by 1939, it was gone in a maelstrom of political struggle. The fear of Communist infiltration sent the whole program into budgetary limbo. The productions produced by the Theatre Project often challenged the mainstream political environment to the point that it lost its base of support in Congress.[4]

The input from such sources as professional associations, consultants, board members, and the media can help shape how an organization will adjust to changes in the political and legal environment. For example, the quote from *The Quiet Crisis in the Arts* given earlier in this chapter spoke of the issue of conservative political and religious movements bringing pressure to bear on funding sources and influencing the freedom of expression. The board and staff of an arts organization, whether they like it or not, face increasing scrutiny when they receive public funds. It is unlikely that the next few years will lessen the pressure on arts groups to justify their needs against the needs for the poor, the ill, and the homeless. The current trend at state and local levels is to cut support for the arts and to allocate the funds to other needs. (See the case study in Chapter 13 for an example of state and local funding being cut in response to legislative pressures to control government expenses.)

Changes in the legal environment are carried out through federal, state, and local enforcement agencies. For example, the Occupational Safety and Health Administration (OSHA) and similar state agencies issue regulations that have a direct impact on the operation of the technical production shops and museum laboratories in the United States. Employees have sometimes reported unsafe conditions and practices to these agencies as a way to force unresponsive arts managers to make the work place safer. The impact of laws that affect the design of public spaces and the work place by mandating access for people with disabilities must be taken into account. Issues relating to smoking, medical and retirement benefits, maternity leave, and so on, have had an impact on arts organizations over the last ten years. In most cases, changes in the legal environment have a price tag attached, and the implementation of new laws translates into expense items that appear in the operating budget.

In the News—The Economic Environment's Impact

This article from the *Chronicle of Philanthropy* shows the impact of the economic environment on one sector of the arts community.

Recession Begins to Pinch Art Museums, Study Finds

Art museums have begun to feel the pinch of recessionary pressures, according to a new study of the 160-member Association of Art Museum Directors.

Among 110 members who participated in the survey, 69 percent said the recession had a negative effect on their operations, and 12 percent said it had been very negative.

Shrinking corporate profits, greater competition for contributions from other nonprofit organizations, and a downturn in government support were among the leading reasons cited by the museum directors.

Board and staff recruiting has been less affected by the economic decline: more than two-thirds of the directors said the recession had no effect on staff recruiting and nearly 90 percent said that economic conditions had not affected board development.

Attendance has not suffered greatly, perhaps because many of the museums don't charge an entrance fee: 62 percent of the survey respondents said economic conditions had no effect on admissions.

The greatest area of concern involved government support for museums. Nearly three out of four of the museum directors said state funds had been cut back, and 57 percent said local financing had been harmed.

The increasing demands on local social service organizations, resulting in greater competition for contributions, were cited by 72 percent of the museum directors as having had a negative impact on their institutions.

SOURCE: "Recession Begins to Pinch Art Museums, Study Finds" *Chronicle of Philanthropy*, (July 2, 1991). Copyright © 1991 by the *Chronicle of Philanthropy*. Reprinted with permisson.

In the News—Laws and the Arts

This article illustrates how changes in laws can have a direct effect on arts organizations. In this case, the political lobbying activities of unions influenced the legislative system, and now arts organizations are attempting to exert their own influence to change the regulations. The implementation of the legislation was delayed in the fall of 1991 by last-minute lobbying by arts-presenting groups and universities.

Law That Would Limit Artists Causes Uproar

John Bennett

WASHINGTON—They are stacked by the hundreds in cardboard boxes—angry letters attacking new restrictions on the number of foreign artists who can enter this country.

The Immigration Act of 1990 caused the outrage by limiting to 25,000 the number of foreign artists who may enter the country each year. The law, to become effective on October 1 after the Immigration and Naturalization Service publishes the regulations, has drawn opposition as the deadline nears.

The letters arrive from opera houses, symphony orchestras and ballet centers in New York City, Boston, Houston, Los Angeles, Denver, Cleveland and Santa Fe, N.M.

Universities from Michigan to Florida have joined the chorus, arguing their stakes in artistic campus tours arriving from Europe.

The law further requires an artist to be affiliated with the institution or performing group for more than a year before applying to come to the United States.

Further, it bars an application for a visa until 90 days before a scheduled arrival and requires those presenting the tour to explain how the event would benefit the U.S.

Then, in a requirement that annoys U.S. cultural centers, labor unions would have to be consulted before bringing the artists into the country.

Lawmakers feeling the heat from across the country are rushing to amend the law. Many of those who backed the idea have switched sides.

"It will all probably be cured in a few months," said an INS official.

The AFL-CIO lobbied for the alien restrictions on ground that the non-immigrant visas threatened the jobs of U.S. workers.

Congress approved the limitations after a report noted that temporary non-immigrant visas issued to entertainers had been climbing rapidly, creating controversy.

SOURCE: John Bennett, "Law That Would Limit Artists Causes Uproar," Scripps Howard News Service, August 27, 1991.

More arts organizations are adopting an active rather than a reactive approach to coping with changes in the political and legal environment. For example, they issue personal invitations to politicians to arts events that are accompanied by high-visibility social activity. Trips to Washington, D.C., the state house, or city hall for one-on-one discussions with the legislators, governor, or mayor have become mandatory. In their regular communication with lawmakers, arts managers stress that the people who attend the arts are also voters. In addition, lobbyists representing arts organizations are now a regular fixture on the national scene. One of the lobbyist's jobs is to keep abreast of pending legislation that may have an effect on the arts.

Cultural and Social

The United States' cultural and social environment in the late twentieth century is heavily influenced by the economic, technological, and political environments. The traditional social structures (family, schools, and religious organizations), though undergoing change, still play major roles in the transmission of social values and beliefs.

Some of the changes affecting the cultural and social environment include two-income households, attitudes about gender roles and race, health care, and leisure time.The potential impact of these changes on arts organizations is far too complex to predict accurately.

The other major force in the socialization process in U.S. society is the broadcast media. Television and radio are the major sources of information and entertainment for millions of people. Unfortunately, the broadcast media do not focus much attention on the fine arts. An interest in opera or symphony is often the source of humor, if it is mentioned at all, in the shows that are watched by millions every night.

The arts, which are a leisure time activity for most people, face difficult times ahead as the cultural and social environment of the U.S. continues to diversify. The generation that created the baby boom and contributed greatly to the growth of the arts, has not been replaced in equally large numbers. In fact, the trend seems to be toward a generally lower standard of living for Americans under 30. A recent article in *Business Week* pointed out that in a typical American family, heads of household who are under 30 are making less money today than their counterparts made in 1973. Families headed by college graduates are the only segment making more money than in 1973. The education level, which consistently correlates with attendance at arts events, also correlates with the income level. In all cases, families headed by individuals with some or no college were making less than their peers made in 1973. The same applies to African-Americans and Hispanics.[5] These lower numbers may make it harder for arts groups to diversify their audiences because these groups will probably spend any extra money on less expensive entertainment options.

The arts manager must also recognize that alternative living arrangements have created new definitions of "families" and have led to new arts consumption patterns. The high cost of housing has meant that many young people have not moved out of their family homes and established their own independent living arrangements. Single-parent families are also common today. In many communities, a greater number of people are living alone. Career pursuits also contribute to fewer leisure hours for a large segment of the highly educated population. Finding creative ways to reach these potential audiences with special discount ticket plans or through the type of programs offered will mean rethinking marketing and fund raising strategies for many organizations.

The arts manager's energies are probably best directed toward developing, maintaining, increasing public awareness of and interest in the performing and visual arts. These objectives can best be met by working with artists to shape the audience of the future.

Artists are often at the leading edge of change in a society. For example, much of the content of the traditional and, of course, popular titles in the repertories of theater and opera companies reflects gender roles that were much different than they are today. Increasingly, the dominant culture is changing to reflect more diverse points of view as directors and producers seek out other perspectives. Women, African-American, Hispanic-American, and Asian-American artists are seeking to change cultural values to reflect a broader vision than that of the white Euro-centric world view. At the same time, the segmentation of audience tastes is continuing to increase to the point where arts groups need to reexamine fundamental choices in titles and programming options.

Demographic

The arts manager must closely monitor the demographic environment, which comprises the vital statistics of a society. Factors such as gender, age, race, income level, occupation, education, birth and death rate, and geographic distribution influence organizations. The more that is known about a community and the surrounding region, the better able the organization's ongoing assessment process will be to address community needs. For example, the baby boom generation will create a large number of elderly after the turn of the century. Together, the boomers and the current elderly population account for a significant portion of today's arts consumers. Trying to anticipate the changing taste and attendance patterns of the aging boomers will be a high priority for arts managers over the next 20 years.

The arts manager must also be concerned about the low birth rate since the 70s. Arts attendance may drop in direct proportion to the population base. This will have devastating consequences for organizations facing what will no doubt be greater financial pressures in the next 30 years.

It is probably safe to assume that with an aging population health care costs will continue to take up a greater portion of the financial resources of society. This demographic trend is already being noticed by arts organizations as state funding for the arts is cut due to the pressure of financing federally mandated health care support.

The AIDS epidemic has a direct impact on arts organizations as well as on society as a whole. Arts managers need to become informed about the complex legal and ethical issues of AIDS in the work place. In addition, arts organizations will face rising medical insurance costs that will make affordable health benefits for performers and staff very difficult. Arts managers will also have to adopt policies and procedures that deal sensitively with the impact of AIDS on the arts community.

Demographic trends in the next 25 years will require arts organizations to adapt to a more ethnically diverse audience that spans a greater age range. One strategy that arts groups can take is to adjust programming choices to the audience, rather than expecting the audience to adjust to the program. For example, the older crowd might have its own special symphony series, while the younger audience might have a series tailored to its tastes. This is the equivalent of the "narrow–casting" of cable television companies.

Technological

As noted in Chapter 2, the invention and distribution of film, radio, and television had a profound effect on the arts around the world. In the United States, the new technologies were quickly adopted for commercial profit-making purposes. The displacement of live performers with film, for example, put thousands of actors, dancers, singers, musicians, and technicians out of work in the 20s and 30s. New jobs were of course created, but many of the older performers in the larger metropolitan areas were permanently put out of work by the movie screen and the sound track. As the job market adjusted to the new technology, the live performing arts have adapted and grown since the end of World War II.

As we have seen, the arts boom was due in large part to the combination of the birth rate, economic growth, and increased levels of education. Through the 60s and 70s, arts centers and performing arts groups came into existence even though television sets and movie screens could be found nearly everywhere. By the 80s, the home videotape recorder (VCR), videodisc, and compact disc player created a new demand for program material. The VCR provided more opportunities for the distribution of films worldwide. It also allowed viewers to rent or purchase videotapes or videodiscs of opera, theater, and dance performances or museum collections. The issuance of thousands of classical titles on compact discs, coupled with the improved sound quality, has helped keep the classical music industry going strong. On the whole, then, the new technologies seem to present opportunities rather than threats to arts organizations. However, the birth rate may have more to say about the future than technological advancements. It is possible that there will be a decline in attendance as the aging baby boomers stay home to be entertained with their own home theaters.

Some of the developments expected in the next decade offer further opportunities for the arts. For example, high-definition television (HDTV) will make the home entertainment center a reality for millions of consumers. Integrating advanced computer systems with these entertainment centers may make possible home versions of the new technology of virtual reality. Virtual reality (VR) is an interactive computer technology that allows the individual to enter into an electronic world that not only appears real to the viewer but also allows direct interaction in an environment. The current technology and the cost of equipment limit the application of VR to the very rich and to researchers at large universities and corporations. However, when the cost drops, which it will, the way people experience events will be transformed radically. According to Howard Rheingold, author of *Virtual Reality*, it is conceivable that it may soon be possible to enter into a performance, rather than simply viewing it.[6]

The opportunities for the performing arts and VR are still uncertain. It seems clear that VR will change the relationship of the performer and the audience even more profoundly than film, radio, and television. The advertisement flyer for a conference on VR entitled "Cyber Arts International" lists programs such as "Theater of Artificial Reality" and "Tomorrow's Choreography Today" among the many sessions and workshops.[7] Given the tendency of U.S. society to use new technologies in ways that maximize profit, VR represents a far greater impact on the commercial entertainment industry than on the arts. Rheingold, who is very enthusiastic about the new technology, does point out that

> One way to see VR is a magical window into other worlds, from molecules to minds. Another way to see VR is to recognize that in the closing decades of the twentieth century, reality is disappearing behind a screen. Is the mass marketing of artificial reality experiences going to result in the kind of world we want our grandchildren to live in? [8]

In the News—Technology Enters the Orchestra Pit

This article illustrates the economic and technological environments converging on the musicians in the Broadway orchestra pit. The desire to keep costs down will put further pressure on the union to fight the advance of a technology that will put members out of their jobs.

At Issue: A Synthesizer in the Pit
Glenn Colling

Can "Grand Hotel" replace the string section of its orchestra with an electronic synthesizer? Local 802 of the American Federation of Musicians contends that the musical is letting go eight string players to do just that. Marvin Krauss, a "Hotel" co-producer, says that only four musicians are at issue, that the synthesizer—played by a union member—is already being used in the show, and that "it's all a tempest in a teapot."

"We're not arguing that he can't use a synthesizer, we're saying it can't play the same parts our musicians used to play," said 802's lawyer, Leonard Leibowitz. The complex contractual dispute will be decided by an arbitrator before September.

SOURCE: Glenn Colling, "At Issue: A Synthesizer in the Pit," *New York Times*, July 26, 1991.

Educational

Studies show that education is one of the most significant factors in developing an arts consumer. Researcher Lynne Fitzhugh notes, "The socio-economic variable most often and most perfectly associated with cultural attendance is, not surprisingly, education."[9] Many surveys have found that more than half of the people attending arts events have college or graduate degrees. When considering how little focus is given to arts education in the United States, these numbers are all the more startling. One can only guess how much greater the attendance would be at cultural events if the arts were more integrated into the educational environment.

Arts organizations stand to gain the greatest long-term benefit from working in cooperation with the local school systems. However, because the schools seldom have the resources to pay for the services of arts organizations, outside funding is required. Foundation and corporate grants to improve the quality of education may provide opportunities for arts groups to establish good community relations and to build future audiences.

The most effective methods for making the arts a significant part of the educational environment usually combine visits to the schools in conjunction with planned lessons throughout the year. Transporting busloads of kids to an auditorium and putting on a show only offers a superficial connection to the arts event, and in many cases, it only acts to further alienate young audiences from the arts. Without a context for the experience, the concert, play, or opera is an isolated incident at best and boring at worst.

In the remainder of this decade, schools will be the focus of much political attention. Performance standards, national testing, increasing budgetary pressures, and parental choice in selecting schools will be among the issues facing the 15,000 school districts across the United States. It will take aggressive action on the part of arts organizations to positively position themselves in this environment.

Information Sources

To effectively manage change and operate a useful evaluation and assessment system, arts managers must identify the sources they will use for gathering information and must develop an ongoing process for evaluating the opportunities and threats facing the organization. Let's examine the type of information each source generates.

Audiences

Smart arts managers want to know as much as possible about the individual who expends the effort to go to a show or an exhibition or who gives an organization money in exchange for a ticket, subscription, or membership. Why? For the simple reason that the organization's survival depends on establishing a long-term relationship with this person. Within the bounds of an ethical system of gathering data, a manager would want to know (1) why this person made the purchase, (2) what he or she liked about the product, (3) what he or she didn't like, and (4) what other related products this person would be interested in purchasing. Members of the audience, patrons, donors, members—whatever they are called—are tremendous resources usually waiting to be asked what they think and feel. Exit surveys for museums, program insert surveys, phone surveys, or small discussion groups of randomly selected arts consumers are viable techniques for gathering information. Some techniques will be more effective than others, but regardless of the method, the arts organization that is able to provide detailed profiles of the consumers of its products will be better able to predict how a planned change will affect the relationship that exists between the individual and the organization.

Does this data-gathering process imply pandering to the audience's tastes? Hardly. The primary purpose of asking people what they think about a product is to learn how to communicate better with them. Arts organizations forget that their audiences don't use the same vocabulary to describe the product and the process of the arts. In the open system, the arts manager designs the communication devices (brochures, letters, posters, and so on) to the outside world to reflect terms and concepts that effectively translate the organization's mission to the widest possible audience. Ineffective communication only raises barriers between the organization and repeat customers or future customers. (Chapters 12 and 13 discuss this topic in more detail.)

To summarize, then, establishing an ongoing communication process with the users of a product is essential to the long-term health of an organization. The importance of knowing as much as possible about *who* is interested in *what* can not be stressed enough. Feedback from the consumers of the product is a resource that will shape the future of an organization.

Other Arts Groups

A community with several arts groups can achieve a synergistic boost from the combination of programs and activities—the whole is greater than the sum of its parts. When the different arts groups recognize that they can benefit from communicating with others about their seasonal or exhibition plans, the local arts scene can flourish.

Strategic thinking and long-term planning should create a mutual understanding among arts groups that there is a complex "arts audience," in addition to individual audiences for ballet, opera, theater, and so on. Research seems to indicate that a segment of the audience can be classified as users of different art forms, while other segments are loyal to one form and seldom go to see other events.[10] One strategy that seems to address these differences is the consortium approach. For example, many cities publish a quarterly arts calendar covering different arts groups and museums. These calendars give potential arts consumers an overview of all events happening in their area. Discount coupons and advertisements are often used to highlight special events. Multipage flyers can be widely distributed through a Sunday newspaper or a mass mailing. The net result can be a piece that enhances awareness of the overall arts scene, recognizes individual audience segments, and promotes cooperation among the arts groups.

In addition to cooperative publications, different arts groups can work together to present new programming combinations that benefit both groups. The symphony and the ballet or the ballet and the opera can pool their resources on occasion to present larger-scale productions than either could mount individually.

If nothing else, a regularly scheduled meeting among the different presenting groups in a community offers an opportunity to share ideas about trends in the different art forms. The sharing of information ultimately helps a manager better understand the overall arts dynamic of the community.

Board and Staff Members

The board of directors and the staff of an arts organization are a vital component in the information-gathering process. The key to success is ongoing input via staff meetings, suggestion boxes, retreats, informal social gatherings, and formal planning sessions. When a board member asks about presenting a particular type of program or a staff member suggests a new procedure, the organization must have a mechanism for responding to the input. An open system depends on these suggestions and works from the assumption that there are always alternatives to what is currently being done. An organization that does not allow for input from the board or the staff will probably become stagnant and dysfunctional over time.

The Media

The print and broadcast media provide the arts manager with up-to-the-minute information about many of the external environments that have an impact on the organization. It is also possible to gain some insight into the general mood of the country or region from polling conducted by the media. Trade publications in the arts as well as national and regional news sources should be part of the arts manager's regular reading list. As noted at the beginning of the chapter, contradictory information is often generated by these sources, but that is to be expected.

Cultivating and sustaining a positive working relationship with the press and the broadcast media can be of obvious long-term benefit to arts organizations. However, arts and nonprofit groups are often naive about the realities of media coverage. Column space or air time is usually an issue of money. For the print media, advertising sales space and articles are always in a complex struggle with each other. For the local commercial television or radio station, ratings determine advertising revenue. Therefore, coverage that will generate ratings is often the focus of attention. Getting a feature story in the arts section of a newspaper or getting 30 seconds of air time at the end of the six o'clock news can be a struggle. Attaining a level of visibility is critical for the arts organization's interactions with the external environments. No matter how good and noble the programs or projects of an organization may be, it is hard to establish credibility in the community without publicity.

One example of the ebb and flow of media coverage was the NEA struggle described in Chapter 2. The media focused on the obscenity issue rather than on the larger questions of government support for the arts because the struggle of Congress with the NEA was simply more interesting than an abstract national policy issue. The net result was a great deal of publicity for the NEA, most of which unfortunately cast it in a negative light. Very few stories mentioned the thousands of grants made each year and the millions of people who benefit from grants and endowment services.

Professional Meetings and Associations

Each of the arts has a professional service organization that provides regular information about issues of importance to its constituency. Many service organizations publish newsletters or magazines, and almost all hold annual conferences. The information-exchange process among members often focuses on current operational problems or topics related to new methods for raising money. The benefit for the arts managers of belonging to these associations or attending these conferences lies in expanding their knowledge of how other organizations are adapting to external forces.

Consultants

Consultants are another source for information about methods of keeping an organization functioning effectively. In theory, a consultant gives the organization a needed outside perspective. Of course, arts managers should never assume that consultants are always right any more than they would blindly trust any other source of information. However, because consultants usually deal with several organizations at one time, they can suggest new ideas and approaches that would not occur to the internal management staff. Consultants can also validate the staff's ideas about how best to manage change in the organization.

Other Sources

Depending on the art form, other input sources may provide valuable information to the manager of an open system. For example, the U.S. government regularly publishes statistical data from the Census Bureau and the Commerce Department that arts managers could apply. This is especially useful when the data profiles a region of the country in which the arts organization resides. Another source of information might be found among the various suppliers of goods and services purchased by the organization. For example, the bank used by the organization could be an excellent source of local economic information. Printers or graphic arts firms could be a source of information about new trends and techniques in advertising. Other useful sources might be the local chamber of commerce. After all, the arts organization is a business in the community, and belonging to local groups that attract other businesses could prove helpful when seeking direct information about the economic health of the area.

The Impact of Future Trends on the Arts

Trying to anticipate the direction that change may take is a difficult task. As we have seen from this brief overview of the major environments affecting arts organizations, complex forces can interact to produce unforeseen results. Some of the issues of today that will continue to have a direct impact on the arts tomorrow include censorship, the AIDS epidemic, and mounting economic pressures.

Censorship

The choices for programming in U.S. arts centers will ultimately be influenced by the arts organizations' perception about the degree of censorship in the community. A fair amount of self-censorship is already occurring in many arts organizations. For example, arts organizations who seek corporate funding and a broad base of community support think twice before tackling controversial topics that may turn donors away. Instead, these arts institutions tend to leave the topical work to the fringe arts groups.

The typical result when an arts organization begins to move onto a path that the majority of the audience isn't willing to follow is a sudden drop in ticket revenue. The reality of keeping a steady cash flow and meeting the payroll provides another type of censorship that most arts groups face: economic. Once an arts group reaches the level of requiring a staff and an organizational structure to operate, the pressure to take fewer programming risks increases. Unfortunately, the economic pressures on arts groups appear to be increasing.

The AIDS Epidemic

In the 1970s, no one foresaw an epidemic like AIDS decimating the arts community. Although several promising medical research projects are investigating cures, the number of cases is expected to rise throughout the remainder of this decade. The personal and economic costs of this epidemic will also rise. The arts community and arts organizations will bear the increasing cost of AIDS. The tragic loss of many talented people and skyrocketing medical costs will influence arts groups through the rest of this decade.

Economic Pressure

As mentioned earlier in this chapter, the authors of *Megatrends 2000* set an optimistic tone about the future of the arts in the United States. Part of their reason for being so optimistic is their prediction that "as we turn to the next century we will witness the linkup of North America, Europe, and Japan to form a golden triangle of free trade."[11]

Although there is no reason to be overly pessimistic, many people in the arts world are a great deal less hopeful than Naisbitt and Aburdene. The escalating costs of operating an arts organization combined with increased competition for donors and ticket buyers is not a mix that promises much relief in the years to come. The fundamental structural problem may be one of costly operational redundancy. It makes little sense for several small arts groups in one community to have their own staffs and operational overhead. Consolidation of management functions and fund raising and marketing activities through cooperative management firms may become a necessity in the next few years.

In some ways, the growth attributed to the baby boom generation may have led to an oversupply of arts organizations in relationship to the demographic changes currently in progress. Ticket sales and donation levels may have peaked, and the market place may now be left with too many arts groups chasing too few dollars. Some combination of regional cooperative productions and cable television distribution may provide the needed revenues to keep the arts an important part of U.S. society. It is possible to envision a whole new subscriber base made up of television viewers watching the show from their home theaters.

Summary

All organizations in an open system interact with changing environments that shape the transformation and output of the product. The economic, political and legal, cultural and social, demographic, technological, and educational environments interact to form a complex set of conditions that influence how well an organization will be able to meet its objectives. The evaluation of the six environments is a function of information gathered from audiences, other arts groups, board and staff members, the media, professional meetings and associations, and consultants. Managers must assume that the environments are constantly changing and must therefore develop a process for continually evaluating input.

The economic environment is the most influential external force. General conditions such as inflation, recession, interest rates, and the taxation system determine the financial health of the operation.

The impact of the political and legal environment on an arts organization extends from the international scene to the local level. Cultivating positive communication and stressing the important part the arts play in the lives of voters can help build support from within the political arena.

The cultural and social environment is a combination of the values and beliefs of the society, as communicated through the family, the educational system, religion, and, increasingly, the broadcast media. The changing family profile, increased racial diversification, expanding career and work choices for women, and gender role differences in U.S. society are creating a different profile of the potential audience member.

The distribution of the people in the United States is changing in terms of age, sex, race, income level, education, and location. The baby boom generation that fueled much of the growth in the arts is aging, and it is not being replaced in equally large numbers. The birth rate has been dropping since 1970. The impact of these demographic changes will have a profound effect on the arts well into the next century.

Technology, once a major threat to the live performing arts, is now helping artists reach a wider audience than at any time in history. New technologies such as videotape and videodisc have helped increase the distribution of the arts in the United States. New technologies may make the experience of the live performance available to consumers in their homes.

The U.S. education system is undergoing tremendous pressure to increase its effectiveness. Because education levels are a strong predictor of later attendance at arts events, arts managers would do well to become part of the education revolution by working to incorporate the arts into the changing educational environment.

Issues of censorship, the AIDS epidemic, and the increasing economic pressures on arts organizations present challenges few organizations have effectively solved. It is clear that new strategies for delivering the product must be developed if the current demographic trends continue.

Key Terms and Concepts

Environments in the open system:

- economic
- political and legal
- cultural and social
- demographic
- technological
- educational

Continual evaluation process
Content analysis
Demographic describers

- sex
- age
- race
- income level
- occupation
- education
- birth and death rate
- geographic distribution

Virtual reality
Information sources:

- audiences
- other arts groups
- board and staff members
- the media
- professional meetings and associations
- consultants

Questions

1. Do the six environments affect the various art forms in different ways? For example, are theater groups more or less influenced by changes in these environments than art museums are? Explain.

2. This chapter focused on the influence of the environments on organizations. What influence do these environments have on the individual artist?

3. Do you agree with the views expressed in *Megatrends 2000*, or do you feel that McDaniel and Thorn's assessment of the arts in crisis is closer to your view?

4. Using the article on the effects of the recession on art museums, detail the various influences being felt by museums.

5. Although it appears that the ruling to limit the number of artists granted entry visas is headed for revision, can you make an argument for defending such restrictions?

6. What combination of demographic describers would you use to outline why you and your family or associates are arts consumers?

7. What opportunities and threats will artists and arts organizations face over the next 20 years?

Case Study

The Future Looks Grim for a Theater in Los Angeles

Robert Reinhold

LOS ANGELES—When it opened in a converted Greek Revival bank building six years ago, hopes were high that the Los Angeles Theater Center would form the nucleus of a Soho-like arts district to revive the seedy Spring Street area downtown. The city pumped more than $27 million into the experimental multicultural company.

But today the nearby storefronts are mostly vacant, drug dealers and homeless men gather in littered doorways and the young company teeters on the brink of financial collapse, a stark example of the risks of using the arts as a tool of urban redevelopment. The city has said no more money, and *The Los Angeles Times* has refused to run the company's advertisements for failure to pay.

By most accounts, the theater center, which has produced nearly 100 plays in its short, busy life, has been a considerable artistic success and a boon to the city's vibrant Hispanic, Asian and black theater scene. On Monday night, the cast of "The Phantom of the Opera" gave a special cabaret-type performance, raising $214,000 for the debt-ridden company, which needs to raise $500,000 by the end of the month to complete a season scaled back to just 7 plays from the 14 it used to produce.

But almost all those in the Los Angeles theatrical world agree that the money would provide only a temporary reprieve from the inevitable and that the company cannot survive much longer. Its probable demise has set off deep soul-searching among those in the arts about why the nation's second-largest city, ranking only behind New York City, and seat of the movie industry, cannot support a public theater or create a downtown arts district.

"This is a major tragedy for Los Angeles: its loss will not easily be repaired," said the president of the California Institute of the Arts,

Steven D. Lavine, whom Mayor Tom Bradley appointed last year to lead a special study of the theater center.

"No other major American theater has such a diverse audience and diverse casting," Mr. Lavine said. "They push the edges of audience taste. Part of the problem is that L.A. has relatively new cultural institutions and less of a history of facing this kind of setback. We are not practiced at organizing ourselves to address an emergency, as New York has had to do often."

Under the leadership of William Bushnell, the 1215-seat theater complex opened in 1985 and quickly became one of the largest of the most productive nonprofit public theaters in the United States, comparable in many ways to Joseph Papp's Public Theater. It supports theater laboratories for Hispanic, black and Asian actors and playwrights. Its provocative, often iconoclastic productions, including "The Illusion" and "August 29," won critical acclaim.

"It was a much needed component of cultural life: it offered a lot of experimentation," said the director of the city's Department of Cultural Affairs, Adolfo V. Nodal, who oversees the theater.

Its current productions include Reza Abdoh's "Bogeyman," which blasts critics of public financing of the arts, and "Bowl of Beings," a biting satire by Culture Clash, a trio of Hispanics signed by Fox Television for a sitcom.

The theater was to have been the centerpiece of Los Angeles' efforts to breathe new life into Spring Street, once the center of the city's business life but now full of low-price clothing stores, residential hotels and drug dealing. Despite an investment of $65 million in the area from the Community Redevelopment Agency, the revival has failed. The restaurants and night clubs that opened soon after the theater center have all become bankrupt, and only two of the many planned developments—a state office building and a condominium—have opened. Though ticket sales are good, many Angelenos are frightened to enter the area after dark.

But the larger reasons for the center's troubles have had to do with its financing and a 34-member board that has been unable to raise sufficient private funds.

"The real costs of operating a complex like that were not thought through from the beginning," said Gordon Davidson, the artistic director of the Center Theater Group, which operates the Mark Taper Forum theater at the Music Center of Los Angeles County. "There was not sufficiently diversified backing. At the same time, other things did not happen on Spring Street."

'He Has Made Enemies'

There was also the issue of Mr. Bushnell, a forceful, frequently abrasive figure who made as many enemies as friends for the theater. "He kind of runs over people in order to get what he wants, and in some cases, he has made enemies," said Daniel S. Lewin, a former stage manager for the Los Angeles Theater Center. "But the place wouldn't exist if he wasn't that way."

For his part, Mr. Bushnell cites a combination of factors, including the recession, which has left the city strapped. "If I were a better politician in the traditional American way of never really saying what you feel, I probably wouldn't have had some of the problems I've had," he said.

But he bridled at suggestions that the rundown neighborhood was a factor. "People in Manhattan don't think anything about going to

Broadway, and that area is no less dangerous than our neighborhood," he said. "What did not go wrong is the art, and our audience has continued to grow."

The turning point came last year when the theater center asked for a $32 million subsidy over 5 years. An infuriated Los Angeles City Council balked, but approved a plan by which the redevelopment agency transferred title of the theater center to the city's Department of Cultural Affairs and paid off $5.3 million in bonds. That still left the company $1 million in debt, and responsible for maintaining the building and the payroll.

Mr. Nodal, the director of the city's Department of Cultural Affairs, said efforts were being made to reorganize the operations so that a consortium of theater companies or another operator, possibly including local universities, could take over the center, with the company remaining only as a tenant.

"They have not had a problem selling tickets," Mr. Nodal said of the center. "What they have been unable to do is generate enough foundation and corporate support to sustain themselves. The problem is they are controversial. That has put a cloud over their heads."

The fate of the center is unclear today, but Mr. Nodal said that whoever ultimately runs it, "the building is not going to get turned into a bowling alley."

SOURCE: Robert Reinhold, "The Future Looks Grim for a Theater in Los Angeles," *New York Times*, August 14, 1991.
NOTE: The Los Angeles Theater Center closed in October 1991.

Questions

1. Do you agree with Mr. Lavine that the theater district has not been able to thrive in the environment in which it is placed because the L.A. theater scene is "not practiced at organizing" itself?

2. Based on what the author says about the issues of bankrupt restaurants and nightclubs, drugs, street crime, and homeless people sleeping in doorways, what conclusions can you reach about how well this plan was conceived in relation to the social, legal, political, and economic environments in L.A.?

3. Gordon Davidson is quoted as saying that the "real costs of operating a complex like that were not thought through from the beginning." What sorts of costs do you think he was referring to?

4. Theater manager William Bushnell admits that he could have been a "better politician" but then goes on to describe a politician as a person who never says what he or she feels. As an arts manager dependent on public funding, do you think Bushnell's attitude will lead to a constructive working relationship with the political community?

5. When Bushnell says that the Spring Street district is "no less dangerous" than Broadway, do you think the frightened Angelenos cited earlier in the article will feel any better about going to his theater?

6. What do you suppose the author means when he says that the request for the five-year subsidy "infuriated" the city council? What could the theater complex have done to more effectively win the support of the political community?

7. Isn't the point of the whole article negated by Adolfo Nodal's comment that the complex did not get foundation and corporate support because "they are controversial"? In other words, perhaps the problem isn't that L.A. cannot sustain a theater district, but that the product being offered doesn't fit within the parameters of the funding community. Short of doing musicals and comedies, what steps could the company take to fit better within the funding environment?

References

1. John Naisbitt and Patricia Aburdene, *Megatrends 2000* (New York: Avon, 1990), 76.
2. Nello McDaniel and George Thorn, *Workpapers: A Special Report—The Quiet Crisis in the Arts* (New York: FEDAPT, 1991), 7–10.
3. Lynne Fitzhugh, "An Analysis of Audience Studies for the Performing Arts in America," Part 2, *Journal of Arts Management and Law 13*, (Fall 1983), 7.
4. John O'Connor and Lorraine Brown, eds., *Free, Adult, Uncensored—The Living History of the Federal Theatre Project* (Washington, DC: New Republic, 1978).
5. "What Happened to the American Dream," *Business Week*, August 19, 1991, 80–85.
6. Howard Rheingold, *Virtual Reality* (New York: Summit, 1991).
7. "Cyber Arts International," November 15–17, 1991, Pasadena Center, Pasadena, CA. The information was taken from a multipage promotional flyer listing Bob Gelman and Dominic Milano as publishers and Linda Jacobson as editor.
8. Rheingold, *Virtual Reality*, 19.
9. Lynne Fitzhugh, "An Analysis of Audience Studies for the Performing Arts in America," Part 1, *Journal of Arts Management and Law 13*, (Summer 1983).
10. Ibid., p. 56.
11. Naisbitt and Aburdene, *Megatrends 2000*, 5.

Strategic Planning and Decision Making

5

If you aren't thinking ahead, you'll be left behind.

As noted in Chapter 1, planning is one of the primary functions of management. In this chapter, we will define planning and look at strategic and operational planning and the decision-making process. We will examine types of plans, the planning process, and the development of a mission statement. The decision-making process will be examined as a tool to assist with the planning process. This chapter borrows terminology and concepts from the business world and adapts these to the needs of arts organizations.

Before we can delve into the topic of planning, we need to step back for a moment and consider an important question: Why are we doing this concert, play, or exhibition? Are we trying to raise money for a cause? Are we trying to make a profit or generate a surplus in our depleted operating budget? Are we trying to bring something new to the audience—a new play, a new choreography, a new composer, a seldom-seen artist?

As we discovered in Chapter 4, there are complex forces at work in the various environments in which an arts organization must function. Artists and organizations, whether they like it or not, have had to adapt to the pressures of the external environments that are an integral part of our society. For example, solo artists unencumbered by a board of directors and an administrative staff may be able to achieve the goal of performing a new work through the sheer force of their energy and drive. A well-established orchestra, on the other hand, may debate for months over the conductor's desire to do a new series of modern music concerts. In the latter case, the various and probably differing attitudes of the members of the board enter into determining why an organization selects a new direction.

After answering the hard questions—the whys—the remainder of the planning process may seem simple.

Planning Basics

Planning, Goals, and Objectives Defined

At the most basic level, *planning* is a process of stating objectives and determining what should be done to accomplish them. Planning involves thinking about the future. It requires imagination, careful thought, and, most importantly, time. Management textbooks often cite these reasons why planning fails: (1) people do not think beyond the immediate future; (2) they are too impatient to work through the details of a plan; and (3) they think planning is part of someone else's job.[1]

This text will define *goal* as a desired outcome and *objective* as the means to achieve the outcome. For example, "It is our goal for the museum to have the widest possible distribution of members. One of our objectives is to increase minority membership by 10 percent in a new membership recruitment campaign by June 1993."

Types of Plans

A *plan* is a statement of intended means for accomplishing stated results. A plan should answer five questions:

1. Why?
2. What?
3. When?
4. Where?
5. Who?

Here is an example of this approach: "In order to fulfill one mission of bringing new music to the community, our marketing and sales staff will expand our subscription audience by 7 percent for next year's concert series in Smedly Hall by contacting corporate personnel departments and offering group discounts." As you can see, we have our *why*—the organization's mission to bring new music to the area, *what*—expand concert subscriptions by 7 percent through group sales, *when*—for next season, *where*—Smedly Hall, and *who*—the marketing and sales staff.

Planning Proverb #1

"Planning is 80% thinking and 20% writing. Then 100% *doing*!"[2]

Now let's look at the various types of plans before describing the planning process in more detail.

Short-, intermediate-, and long-range plans A short-range plan covers one year or less, intermediate-range plans cover one to four years, and long-range plans cover five or more years. Long-range plans that exceed five years are of limited value because there are too many unforeseen variables.

It is important to consider how people within the organization perceive of time. Research on planning points out that most people are comfortable with thinking three to six months ahead. Once you get past one year, people are only able to think in the most general terms. The age of an organization also determines perceptions of time. When you are first getting an organization up and running, four months can seem like a long time. However, if you are part of a long-standing arts organization, three- to five-year plans might not be so difficult to comprehend.

Strategic and operational plans The phrase *strategic planning* has an impressive "corporate" ring to it. The original use of the phrase was to describe the planning and direction of large-scale military operations to maximize forces before engaging the enemy. The business world regularly transposes military concepts into their daily operations. We will define *strategic planning* as a set of comprehensive plans designed to marshal all of the resources available to the organization for the purpose of meeting defined goals and objectives derived from the mission statement.

Operational plans are usually more limited in scope. They are the tactical plans in our military model. In this case, we are talking about marshaling the resources to support the strategic plan. For example, if a museum wants to achieve a larger membership base, it would need an operational plan that might include a specific marketing plan. At the same time, to undertake the strategic plan, the museum might need a personnel plan that calls for hiring new people with the expertise needed to run an effective marketing campaign.

Single-use and standing-use plans These plans are often the most common ones found in arts organizations. Examples of *single-use plans* include a budget, a schedule, and a project. Examples of *standing-use plans* include a policy, operating procedures, and rules.[3]

A *budget* is a plan for distributing resources. A budget that allocates more money for costumes than for scenery says something about the idea driving the production. An exhibition budget may allocate 80 percent of the money for a full-color book and only 20 percent for mounting the exhibition. These choices ideally represent a plan agreed to by all of the people involved in putting on the show or setting up the exhibition.

Another single-use plan would be a schedule. A *schedule* is a list of deadlines for completing specific tasks designed to meet the overall objective. For example, when an opera company sits down to plan a season, it works from a single-use plan: the season production schedule. If the company knows it will perform *Aida* next season, it can plan for accommodating the animals for the triumphal march scene. The logistics related to this part of the plan can be arranged far in advance. Budgeting and scheduling are discussed in more depth in Chapter 9.

A standing-use plan is designed to be used over and over again. For example, an arts organization should have a standing-use plan dealing specifically with how the administrative offices will operate. A theater box office should have a standing-use plan detailing the day-to-day operational procedures for processing orders and accounting for all revenue. Standing-use plans are also found in policy books, employee regulations, or posted rules.

On the surface, these two planning components may seem less weighty than the grand strategic plans, but they are often critical for the success of an organization. As we will see in Chapter 7, employees depend on well-designed single-use and standing-use plans to do their jobs.

The Planning Process

Most of us use a planning process of one sort or another to get through the day. You start your day by saying, "After class I have to go to the bank, then lunch with Fred, and over to the library at 2P.M." Let's take a look at a more formal process.

Five Steps in Formal Planning

1. Define your objectives. This key first step defines what you want to achieve. For example, "I want to have 40 bookings for my touring concert group by November 1." You must be specific in this first step. Specifying a quantitative achievement by a fixed date is one way to define your objectives.

2. Assess the current situation in relation to your objectives. You must clearly assess where you are and just how far you have to go. "I have 15 bookings, and it is September 1. I still have 25 to go in two

Business Insight

The following excerpt from the *New York Times* shows one corporation's strategic planning in the entertainment world. The question is whether the public is ready for Sonyworld.

Columbia Pictures to Enter the Theme Park Business
Michael Lev

LOS ANGELES—Columbia Pictures Entertainment, the Hollywood studio owned by the Sony Corporation, plans to enter the theme park business and may build one or more attractions in the United States and abroad.

Columbia said it had received permission from Sony to move ahead with plans for Sonyland, but that details on the size, scope and location of any proposed project would not be ready until later this year.

One option it is considering is building an attraction in Southern California to compete against Disneyland and the Universal Studios' Tour in Hollywood.

Theme parks are crucial to the business of several of Columbia's competitors in entertainment, including the Walt Disney Company, and MCA Inc., now a subsidiary of Matsushita Electric Industrial Company. The theme parks are highly lucrative and considered valuable accompaniments to the film and television businesses of the two large entertainment companies.

SOURCE: Michael Lev, "Columbia Pictures to Enter the Theme Park Business," *New York Times*, January 10, 1991.

NOTE: A change in management leadership at Columbia Pictures led to the abondoning of this plan in 1992.

months. It took me 6 months to book the first 15. I'd better be more aggressive in seeking out bookings, or I'll never make my target."

3. Formulate your options regarding future outcomes. Now you must design specific options to choose from to reach your objective. "I will need three to four bookings per week over the next eight weeks. I can only devote two more hours per day to this project after I adjust my schedule. I could hire someone to help me, but I'll have to pay them. I could lower my target figure to 30 bookings, but my portion of the booking fee will go down. I could extend the deadline to January 30 in the hope of reaching my original target figure."

4. Identify and choose among the options. After creating and reviewing all of your options, you must select the option you assess as the most effective. For example, you might decide to hire a temporary assistant to help you secure bookings.

5. Implement your decision and evaluate the outcome. If this plan is to work, it will be critical for you to set up short-term measuring points to mark how well you are doing. You may find that you need to implement other options if the outcome still seems questionable.

These five steps may seem simple and straightforward, but more often than not, people and organizations fail to even make a plan. It takes work to keep on top of this process. One of the most important skills you can develop as a manager is to master the planning process and effectively put it to use.

The business world is filled with different approaches or combinations of approaches to take when planning. Arts organizations are usually not as bureaucratic in the planning process as businesses are.

Other Planning Approaches

Top-down and bottom-up planning *Top-down planning* simply refers to a process where the upper-level management sets the broad objectives, and then middle- and lower-level management work out detailed plans within a limited structure. *Bottom-up planning* begins with lower and middle management setting the objectives; upper management responds with final planning documents that reflect the input. Mixtures of these approaches make sense for most organizations.[4]

Top-down planning can fail if upper management does not consult with middle and lower management and labor when setting objectives. For example, suppose that the board and artistic director of a theater company plan to expand the season from 24 to 36 weeks. Before trying to implement this plan, they should ask the people in the other levels of management (design, technical production, and marketing) to evaluate the impact of this change. Middle- and lower-level management will be asked to prepare reports showing the increased costs and increased revenue anticipated as a result of this plan. Upper management will then have the information it needs to assess the consequences of expanding the season. Modifications can be made in the plan before final implementation. (Given the number of anecdotal stories by staff people about never being consulted when sweeping changes are being made in their arts organizations, I surmise that an effective top-down planning process is an ideal in some settings.)

Pure bottom-up planning is fairly rare because it is usually a very cumbersome process. Too much staff time is spent meeting and reviewing every detail of the planning documents. More typically, the process might begin with upper management requesting that middle and lower management draw up planning documents for their areas or departments. Difficulties may arise when middle and lower management are not well informed about the overall organizational goals and objectives.

Contingency planning As the name implies, the *contingency planning* approach sets alternative courses of action that depend on different conditions. Contingency planning is most effective when trigger points are built into the process. For example, suppose that your season subscription campaign began in March. You expected to have a 70 percent renewal rate by July, but the box office reports only a 40 percent renewal by July 15. You would now activate your contingency plan for another mailing and a media blitz.

Crisis planning *Crisis planning* is an offshoot of contingency planning. Unfortunately, many arts organizations adopt this planning process as a standard operating procedure or management style. The key reason why many arts organizations operate in this crisis mode is because of the poor management skills of the staff. When a ballet company plans a season based on selling 80 percent of capacity and has no formal plan when only 50 percent of the seats are sold, there will be an atmosphere of constant crisis. Crisis management often leads to middle- and lower-level staff saying, "We seem to be making it up as we go along."

Plans for dealing with a crisis do have a place in arts organizations. This is especially true when the organization must deal with the media, supporters, subscribers, or the general public. For example, an arts organization should have a plan ready to activate in the event of the death of a key person like a founder-director. Arts organizations all too often go through months and years of chaos because no one took the time to map out a plan before a crisis strikes.

Strategic Planning

To engage in strategic planning, which was previously defined as a set of comprehensive plans designed to marshal all of the resources available to the organization for the purpose of meeting defined goals and objectives derived from the mission statement, the management must start at its source: the mission statement.

Figure 5–1 depicts a flow chart for the strategic planning process. The cycle begins with a mission statement that expresses a general approach to solving a problem or meeting a particular need in a society. The organization then formulates goals and establishes specific objectives designed to fulfill the mission of the organization. The specific plans are implemented to influence the input environments that created the mission of the organization. Let's look more closely at each phase of this process.

The Mission Statement

A clear *mission statement,* which defines the organization's "reason to be," is the source from which all plans should spring. It makes no difference whether the organization is a small modern dance company or a large regional arts center. Groups of all sizes need a concise state-

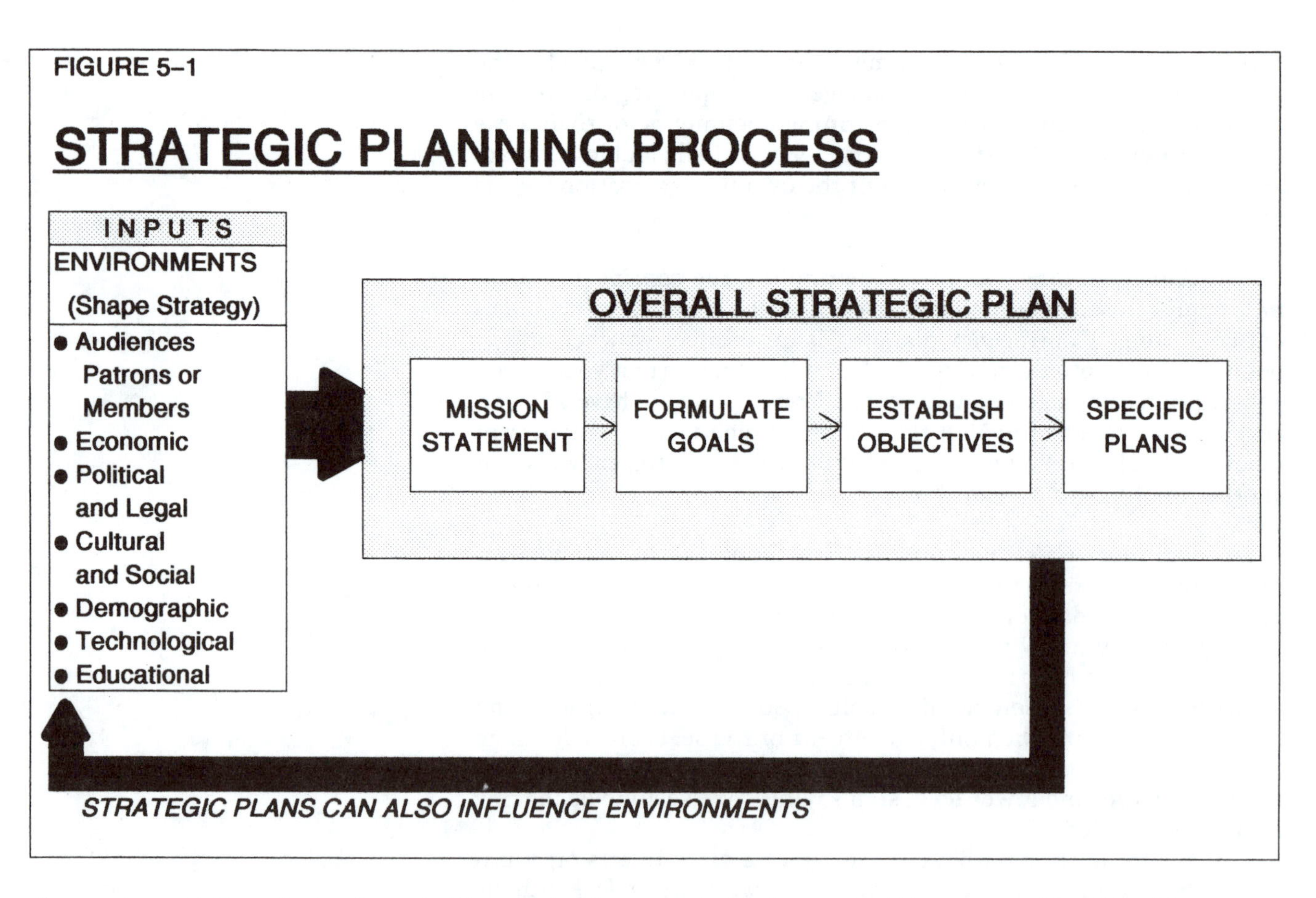

Source: James H. Donnelly, Jr., James L. Gibson, John M. Ivancevich, Fundamentals of Management, 6th Edition (Plano, TX: Business Publications, 1987), Copyright (c) 1987 by Business Publications, Inc. Used with permission.

The Effective Nonprofit Executive Handbook,[5] an excellent resource book published by the Public Management Institute of San Francisco, provides a check list to help arts managers organize their thoughts about the mission statement. They suggest that arts groups ask these nine questions:

1. Why do we exist?
2. What "business" are we in?
3. What is our most important product or service?
4. Who are our clients, volunteers, and donors?
5. Why do they come to us?
6. How have we changed in the past five years?
7. What are our organization's unique strengths and major weaknesses?
8. What philosophical issues are most important to us?
9. What would be lost if we ceased to exist?

Mission Analysis

An organization's mission statement should broadly describe its "reason to be." For example, a theater company might be dedicated to presenting new works, or a ballet company might be committed to performing classical works. The following are some real-life examples. The Guthrie Theatre "exists to celebrate the shared act of imagining between audience and actors."[6] The San Diego Symphony Orchestra

In the News—Theater Company and Strategic Planning

The following article demonstrates how an arts organization can make a fundamental change in its season schedule as part of an overall strategy of finding a better fit within the local presenting environment.

GLTF Season to Run September to May

Marianne Evett

The Great Lakes Theater Festival announced its season will run September to May, like most of the city's performing arts organizations. The 1991 season will open on Sept. 14 at the Ohio Theater in Playhouse Square.

"People haven't been clear about when the season opens or when it closes," said managing director Mary Bill of the festival's split season of May–June and September–October. "And our mid-summer hiatus has caused us to lose momentum in the middle of the season."

Originally a summer Shakespeare theater in Lakewood, the Festival moved to Playhouse Square Center in 1982. It retained a July–October schedule (sometimes adding a holiday show) until 1986, a year after Gerald Freedman was appointed artistic director.

At that time over $700,000 in debt, the festival split its season, eliminating performances during July and August. "Our competition was the green spaces and good weather in Cleveland," said Bill. "Lakewood was perceived as a suburban green space, more conducive to summer theater. Our move downtown was good on all levels, but it positioned us differently in the city. The Ohio was recognized as a 'real theater,' and it became more difficult to draw audiences for our (classical) repertoire to that space in summer."

Bill said that the festival's commitment to the classics made educational institutions a natural audience segment. "If we perform when the schools are not in session, we are defeating our purpose," she said. The new schedule also will make collaborations with other theaters easier and help with cash flow problems.

The change does mean that the Festival will have the same season schedule as the Cleveland Playhouse. The Festival mounts six productions to the Playhouse's eight, and has a subscriber base of 4,100 in 1990 as compared with the Playhouse's 1989–90 base of 9,100.

Neither theater looks on the change as putting them in competition. "We overlapped with four of our productions this year," Bill said, "and our missions are entirely different."

"I think it's terrific," Playhouse artistic director Josephine Abady said about the change. "I hope we can develop an appetite for theater in Cleveland that will mean larger audiences for us and for them. We've been careful to choose a direction that doesn't compete with their commitment to classical theater. Together we can offer a well-rounded look at theater for the community.

"But, I'd have to say I'd be disappointed if they duplicate the direction we have chosen and start programming new works and American plays."

SOURCE: Marianne Evett, "GLTF Season to Run September to May," *Cleveland Plain Dealer*, March 5, 1991. Copyright © 1991 Reprinted with permission by the *Cleveland Plain Dealer*.

"exists to develop and maintain a full orchestral ensemble of the highest artistic quality presenting on a regular basis a variety of work from the entire orchestral repertory."[7] "The purpose of the Milwaukee Art Museum is to enrich life by advancing the appreciation and understanding of visual arts through collection, preservation, display, research, education and interpretation."[8]

Analysis of external environments The mission statement must be based on the organization's interaction with the world around it. For example, the Milwaukee Art Museum incorporated into its planning process an "Environmental Scan Report, . . . which examined local and national issues affecting the Art Museum, encompassing demographic, economic, social and art-related data."[9]

Analysis of strengths and weaknesses The process of looking at yourself and frankly assessing your good points and shortcomings is as difficult for an organization as it is for an individual. The arts consulting business thrives on bringing the outsider's viewpoint into what too often becomes a self-congratulatory process. Organizations, like people, have a hard time seeing their flaws.

One approach to this analysis is to break down the review into an evaluation of the organization's human, material, and technological resources and its operating system. Questions such as the following must be posed:

1. Do we have the people with the skills we need to realize the plan?
2. Do we have the facilities, money, equipment, and so on, to make the plan work?
3. To what extent can we apply technology to our plan?
4. Do we have the systems in place for supporting additional programs of activity?

An organization's entire planning process should be directed at creating objectives and plans of action based on fulfilling its mission statement. This may seem like an obvious starting point for planning, but I have spent many hours in meetings where conflicting views of an organization's mission statement were never resolved. The planning went ahead, but conflicts always came up when the planning process led to the stage of deciding what was most important to the organization. A contradictory mission statement is like an out-of-focus photograph. When people view the picture, they often see different things. They reach conclusions and make assumptions based on imprecise information.

Formulating an Organization's Strategy

Strategy defines the direction in which the whole organization intends to move. It also establishes the framework for the action to be taken to achieve the goals outlined in the strategy. We have noted that the strategy should relate to the environment in which the organization must function. In Chapter 4, six environments were outlined: economic, political and legal, cultural and social, demographic, technological, and educational. Depending on any number of conditions, one or more of these environments could be stable, undergoing change, or even be uncertain.

Strategic planning usually draws on one of the following approaches: stability, growth, retrenchment, or some combination of these.

Stability strategy The basic thinking behind this strategy is, "We are doing pretty well with our current operation, and there is no reason to make any big changes." This does not mean that the organization is doing nothing about meeting its stated goals and objectives, but it does imply that there is no reason to move off into new directions. Many arts organizations would probably feel comfortable adopting this strategy. The major arts organizations in a community are often seen as institutions that are part of the basic fabric of the area. People can't imagine not having the museum, the symphony, and so on.

Growth strategy In the business world, this approach makes sense when expanding operations into new markets. With this strategy, a company may diversify its product line or actively seek a bigger share of the market. Arts organizations may adopt growth as an overall strategy by doing such things as increasing the numbers and types of events that it produces. Another example of a growth strategy is to deliberately push for greater community involvement by adding a ballet school or an art school to the dance company or museum. With growth comes increased costs and, it is hoped, increased income. These elements should be carefully calculated in the overall strategy.

Retrenchment strategy The third strategy describes a slowdown, cutback, or elimination of some portion of the organization's activity. Because this process is often viewed as retreating, many organizations will go to great lengths to describe it as something else. For example, a music group might say, "We are engaged in a planned phaseout of our Tuesday night concert series." In other words, the group is retrenching and cutting back on its programming to save money. As we saw in Chapter 4, the early 1990s have been marked by a great deal of cutting back and retrenching among many arts organizations.

Combination strategy An organization might use all three of these strategies at any given time. Again, the influence of the external environments will determine to what degree various strategies must be adopted. If the community is experiencing an economic slump combined with an uncertain political environment, the organization might need to retrench in some areas and expand in others.

To answer these questions effectively, an honest appraisal of the organization's mission statement, threats and opportunities in the external environments, and the organization's strengths and weaknesses must be made. Here, again, the services of an outside consultant can help the board of directors and staff to keep a sense of perspective about what the organization will really be able to accomplish through its strategic plan.

Limits of Planning

In *Management for Productivity*, John Schermerhorn cites seven reasons why organizational plans fail:[10]

1. The upper management fails to build a formal planning process into the general operating routine.
2. The people involved in planning are not very skilled in the planning process.
3. The data used in making the plans are incorrect.
4. The resources needed are not made available to execute the plans.
5. Circumstances change due to unforeseen events.
6. Staff members do not want to change, and they hold to plans that do not work.
7. Staff members become bogged down in the details and fail to reach the broader objective of the plan.

Planning Proverb #2

Once you become committed to a plan, it can be hard to back out, even if it isn't a good plan.

Bachman's Inevitability Theory
"The greater the cost of putting a plan into operation, the less chance there is of abandoning the plan."[11]

Relationship of Planning to the Arts

The creation and ongoing use of a strategic plan can be an excellent way to provide the overall framework for keeping an organization headed in the same general direction. All too often, though, the strategic plan and the mission statement are put away in an attractive folder and only pulled out once a year at the annual board meeting. However, for these ideas and plans to be effective, they should be integrated into the daily operation of the organization. Does this ensure that the organization will be a success? No. In fact, many founder-driven arts organizations came to life and struggled to national prominence without any of this documentation. However, it would be very difficult today for an arts organization to get any long-term support from foundations, corporations, or government agencies without a published mission statement and strategic plan.

The Necessity of Planning

Over the last 25 years, arts organizations and artists have had to deal with ever-increasing accountability, especially when dealing with public money, corporate donations, and foundation support. It is typical for arts organizations to provide five-year plans in their grant applications. However, the need for arts organizations to remain flexible and open to change is also important. Planning that locks an arts organization into rigid thinking can be deadly to the whole enterprise.

The management of any arts organization must assume that change is a given. Opportunities and threats to the organization will constantly present themselves. Therefore, there is no choice other than to draw up plans detailing how the organization will respond to change.

The Organization's Map and Leadership

Most of us have had to read a map at some point in our lives. In effect, the strategic plan for an organization is the map it plans to use to get to its ultimate destination. Although most of us have read maps, few of us regularly create them. Arts organizations need the skilled assistance of managers who know how to use the techniques of strategic planning to help make these maps. Ultimately, all of the planning in the world will not ensure success. Without dynamic and articulate leadership, an organization will suffer and probably fail.

Decision Making

For any planning process to succeed, the organization must have a well-defined decision-making process in place. A good arts manager (or any manager for that matter) locates problems to be solved, makes decisions about appropriate solutions, and uses organizational resources to implement the solutions. Our discussion of planning was based on the assumption that the ability to make decisions was an integral part of the manager's background. Let's take a closer look at this key part of the entire planning process.

Choices, Decisions, and Problem Solving

You make hundreds of decisions every day. For example, you make a choice (to wear the long coat) based on a decision arrived at after (1) identifying a problem (it's cold), (2) generating alternatives (wear no coat, wear two sweaters, wear short coat), and (3) evaluating the alternatives (no coat and freeze, wear two sweaters and look bulky). This process lead to a problem being solved (keeping warm). *Problem solving,* then, is "the process of identifying a discrepancy between an actual and desired state of affairs and then taking action to resolve this discrepancy."[12]

Schermerhorn identifies three styles of problem solving: problem avoiders, problem solvers, and problem seekers.[13] The first two styles need little explanation. The third style describes the rare person who actively goes out and looks for problems to solve. At any given time, all of us have probably exhibited a little of each of these styles. You should give it some thought if one of these styles dominates your problem-solving approach.

When approaching problems, it is helpful to define whether you are dealing with expected or unexpected problems. For example, you should expect that from time to time an audience member will appear on Saturday night with a ticket for Friday's show. You should have a solution to this problem ready to be activated when the situation arises. An unexpected problem might be smoke pouring into the lobby from an overheated motor in the air circulation system. In this case, you must quickly assess your alternatives without creating a panic.

Steps in Problem Solving

The following example illustrates one way of proceeding through the problem-solving process:

1. Identify the problem. What is the actual situation? What is the desired situation? What is causing the difference? For example, suppose that your interns are always late, and you want them to be on time. You must try to determine why they are late. Is it inadequate transportation, inappropriate work schedules, an unsafe work place, the workload and expectations, or their supervisor?

2. Generate alternative solutions. This step is critical and requires some imaginative thinking. Your investigation of the situation should allow you to gather as much information as possible to evaluate various courses of action. For example, you may discover that the interns are late because the supervisor is disorganized and sometimes verbally abusive. Further investigation reveals that one of the interns has an attitude problem. He has been rallying others to stage a work slowdown by showing up late every day.

3. Evaluate alternatives and select a solution. You consider replacing the supervisor, the intern, or both. You also assess the workload expectations and any other relevant information you gathered about the situation. Your solution is to dispense with the assistance of the intern and reassign the supervisor. You also enroll the supervisor in a two-day workshop on human relations skills.

4. Implement the solution. After consulting others within the organization (there could be some legal or interpersonal problems you had not foreseen), you implement your solution.

5. Evaluate the results and make adjustments as needed. You monitor the new supervisor and interns on a regular basis, conduct formal and informal talks with all concerned, and monitor the former intern supervisor.

Problem-Solving Techniques

If problem solving were as easy as these five steps imply, then managing would be a much simpler task. In reality, problem solving is a difficult and demanding part of the manager's job.

Defining problems, making hasty decisions, and accepting risk One of the many difficulties in problem solving is accurately defining the problem. People often misidentify the symptom as the cause of the problem. For example, the lateness of the interns was the symptom for which we later identified the causes. In the planning process, you may create many extra difficulties if you formulate objectives based on incorrectly identified problems. For example, a drop in subscription sales is a warning symptom of a whole host of possible problems. The ultimate cause may be show titles, prices, schedule, or even sales staff, among other things.

Another difficulty in problem solving is jumping to a solution too quickly. The first solution is not always the best solution. This is where trying out ideas on others can be helpful. A group brainstorming session may give you the added dimension you need to solve the problem.

Management texts frequently note that problem solving can take place in environments that are uncertain and risky.[14] The way you go about implementing the five-step problem-solving process depends a great deal on factors that you may have little control over. For example, if the intern supervisor happened to be the spouse of the artistic director of the theater company, there would be an added element of risk in your decision.

Analyzing alternatives Probably the best approach to analyzing your alternatives is to write them down. You can make an *inventory of alternatives*[15] by simply listing all of the alternatives you have and writing out the good and bad points of each choice. Figure 5–2 demonstrates the use of this process to decide whether to buy new computer equipment or to upgrade the old equipment. By forcing yourself to write it down, you may see other alternatives or ramifications of a decision.

Making the final choice After you have written out all of the alternatives, you have reached the stage of making the decision. After all is said and done, you need to ask yourself, "Is a decision really necessary?" The intern may quit out of frustration, or the supervisor may ask for a transfer to some other part of the operation before you have finished gathering all of the evidence you need for a decision.

Decision Theory

In reality, the *classic decision theory* situation (clear problem, knowledge of possible outcomes, and optimum alternative) seldom exists.[16] Arts managers more commonly find themselves operating in the realm of *behavioral decision theory*. This theory assumes that "people only act in terms of what they perceive about a given situation. Because such perceptions are frequently imperfect, the behavioral decision maker acts with limited information."[17] According to this theory, people reach decisions based on finding a solution they feel comfortable with given limited knowledge about the outcome. For example, when faced with the problem of the difficult intern, you may opt for firing the intern and tolerating the obnoxious supervisor. You may assess the risk of transferring the supervisor and in turn alienating the artistic director and find that it is too high.

Conclusion

Planning, as described in this chapter, is a series of logical steps that can lead to creative solutions to problems. One of the manager's most important functions is to solve problems. An excellent way of solving

Figure 5–2

Decision Inventory

Decision: Three Box Office Computers – Buy New 486 or Upgrade XT?

Alternatives	Time Required to Implement	Estimated Costs	Pros	Cons
UPGRADE? 1. Replace motherboard, power supply and disk drives	10 days	$1500	Newer, faster processor inside	Need new hard disk drive, add $400 per machine
2. Replace as above and add new 200 megabite hard disk	3 days	$1200	New hard disk is faster, accesses data faster	Still have old keyboard and limited warranty
3. Replace with used 286 12mhz machines	5 days	$1800	Only $600 each, comes with hard disk and is ready to go	Used, obsolete machine, slow, limited warranty
REPLACE? 4. Replace with 486 33mhz machines	3 days	$6000	In stock, fast, full warranty, upgradeable	Higher cost, not newest technology

problems is to ensure that planning is integrated into all phases of an organization. For an arts manager, the organization's mission statement is a fundamental element in the planning process. The mission statement is not some historical relic to be taken off the shelf once a year and dusted off for a board meeting. Rather, it is a statement of the purpose of the organization, and at the same time, it is the force behind all decision making. The distribution of resources to performance, production, marketing, fund raising, and administration should be traceable back to the mission statement. When this link is broken, an organization finds itself in a struggle to make sense of why it is doing what it is doing.

Planning is a tool that any organization can put to good use. In his introductory chapter in *No Quick Fix (Planning)*, Robert W. Crawford writes, "Planning is, in reality, a commonsense way of defining what it is that one wants, when one would like to attain it, and how one goes about attaining it."[18] Crawford makes an excellent point about how people misconceive what planning really involves. He observes:

> It is fascinating how difficult it often is for individuals to transfer their understanding of planning in their own lives, and its flexibility, to organizations of which they are a part. More often than not, when organizational planning is brought up or initially discussed, psychological blinders appear. It often is assumed that planning is a restrictive process; that the organization and its creative leadership will be locked into a plan which may well not be good for either; that a plan must be adhered to rigidly once it is formulated and approved; that change is impossible, or at the very best, difficult; that it forces people to do things when they realize from further experience that doing something else would be better; that because one doesn't know what is going to happen in the future, one is precluded by a plan from taking advantage of opportunities which may arise unexpectedly. To put it succinctly, such perceptions of planning are ridiculous.[19]

Later in this text, we will focus on planning as it relates to the areas of finance (Chapter 11), marketing (Chapter 12), and fund raising (Chapter 13). All three areas rely on the work done in the strategic planning process.

Summary

Planning is a primary function of management. For arts organizations, creating a mission statement that defines their "reason to be" is an important first step in the planning process. A plan is a statement of means to accomplish results. The entire process of planning should clearly state the organization's objectives and help determine what should be done to achieve those objectives. Short-range plans (under one year), intermediate-range plans (one to four years), and long-range plans (five to ten years) are used to reach the stated objectives.

The overall master plan, called a *strategic plan*, supports the mission of the organization. Strategic plans may stress stability, growth, retrenchment, or some combination of these. The strategic planning process analyzes the organization's mission, reviews external environments, and examines the organization's strengths and weaknesses. Within the strategic plan, various operational plans are designed to achieve specific objectives. Operational plans include single-use plans (budgets and schedules) and standing-use plans (policies, rules, and regulations).

There are five steps in formal planning: defining objectives, assessing the current situation, formulating options, identifying and choosing options, and implementing the decision and evaluating the outcome. Planning approaches include top-down and bottom-up planning, contingency planning, and crisis planning. Organizations can benefit from formulating plans in case a crisis occurs.

Planning can fail if it is not integrated into normal operations, planning skills are poor, inaccurate data are used, resources are scarce, unforeseen events intervene, staff members resist change, or plans become bogged down in excessive detail.

For the planning process to be effective, an organization must have a decision-making system in place. Problem solving is the process of identifying a discrepancy between an actual and a desired state of affairs and then acting to resolve this discrepancy. There are five steps to the process: identifying the problem, generating alternative solutions, evaluating the alternatives and selecting a solution, implementing the solution, and evaluating the results. You must assess the risks involved in your decision and carefully analyze alternatives.

Key Terms and Concepts

Planning
Goals
Objectives
Short-, intermediate-, and long-range plans
Strategic plans
Operational plans
Single-use and standing-use plans
Top-down and bottom-up plans
Contingency and crisis plans
Mission statement
Decision making
Inventory of alternatives
Decision theory

Questions

1. What would be a good strategy for an arts organization to adopt if the national economy is in a recession?
2. According to "Columbia Pictures to Enter the Theme Park Business," what is the strategy Columbia Pictures is taking? Do you think this strategy will work? Explain.
3. Use the five steps of the formal planning process to plot out your own personal short-range plans (within the next year) and intermediate-range plans (within the next two to three years).
4. Summarize the strategy behind the change discussed in "GLTF Season to Run September to May." Will there be a more competitive theater environment in the community with the two groups running at the same time? Based on the information in the article, how do the missions of the two organizations differ?

Case Study

Karamu Renaissance: Troubled Theater Company Again Getting Its Act Together

Marianne Evett

"The renaissance on E. 89th St." blazes across the new season brochures at Karamu House. In the theater, the season opens with an ambitious production of the hit musical "Dreamgirls."

The promise of rebirth comes after a year in which financial and administrative problems nearly proved terminal for this 73-year-old Cleveland institution, where generations of black theater artists have been bred.

But there are signs that "the renaissance" is not just a public relations slogan; that an effort has begun not simply to stabilize Karamu but to rebuild it as an arts center primarily expressive of black culture but open to all. New executive director Margaret Ford-Taylor, who has worked at Karamu in many capacities over nearly 20 years, sees its mission "as a bridge for cultural understanding and a place for black people to go. It fulfills a great need."

Ford-Taylor has already begun to make changes.

Mike Malone, who left as theater director in 1982 amid considerable controversy has returned to direct "Dreamgirls" and "Black Nativity," the holiday gospel celebration by Langston Hughes, which was one of Malone's great successes here. William Lewis, Karamu's former executive director . . . returns in January to direct Athol Fugard's "The Bloodknot."

The theater administration has been reorganized. Don Evans, appointed theater director in 1984, remains this year as a consultant and artist-in-residence, involved in raising money, contacting Karamu's far-flung alumni and "searching for new scripts," he said. Reyno Crayton, a Karamu alum whose New York career has included work with the Negro Ensemble Company and the New York Public Theater, has become the theater manager.

Karamu's new Theater for Young Audiences program begins Saturday. A company of five professional adult actors, directed by Jeff Gruszewski, . . . will perform a series of children's plays on the weekends at Karamu. The first, "The Emperor's New Clothes" by award-winning playwright Max Bush, will be available for touring to schools, along with workshops lead by the actors.

Karamu will also give a home this year to three productions of the Cleveland Cosmopolitan Theater, an interracial group presenting original works.

Jane Palmer is new coordinator of the visual arts program, which is known locally for its offerings in ceramics. Palmer will add a course in mask-making and a daytime course for parents.

Other changes will address more deep-seated needs. Ford-Taylor reported that the board is being reorganized, and Karamu has been given financial breathing space to develop long-term funding. The chaos of the past year—when payrolls were sometimes not met, utility bills went unpaid and morale sank to new depths—seems far away.

Karamu's woes partly stemmed from conflicts, especially during the past 12 years, about its proper roles—as both an arts institution and social service agency; as a place to empower blacks and explore black culture, and one to exemplify black interracial cooperation.

Founded as an interracial institution by Russell and Rowena Jelliffe in 1915, Karamu had mostly white leadership until 1976. When a black administration took over, it was perceived as militant, alienating some supporters of both races. The divisions that erupted about Karamu's purpose and mission resulted in rapid changes in leadership over the next dozen years. Ford-Taylor is the fifth director since 1976; board leadership has lacked continuity and experience, and community support dropped.

Although the twin problems of leadership and financial stability continued to plague Karamu, things improved gradually under executive director Milton Morris. When Morris died in 1987, however, the crisis became acute.

Ford-Taylor, a doctoral candidate in theater at Kent State University, Emmy-nominated playwright and former director of the Afro-American Cultural Center in Buffalo, was appointed interim director by the board last fall and executive director in June. At Karamu, she has been volunteer and employee, cultural arts director, actor, playwright and play director.

Ford-Taylor explained in a recent interview that the financial crisis led board representatives to ask help from Karamu's major supporters, the Cleveland and Gund Foundations and the United Way. Under their sponsorship, an ad hoc committee was formed to study the situation and make recommendations about Karamu's continuing mission, organization and long-range financial planning.

The mission statement that emerged has Ford-Taylor's wholehearted endorsement. It reaffirms Karamu as a metropolitan center for the arts for people of all ages and races but defines its "unique responsibility" as "providing avenues and arenas for black artists to demonstrate, practice, share, communicate and further develop their skills and talents."

"The mission is clear and reflects what I feel about where Karamu is going," Ford-Taylor said. "It's a bridge, but on a bridge the traffic goes both ways or it doesn't work." She sees the "bridge" extending across age lines into the schools, and, she hopes, eventually into Cleveland's Hispanic community.

The mission statement ends, however, by stressing that Karamu should "adhere to and promote high standards of excellence. Since it was organized as the Gilpin Players in the 20's, the Karamu Theater has boasted alumni like Langston Hughes, Minnie Gentry, Ron Neal and Bill Cobbs. Under directors like Benno Frank and Reuben and Dorothy Silver, it acquired a reputation for quality work.

Shows at Karamu will no longer be double cast, except certain roles in musicals, which are vocally demanding. "That means (the production) will be more disciplined," Reyno [the newly appointed theater manager] said.

Ticket prices have also been lowered in an effort to draw a wider audience. Once a standard $12, they are now set at $5.50 on Thursdays, $7 on Fridays and Sundays and $8 on Saturdays. Runs have also been shortened from eight weeks to four for plays and six for musicals, making it somewhat easier to recruit actors.

At the start of her first year as executive director, Ford-Taylor is full of hope. "I have a vision, and a chance to try it," she said. "It's not that we haven't any problems, but I am excited by the possibilities. I love the place, and that makes me strong.

"I say, 'Just hold on; we're almost there.'"

SOURCE: Marianne Evett, "Karamu Renaissance," *Cleveland Plain Dealer*, September 26, 1988.

Questions

1. Identify three general strategic planning objectives.
2. Name at least five of the specific planning objectives that Karamu envisions.
3. Do you think the revised mission statement and the overall strategy will work? Explain.

References

1. John R. Schermerhorn, Jr., *Management for Productivity*, 2d ed. (New York: John Wiley and Sons, 1986), 98.

2. Harold R. McAlindon, *Management Magic* (Lombard, IL: Great Quotations, 1989).
3. Schermerhorn, *Management for Productivity*, 100.
4. Ibid., 105.
5. *The Effective Nonprofit Executive Handbook* (San Francisco: Public Management Institute, 1982).
6. Guthrie Theatre, *Long Range Plan* (1988), II.2.
7. San Diego Symphony, *Five Year Plan* (1990), 2.
8. Milwaukee Art Museum, *Second Century Plan* (1990).
9. Ibid., 3.
10. Schermerhorn, *Management for Productivity*, 114.
11. U.S. Institute for Theatre Technology, Computer CallBoard.
12. Schermerhorn, *Management for Productivity*, 64.
13. Ibid., 65.
14. Ibid., 76.
15. Ibid., 77.
16. Ibid., 79.
17. Ibid., 80.
18. Robert W. Crawford, "The Overall Structure and Process of Planning," in *No Quick Fix (Planning)*, ed. F.B. Vogel (New York: FEDAPT, 1985), 14.
19. Ibid.

Additional Resources

The following sources were also used in writing this chapter.

Bedeian, Arthur G. *Management*. New York: Dryden Press, 1986.

Donnelly, James H., Jr., James L. Gibson, and John M. Ivancevich. *Fundamentals of Management*, 7th ed. Homewood, IL: BPI/Irwin, 1990.

Fundamentals of Organizing and Organizational Design

6

Whether we like it not, we spend the greater part of our lives in organizations. Our contact with organizations may start with a day care center, then move to a series of educational organizations, then to a work organization, and finally, we may live out retirement in an organized elder care system. The family, which is also an example of an organizational unit, can be a powerful force in shaping how we interact with others and with organizations. Our ability to relate to the numerous complex organizations in our society determines how successful we are in achieving our personal goals and objectives. The powerful myths of the individual going it alone in society are offset by the reality that we need the support of people in an organization to achieve maximum results. One person can make a difference, but many people working together can create permanent change.

In this chapter, we will analyze many of the basic concepts pertaining to organizations and organizational design. Then we will apply these theories to arts organizations. We will also examine the importance of matching the organization's structure to the task at hand. Finally, we will review the phenomenon of organizations as cultures.

The Management Function of Organizing

In the study of management, organizing is usually listed as the second basic function.[1] If you are to implement effectively the strategic plans formulated in Chapter 5, you will need a way to organize your resources to realize your objectives.

A good starting point is to return to our earlier definition of an *organization* as "a collection of people in a division of labor working together to achieve a common purpose."[2] The term *organizing* was defined as "a process of dividing work into manageable components and coordinating results to serve a specific purpose."[3] We previously defined a *manager* as a person in an organization who is responsible for the work performance of one or more other people, and we defined *management* as a process of planning, organizing, leading, and controlling.

Four Benefits of Organizing

No matter what project or production you plan to undertake, four benefits can be derived from organizing:[4]

1. making clear who is supposed to do what
2. establishing who is in charge of whom
3. defining the channels of communication
4. applying the resources to defined objectives

It is part of the arts manager's job as an organizer to decide how to divide the workload into manageable tasks, assign people to get these tasks done, give them the resources they need, and coordinate the entire effort to meet the planning objectives.[5]

Organizing for the Arts

The task of organizing to achieve results should always be the arts manager's objective. It is the underlying assumption of this text that an effective arts manager is functioning in a collaborative and cooperative relationship with the artist. People outside the arts sometimes erroneously assume that artists and arts organizations, by their very nature, are less structured than other organizations or that they function best in a disorganized setting. Nothing could be further from the truth. In fact, artists and arts organizations often foster what is popularly known as a *workaholic* attitude among employees. (See the "Books of Interest" for a brief discussion about the addictive organization.)

Although there are different ways to approach the process of putting on an exhibition or presenting a theater, dance, opera, or concert performance, each art form shares an inherent organizational structure that bests suits its function. For example, theater is rooted in developing a performance based on many hours of text study, blocking and line rehearsals, and technical and dress rehearsals. The organizational support required to prepare, rehearse, design, produce, and find an audience for the play need not be discovered each time a new show is put before the public. There are standard ways of pulling together a production. When the support system is in place and functioning correctly, it is almost invisible. However, when something goes wrong with this system, it becomes the hot topic of discussion. As we will see in this chapter, there are various ways to go about organizing any enterprise. It is not an issue of a right way or a wrong way. As previously noted, organizational design and organizing should be aimed at achieving the desired results.

Before exploring the structural details of various arts organizations, let's take a look at the overall concept of the organization as an open system.

Organizational Design Approaches

Management theory approaches organizational design by using concepts such as mechanistic versus organic organizations,[6] the relationship to external environments, and the degree of bureaucracy within an organization. The mixture of these concepts may be outlined in a model of an open system. Figure 6–1 depicts how an organization transforms inputs to outputs. Within the overall environment, a constant feedback loop exists back to the input stage. For example, if the output is exhibitions and the stated mission is education, you will want to monitor the input from the people using your product to see if you are fulfilling your mission. You would then be sure to design into the organization a way to transform the data you received from the people who used your service. This translates into a department that gathers survey data, tabulates the results, and publishes reports informing management of how effectively the mission is being met.

Another way to turn this model into a practical tool is to examine how an arts organization might go about more effectively reaching beyond a small segment of the population. For example, if an arts organization's objective is to build a larger subscriber or membership base from minority groups in the community, how should the organization's input, transformation, and output stages be designed? One obvious structural element that could be put in place would be a staff member heading up a minority development project. This position would function with horizontal authority. In other words, the staff member in this area would be given sufficient authority to cut across departments in an attempt to increase the number of minority employees within the organization. He or she could work with the board of directors to get board seats for minorities. Over time, the organization's overall output should begin to create a feedback loop that signals to the community (input stage) that this is an organization that is trying to address minority concerns.

Mechanistic and Organic Organizational Design

Arts organizations tend to be found at the organic end of the continuum of mechanistic versus organic organizational structure. The distinguishing features of an *organic organization* are a less centralized structure, fewer detailed rules and regulations, often ambiguous divisions of labor, wide spans of control or multiple job titles, and more informal and personal forms of coordination. A *mechanistic organization* tends to have a great deal of centralization, many rules, very precise divisions of labor, narrow spans of control, and formal and impersonal coordination procedures.[7] Of course, organizations vary in degree when it comes to classifying them within this continuum. Some arts organizations may adopt aspects of a mechanistic organization. The size and complexity of the operation usually dictates these attributes. For example, the Metropolitan Opera is more likely to adopt aspects of a mechanistic organization than would a smaller organization like the Opera Theatre of St. Louis.

Bureaucracy

When an organization adopts a mechanistic organizational structure, we often refer to it as a bureaucracy. Ideally, a bureaucratic organization has clear lines of authority, well-trained staff assigned to their areas of specialization, and a systematic application of rules and regulations in an impersonal manner.[8] This form of organization, originally found in government by bureaus or unelected civil servants, was created to overcome the excesses of nepotistic control of social infrastructures. Today, we tend not to associate bureaucracy with democratic principles. Instead, we have very negative perceptions of bureaucracy. Anecdotes abound regarding bureaucratic structures bringing out the worst in an organization. We have all seen a rigid and unwieldy organization where, like a black hole in space, things are sucked in and are never seen again. Governmental agencies are often cited as prime examples of bureaucracy at its most entrenched. However, James Q. Wilson's insightful book, *Bureaucracy: What Government Agencies Do and Why They Do It,* argues that bureaucracies are not black boxes into which input is impersonally converted to output, but rather changing complex cultures.[9] Wilson also argues that the perceived mission and the people within the government bureaucracies are significant forces in shaping interactions with the society at large.

Modern views of bureaucratic structure suggest that it would be best to adopt a contingency approach in organizational design. In other words, the organization adopts only the amount of bureaucratic structure necessary to accomplish its objectives. Does this mean that an arts organization needs to have some bureaucratic structures? Yes, to some degree. For example, an arts organization must establish ticket refund, crediting, and billing policies and procedures. It needs a consistent structure that handles most, if not all, transactions in the same manner. You cannot operate a box office with every employee setting his or her own rules or procedures for routine tasks. The patron who comes to the ticket office to request a refund or exchange should be given fast and efficient service. Staff members should not have to consult one another on the proper procedure.

Another area in which arts organizations need a great deal of structure is payroll. Employees may complain about the bureaucracy, but because the payroll department must interface with local, state, and federal agencies, there is no choice in the matter. Because of the complexity of payroll regulations, you cannot make it up as you go along.

The idea of an open framework of policies, rules, and regulations makes a great deal of sense for an arts organization when it comes to adopting bureaucratic structure. As we have seen, different areas within the organization need more rigid structure than others. The box office and payroll both need a clearly defined framework with rules, regulations, and policies. However, a resident scene designer in a regional theater will not find a highly structured framework very helpful if she has to fill out a purchase requisition every time she wants a new pencil.

Probably the most useful tool for an arts manager to use in managing the degree of bureaucracy in the organization is testing, that is, walking through the procedures that are in place. The manager may discover procedures that are confusing, contradictory, or nonessential to the mission of the organization.

Organizational Structure

Let's delve now into a more detailed examination of how to structure an arts organization. When we talk about *organizational structure,* we are referring to the "formal system of working relationships among people and the tasks they must do to meet the defined objectives."[10] These relationships and tasks are usually shown in an *organizational chart,* which is "an arrangement of work positions in an organization."[11]

In the business or arts world, adherence to the organizational chart must be tempered with a healthy dose of reality. It is important to establish the organizational chart to help clarify how things work and whom to contact about getting something done. The organizational chart should help, not hinder, operations. If an organization finds itself unable to accomplish its objectives because the organizational structure frustrates action, it is time to reexamine how it is organized.

Organizational Charts and Formal Structure

A typical organizational chart should clearly show six key elements about the organization: divisions of work, types of work, working relationships, departments or work groups, levels of management, and lines of communication.[12]

Figure 6–1

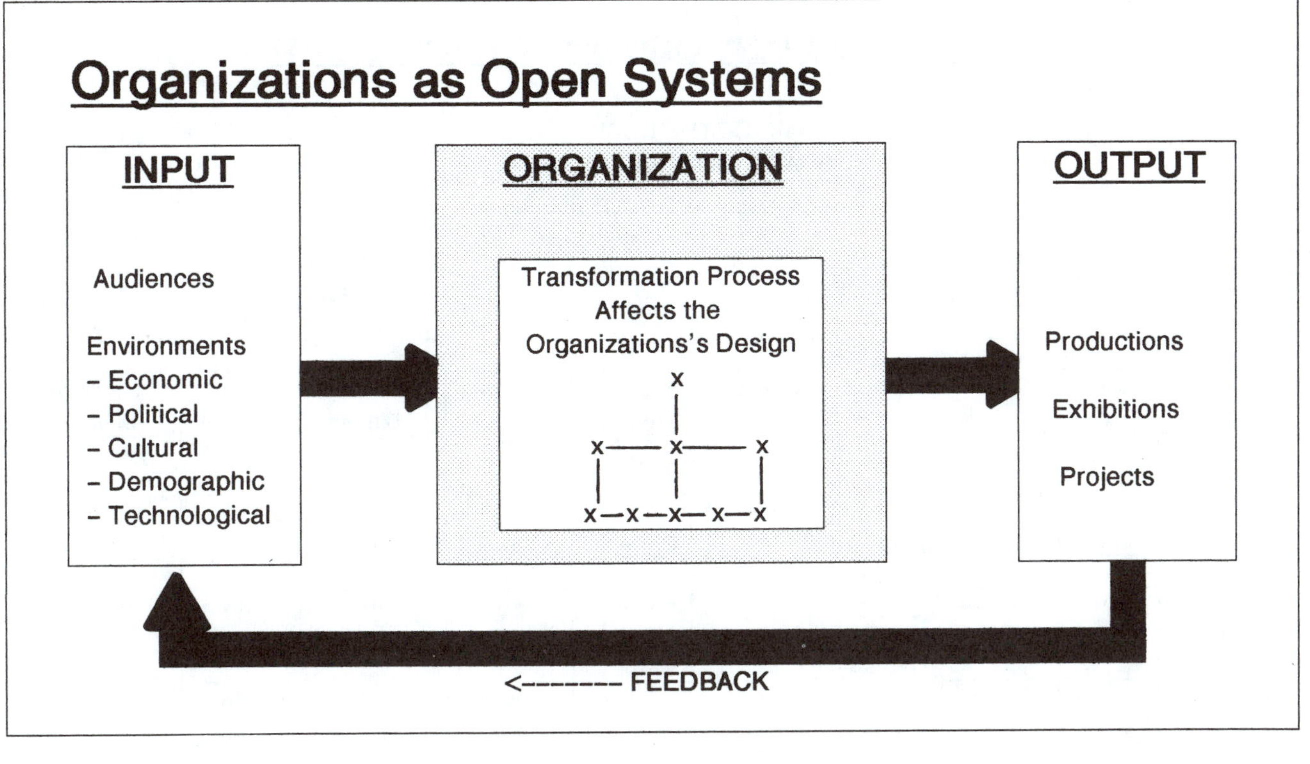

Divisions of work Each box represents a work area. Each area should designate an individual or group assigned to complete the organization's objectives. Figure 6–2 shows divisions of work in a theater company. Under each division are specific work areas assigned to complete the task. For example, the production manager is responsible for the production area. The technical director is assigned the specific task of completing the scenery. The technical director in turn has designated staff supporting the work area: the scene shop manager and the construction crew.

Type of work performed The title you use for the work area—for example, director of marketing, box office manager, costume shop manager—should help describe the kind of work the person or group will do. Care should be taken to avoid obscure work area titles. Vague or misleading titles often indicate that an organization is carrying staff positions that serve marginal functions.

Working relationships The organizational chart should show who reports to whom. The solid line between the production manager and the technical director indicates a supervisor–subordinate relationship. The production manager and director of marketing are on the same level, and a communication line connects them to other supervisors.

Departments or work groups The grouping of job titles under a work group or area should be communicated by your organizational chart. A performing arts organization may be large enough to warrant separate departments for work areas. For example, a highly structured organization like the Metropolitan Opera lists fifteen different departments in the ad-

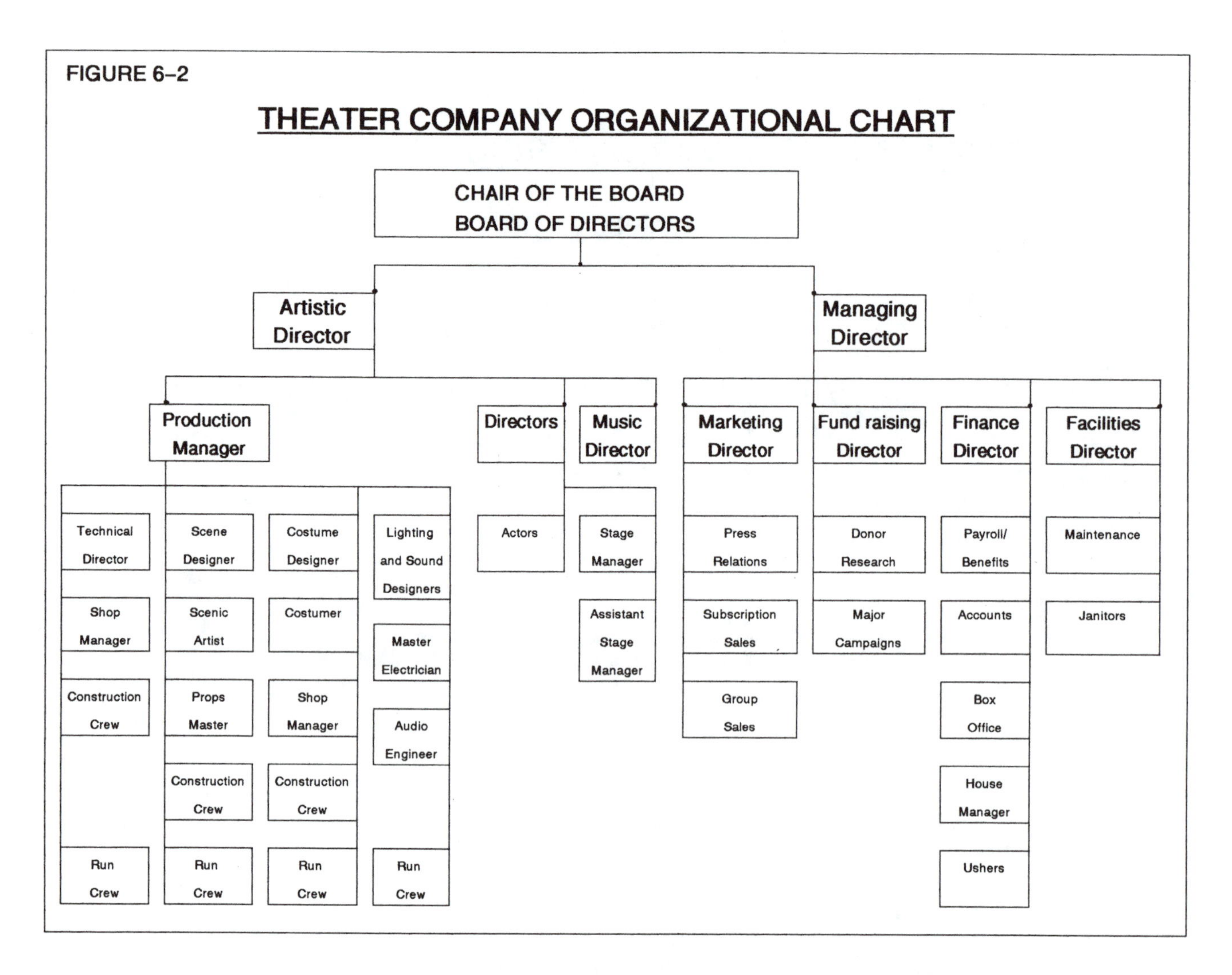

ministration area alone.[13] It is possible to identify four subunits of the Met production department: technical direction, design, stage management, and stage and shop operations. The marketing and annual giving department at the Met is broken down into nine separate departments, including box office, Met ticket service, and subscriptions.

Smaller organizations often combine departments. For example, the box office might include all single and subscription sales. Group sales and telemarketing might also be included in this department. It is not unusual in a small operation to find one person heading up four or five different subdepartments. With growth comes the creation of separate departments.

The levels of management The organizational chart should act like a map in depicting all of the levels of management (upper, middle, and lower). In Figure 6–2, the upper level is represented by the artistic director and the managing director. The production manager and the technical director can be identified as middle management, and the scene shop manager as lower-level management. This hierarchy theoretically reflects how information and work objectives are carried out in the organization.

As an organization grows, the levels of management also tend to grow. When an arts organization first begins to operate, there are usually two levels of management. For example, if you call the artistic director of a new dance company, the phone may be answered by the director or the secretary. In five years, there will be three or more levels of management. Your call will be taken by a receptionist, who in turn transfers you to the appropriate secretary, who then connects you to the associate artistic director. The associate artistic director screens the call to see if you really need to speak to the artistic director. When this scenario actually exists, it might be time to reexamine how many levels the organization really needs.

Lines of communication Finally, the organizational chart should represent the lines of communication throughout the hierarchy. For example, the scene shop manager tells the technical director that they will have to go over budget to complete the set as designed. The technical director informs the production manager of the situation. The production manager informs the artistic director, who in turn communicates to the managing director. In theory, the upper managers decide what they want to do, and that decision is passed back down through the hierarchy.

Informal Structure

Every organization has an informal organizational structure. Good managers remain aware of this underlying framework and use it to their advantage. At the same time, they must discern when this informal structure is damaging the organization and hindering the achievement of its overall objectives.

One of the reasons why an informal structure exists is to fill in the gaps in the formal structure. Interactions in an organization involve people, and that translates into a complexity and a subtlety no chart could capture. Inventive employees will always find a way to get things done in the organization with or without using the formal structure. For example, suppose that the scene shop wants a new table saw, but the technical director has refused to act on the request for months. At a cast party, the shop manager casually lets the production manager know that his department could speed up the set construction process if only it had that new saw. If the crafty shop manager uses this informal communication system properly, he may end up with a new saw before the technical director knows what has happened. He will also enhance his status with the shop staff because of his ability to get what is needed despite the system. In so doing, he may unwittingly send a message to the staff that it is all right to bypass your supervisor to get things done.

Problems inherent in the informal system This example points out some of the problems with the informal structure. One major fault is that the informal structure diverts efforts from the important objectives of the organization. This shadow organization may be more concerned with personal status.

Another difficulty with the informal organization is its resistance to change. You may define new objectives and marshal your resources to put a new procedure in place, but if the informal organization rallies around the "old way," your efforts may be in vain.

As we saw in the example, an alternative communication system usually accompanies the informal structure. One key element of this informal system is the rumor-spreading mechanism. Arts organizations, despite the common perception, are no more or no less prone to rumor spreading than other businesses. The rumor mill is usually a prime source of informal communication in a performing arts setting. Good managers will determine how the informal communication system works in their organizations in order to track what employees are really saying about the work place. From time to time, it can be useful to feed information into the rumor mill. For example, you might let it leak that upper management is not happy with the lax compliance with the new no-smoking policy. If carefully leaked, the rumor might create greater adherence to the new rule if staff members think there will be consequences if they continue their current behavior.

All organizations have these shadow structures. The sooner managers become informed about these informal systems, the better they will be able to monitor and influence them.

Structure from an Arts Manager's Perspective

General Considerations

The complexity of the organizational structure should be directly related to the size and scope of the operation. A good guideline to follow is to keep the structure to a minimum. There are no rules about how much structure is required. If an arts organization's mission is to further its art and serve the public good, creating elaborate organizational charts is not the ideal path. To help keep organizational design in perspective, consider the five elements that are often listed as influences on the final design:[14] strategy, people, size, technology, and environment. Of these five influences, strategy, people, and size have the most direct applications to the arts.

Strategy The organizational structure should be designed to support the organization's overall strategy. For example, if you are starting a new regional opera company, your objective might be to build a subscription base as quickly as possible. Your strategy in building that base might be to select your season around well-known titles with famous guest singers. Your organizational design would therefore stress more staff to take care of the guest artists and to carry out the organization's marketing, public and press relations, and ticket sales activities. To maximize resources, your strategy would also include keeping your operating costs as low as possible. One way to do that would be to rent all of the sets and costumes for the season. You would therefore need only a relatively small production department and no resident design staff.

A few years later, you might expand your audience development strategy by adding a community outreach program. Now your objective is to hire a tour director and a small staff to promote and support bringing opera scenes into schools. You might consider reorganizing another area within your existing organization to save on the expense of adding staff. Either way, you must make changes in the organization's design.

One of the reasons why new programs of activity sometimes suffer in an organization is that no one has thought about what was going to be done by whom. Without careful thought about job design, an organization can quickly overload an employee with too many tasks.

People The most important element in any organization is the people who work within the overall structure. Realistically, there must be flexibility between the structure and the person in an arts organization. The military is a prime example of a rigid organizational structure designed to mold its "employees" to the specific jobs at hand. A strict hierarchy and adherence to numerous rules and regulations are all focused toward a set of specific objectives. People usually do not join the staff of an arts organization because they want a rigid and highly controlled work environment. In fact, people who work in arts organizations are often highly self-motivated, and they vigorously resist regimentation.

Care must be taken when applying organizational theory to real organizations. In an arts organization, for example, the degree of structure varies with the type of job. For example, the director of a museum would probably be wise to allow for a degree of creative independence among the department heads of the curatorial staff. The security guards, on the other hand, would have a rigid work schedule with little independence.

Size When an arts organization is first established, there may be no more than three or four people doing all of the jobs in the organization. The artistic director may direct the operas, write the brochure copy, hire all of the singers and the artistic staff, and do all of the fund raising. As the organization grows—a board of directors is added, staff specialists are hired to do the marketing, scheduling, advertising, and so on—the simple organizational chart and lines of communication suddenly become much more complicated. In management theory, the organization would be said to be moving from an "agency form" to a "functional department form."[15] An *agency organization* refers to a structure in which everyone reports to one boss, and this boss provides all of the coordination. Each staff member is in effect an extension of the boss. When an arts organization decides to hire a marketing director, it may be because the artistic director can no longer supervise all of these activities. A department with a specific function and the support staff to do the marketing would be supervised by this new person in the organizational structure. Problems and conflicts will arise if the artistic director attempts to give direct orders to the staff of the marketing director instead of going through the new structure.

Technology and environment The other elements that influence organizational design are technology and environment. I did not list these as primary influences because of what I perceive to be the unique place of arts organizations in the overall business world. When speaking of the influence of technology in the business world, it is easier to see how new systems and methods can affect how products are produced. There is usually a direct connection between how an organization is structured and how technology may help it become more productive. For example, a large company may add a whole department to do nothing but assess new technologies and advise about their application to that business. Arts organizations, on the other hand, have used new technologies in office and information management and, in limited ways, applied new approaches to the technical production aspects of their operations. However, the technological changes do not radically transform what the arts organization does or how it goes about preparing or delivering its product.

External environments also have some impact on the structural design of arts organizations. The political and legal environments may legislate new laws that affect hiring in the organization. For example, there were virtually no affirmative action programs in arts organizations 20 years ago. Today, a staff person may be designated to monitor all of the activities of the organization to examine their impact on minorities.

Departmentalization

To departmentalize, or to set up departments in an organization, simply means "grouping people and activities together under the supervision of a manager."[16] Departments may be structured in three ways: by function, by division, and by matrix.

Function Most arts organizations use a structure defined by functional departments. It makes sense to group people by the specialized functions they perform within the organization. Figure 6-2 shows how the production manager of a theater company supervises the functional departments of scenery, costumes, lighting, and sound.

Division Departments can be organized around a product or a territory. Figure 6–3 from Langley and Abruzzo's excellent book, *Jobs in Arts and Media Management*,[17] depicts a major film studio organized around six divisions in a large corporation. An example on a smaller scale is a major arts center that not only hosts touring productions, but also produces its own shows, runs a gift shop and an art gallery, and operates a restaurant. An organization may decide to establish a divisional structure that keeps booking, production, exhibition, and food services separate. The logic behind this choice is that each of these activities involves very different operating conditions with specialized supervision and staff needs. The division in charge of production might include a marketing person to supervise the subscription series for the regular events. The division in charge of touring might employ another person to market their shows to other arts centers and producers. Both employees are marketing specialists, but they market their products from very different perspectives.

Another divisional structure would be by territory. A dance company decides to pursue a strategy of dual-city operations. One of the first steps would be to establish an organizational structure to staff two different geographical sites. There would have to be some staff duplication. You cannot expect the city A marketing staff to do the marketing for city B without local staff designated for each campaign.

Matrix The most complex structure is a matrix organizational structure. The matrix is created by overlaying the departmental and functional organizations in a vertical and horizontal pattern. The matrix system was created in the late 50s by the cofounder of TRW Inc., Simon Ramo.[18] The department structure proved to be inadequate when TRW tried to manage several technologically complex projects for the defense industry. His scheme was to use a department structure for important activities like research and development (R&D) and place the department under the control of a department head. However, within the R&D department were smaller groups of people working on different projects under the supervision of a project manager. The vertical matrix in the structure is thus the department head to the

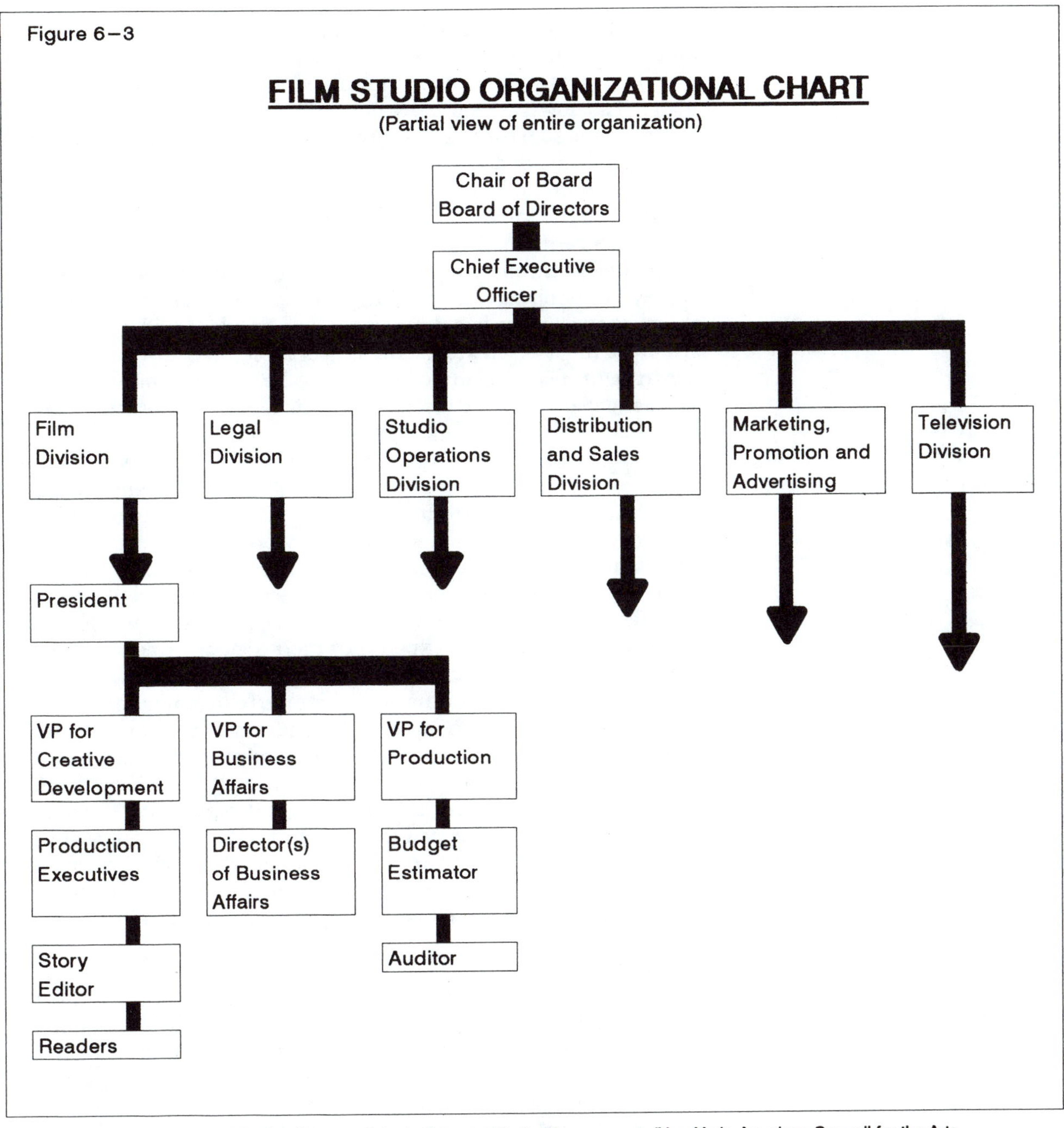

Source: Stephen Langley and James Abruzzo, Jobs in Arts and Media Management, (New York, American Council for the Arts, 1990). Copyright (c) 1990, American Council for the Arts. Used with permission.

various R&D groups. The horizontal matrix consists of the individual project managers working with the separate R&D groups.

In an arts organization, a matrix structure often evolves, although no one sits down and actually plans a change to this type of structure. For example, suppose that a museum is organized around a department structure. The six departments are responsible for various sections of the collection, and there are other departments for marketing, fund raising, operations, maintenance, accounting, and payroll. The museum's centennial is coming up in three years. A staff member is designated as the director of the centennial. If this project director is to achieve the objective of creating a successful celebration of all the things the museum has done in the last one hundred years, a matrix structure must be created. The project director will require that each department head designate a person to be the centennial coordinator for that department. In addition to their regular duties, the marketing staff will also have to work on this special project.

Arts organizations often find themselves involved in special projects. However, without recognition of the need to shift to a matrix organizational structure, trouble may occur. For example, if the added staff required to make a special project work are not hired, the project coordinator will have to work horizontally through the organizational structure. She will discover that the overworked staff in the various departments do not have time to give to the project, or worse, the staff will find the time at the expense of their regular responsibilities, and the effectiveness of the entire organization will suffer.

There are other examples of the matrix organizational structure in arts organizations (see Figure 6–4). There is a matrix between a resident design staff and the guest directors or choreographers that may come into a regional arts organization. The staff is hired by the organization and may work within their separate departments. When a guest director or choreographer arrives, this staff now has a new boss specifically for a particular show. A good production manager will recognize this matrix structure and will establish the needed lines of communication to keep all of these overlapping projects on track.

Coordination

A principal concept in organizing any enterprise is coordination. Coordination can be divided into vertical and horizontal components. The first area of concern, *vertical coordination,* is defined as "the process of using a hierarchy of authority to integrate the activities of various departments and projects within an organization."[19] Vertical coordination is split into four areas: chain of command, span of control, delegation, and centralization-decentralization. Each of these has applications to arts groups.

Horizontal coordination simply refers to the process of integrating activities across the organization. Many arts organizations use this structure to promote interdepartmental cooperation.

Vertical Coordination

Chain of command Classic management theory, as noted in Chapter 3 in Fayol's Fourteen Principles, states that "there should be a clear and unbroken chain of command linking every person in the organization with successively higher levels of authority."[20] This is known as the Scalar Principle. The military is a good example of this concept of chain of command applied completely. Your common sense should tell

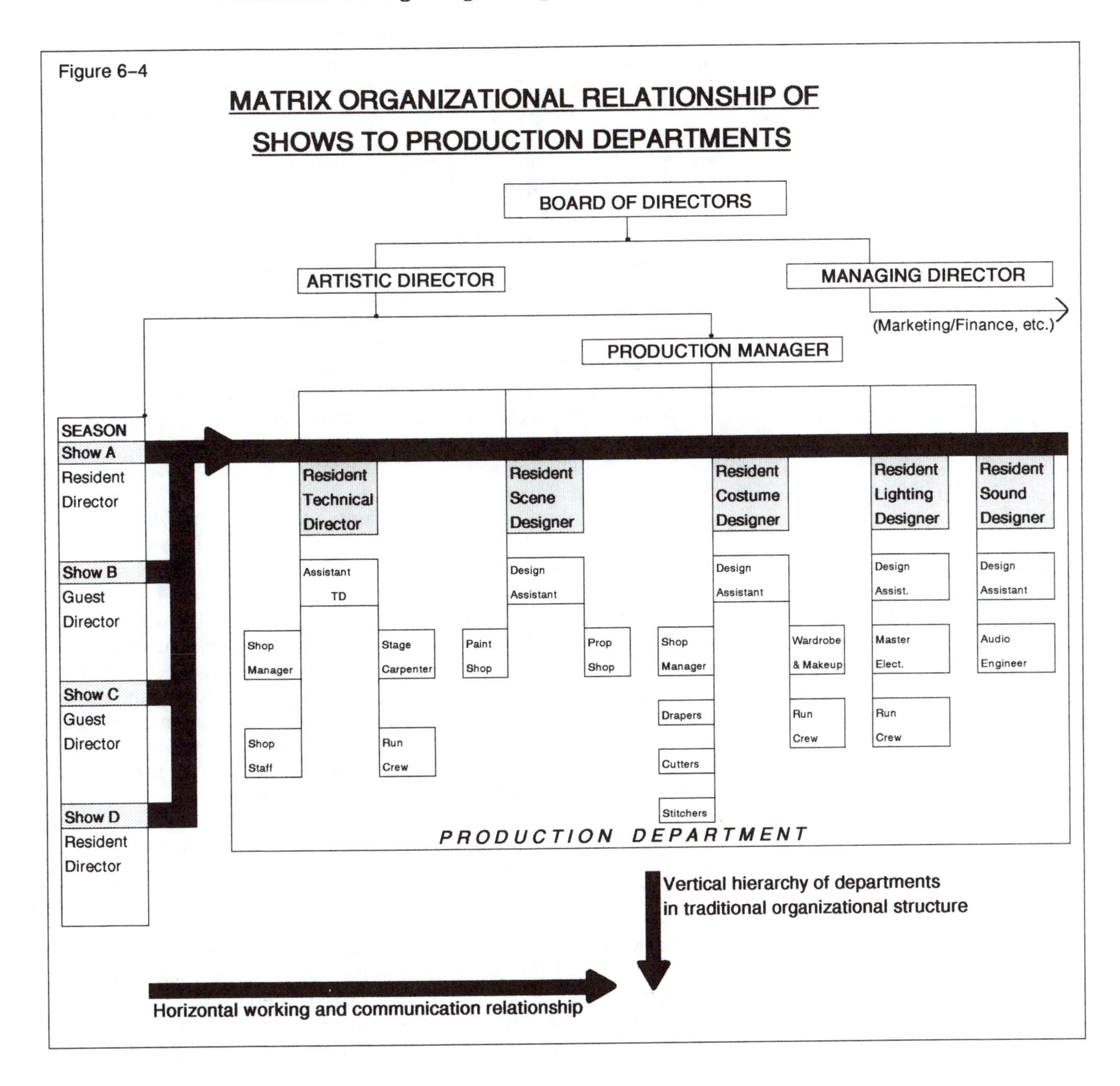

you that problems will develop in an organization that lacks a clear authority and clear lines of communication. On the other hand, not all situations allow for a clear "unbroken chain of command" within an organization. A small arts organization may hire a general secretary to work for several people and departments. On paper, the job description says the secretary is supervised by the managing director. However, as the organization grows, the secretary is asked to complete work for the artistic director, the new marketing director, the newly formed fund raising department, and others. Unless the secretary is very good at time management and setting priorities, work may become backed up or simply not done. In this example, too many people have authority to ask the secretary to work for them.

Span of control The span of control describes how many people report to one person. There are no fixed rules about how much or how little span of control one person should have in an organization. The factors often cited in determining this are "(1) Similarity of functions supervised, (2) physical proximity of functions supervised, (3) complexity of functions supervised and (4) the required coordination among functions supervised."[21]

If the marketing director is asked to take over the supervision of the fund raising and ticket service operation, there is a reasonable match between functional areas and a limited span of control. However, if the marketing director tries to take over the management of the acting company—that is, becomes, in effect, the company manager—there may be problems. You are starting to exceed the marketing director's expertise. The marketing director could possibly supervise the company manager—the person responsible for seeing to all of the needs of the acting company. However, the marketing director may not have enough understanding of the job to know if it is being done properly.

An example of proximity in span of control occurs when the production manager is asked to supervise the activities in theaters in three different locations in the city. The production manager soon discovers that it is impossible to be in three places at once. Three staff assistants are needed: one for each theater. Similarly, if the functional areas become complex, as in the data-management system for an organization, you may need to reduce the span of control. The staff member in accounting who became the resident computer guru can no longer keep control of the expanded system used in the marketing department, the box office, and fund raising. Either someone must be hired to coordinate the organization's computer needs, or the accounting work of the existing staff member must be reduced.

Delegation As a manager, you must decide how much day-to-day work you should do yourself and how much should be assigned to others. *Delegation* is "a distribution of work to others."[22] The process involves three steps: assigning duties, granting authority, and establishing an obligation. When you assign duties, you must have a solid understanding of the work to be done, and you must spend time in analyzing what you expect the employee to do.

Problems sometimes arise with the second step: granting authority. If the delegation process is to be effective, you must give up some of your authority and transfer it to the employee. If you do not grant that authority, it is difficult to establish an obligation on the part of the employee to assume the responsibility. For example, suppose that you delegate the responsibility of providing daily sales summaries for the organization to your box office manager, but you go down to the box office the next day and run the numbers yourself. The box office manager will wonder what happened to this delegated responsibility if you duplicate her work. She will assume that you do not trust her work. A key principle to remember is that the "authority should equal the responsibility when you delegate work."[23] For example, if you give a staff member responsibility for a budget but then take away the authority to expend the funds, you have undermined the delegation process.

No area creates more bad feelings than delegation. Overprotective managers tend to create employees who are bored with their jobs and not particularly committed to meeting the organization's objectives. Because they have not been given any real responsibility, they develop the attitude that is expressed in the phrase, "Don't ask me; I just work here." Meanwhile, the whole organization suffers because the manager is spending time doing work that would be more effectively done by others.

Centralization-decentralization This last area of vertical coordination simply refers to how the organization will concentrate or disperse authority. Colleges or universities are often cited as examples of decentralized organizations. The individual departments (history, government, physics, and so on) often have autonomy over personnel and course offerings. They report to a central authority—that is, the dean—but the faculty members of the department exercise a great deal of control over their own day-to-day activities. On the other hand, a large corporation may have a rigidly organized authority structure in which only designated managers have the authority to make decisions. In an arts organization, the degree of centralization or decentralization depends on the functional area. For example, the process of running a subscription or membership campaign requires a centralized authority structure. You cannot have several people making key decisions about the campaign autonomously. To control the look and language of the campaign, one person must make the final decisions. In other areas, it may be more efficient to decentralize the authority. For example, the production manager delegates the purchasing of production supplies (lumber, fabric, steel, paint, and so on) to the technical director. The technical director then delegates the purchasing to the individual department heads.

Horizontal Coordination

Horizontal coordination is a key function in the matrix organizational structure. In the business world, the function of horizontal coordination is most often identified with such areas as personnel and accounting. All of the departments in the organization use the personnel department as a resource for hiring, firing, and evaluating employees. Payroll is a good example of another department that has a horizontal relationship to a vertical structure. Everyone in the organization is affected by the payroll department. Organizations could not function effectively if every department handled its own payroll.

In an arts organization, similar personnel and payroll functions may cut across departmental lines. As already noted, the project orientation of many arts organizations creates the need for horizontal coordination. A successful production, concert, or exhibition often requires that different departments cooperate and communicate over an extended period of time. For example, a stage manager must be able to coordinate with what is called *functional authority*. This authority allows the stage manager to cut across the formal chain of command. When a problem backstage needs to be solved instantly, the state manager can issue orders that bypass the traditional crew head hierarchy.

Organizational Growth

Organizations seem to have a way of growing beyond anyone's original expectations. As one sage wit once noted, "The number of people in any working group tends to increase regardless of the amount of work to be done."[24] Almost constant attention must be paid to the proliferation of staff with each new cycle of strategic planning. As new plans are implemented, the workload seems to increase, so new staff members are hired, and more levels of management are put into place. The production manager now has three assistant production managers. The secretary is now the director of office operations and has three secretaries working under her. This growth carries with it increased costs to the organization. The cycle of overloading the current staff, hiring new staff to help reduce the overload, overloading the budget, laying off the new staff, thus overloading the original staff, and so on, can be avoided if the manager remains in control of the planning function. One of the elements of planning is keeping accurate records of where you have been. Tracking the growth of a department over a period of time can be a helpful way to monitor growth. In my comparisons of staff growth in arts organizations over the last 20 years, I found that increases of 25 to 125 percent were not uncommon. This is all the more significant when it is realized that many of these organizations were only doing 5 to 10 percent more programming.

Managers must think of the design of the organization as a creative challenge and not a burden. As we have seen, the manager's job is to anticipate and solve problems. Lack of attention to problems with organizational design can lead to fundamental flaws in the operation of the enterprise, and these flaws could lead to the demise of the organization.

Corporate Culture and the Arts

Theory Versus Reality in Organizations

In the last 15 years, management experts have found that organizations behave in ways that the theories cannot always explain. As you would expect, no theory of organizational design can take into account all of the variables that affect businesses. One of the more interesting approaches to analyzing how organizations behave may be found in Terrence Deal and Allen Kennedy's *Corporate Cultures: The Rites and Rituals of Corporate Life.*[25] The basic premise of the book is that organizations are social systems, and these social systems are based on the shared values, beliefs, myths, rituals, language, and behavioral patterns of the employees. All of these factors are carried on from year to year over the life of the organization. A manager may find comfort in the structure, policies and procedures of the company, but in reality, these things are not necessarily why people choose to work in the organization. Deal and Kennedy list five elements in the corporate culture: (1) the business environment, (2) values, (3) heroes, (4) rites and rituals, and (5) the cultural network. As you will see, there are several points at which this study of corporate for-profit organizations intersects with the arts world.

Application to the Arts

Because many arts organizations are going through the transition from a strong founder-director orientation to a system with professional staff and volunteer board management, it is worthwhile to take a brief look at the cultural aspects of an organization. It is also important to gain insights into the rites and rituals that may affect your ability to function effectively within the formal and informal structure of an organization. Let's look briefly at each element of the corporate culture and examine how arts managers may benefit from a keen awareness of this often-overlooked aspect of organizational life.

Five Elements of Corporate Culture

Business environment As previously noted, organizations function in relation to various political, educational, technological, cultural, and economic environments. Deal and Kennedy found that "to succeed in the marketplace, each company must carry out certain kinds of activities very well. . . . This business environment is the single greatest influence in shaping a corporate culture."[26]

The authors identify four types of organizational cultures. As with all classification systems, the reader should keep in mind that creative combinations of some or all of the types may exist in complex organizations. Deal and Kennedy note that companies in highly competitive markets tend to develop approaches that stress a *work hard/play hard culture* to keep the sales force motivated. Companies like McDonalds, Xerox, and Mary Kay Cosmetics are cited as examples. Other cultures present the *tough-guy* or *macho culture*, which includes people in high-risk businesses who get quick feedback on their decisions. Such businesses include construction, advertising, and entertainment. The *bet-your-company culture* functions in the world of high risk and slow feedback. Large oil companies and the aircraft industry are examples of companies that must wait a long time before their decisions pay off. Finally, there is the *process culture*, a low-risk, slow-feedback business world. Government, heavily regulated industries like utilities, and financial service organizations tend to have this type of culture. The very low level of feedback "forces employees to focus on *how* they do something, not what they do."[27]

Arts organizations may take on some combination of these cultures, depending on the circumstances. For example, a newly formed dance company may have a very driven leader who creates an ongoing culture of risk-taking. As the organization establishes itself—a board of directors is formed; foundation, corporate, and public grants are found—another culture with a more process-orientated approach may evolve. The organization shifts from operating on the edge to focus on a more stable and long-term perspective. As a matter of fact, there is an inherent conflict between the risk-taking and process cultures. This may explain why many arts organizations go through a great deal of personnel turmoil. For example, a conservative board may want to move the organization in a different direction than the adventurous artistic director wants. Employees within the arts organization become caught in the conflict. A strong artistic director, creating a culture in the organization that stresses taking risks, will want to be surrounded with like-minded people.

Values The values of the organization define the criteria for employee success and tell the employee what is important to the organization. These values are usually communicated through a corporate slogan (Ford Motors: "Quality is job one"; General Electric: "Quality is our most important product") or through glossy annual reports. Arts organizations usually produce annual reports, and the comments by the artistic director or chair of the board illustrate the values stressed by the leadership. For example, a recent annual report by the American Repertory Theatre in Cambridge, Massachusetts, places a high value on "the presentation of challenging and innovative works."[28]

Although it is important to communicate values to the public, it is more important that the employees of the organization know how their work is guided by these values. A strong culture will reinforce the organization's values at every opportunity. This is not easy in a complex arts organization. For example, the artistic director may not direct every production in a season. Guest directors may bring into the organization values that are at odds with the organization's leadership. This requires that the staff constantly adapt to changing values. The entire strength of the organization's culture can be eroded if the problem is ignored. On the other hand, contrasting cultures could be used to advantage by deliberately creating the value, "We stress diversity."

Strong artistic leadership should bring with it a strong culture with clear values to which everyone in the organization can subscribe. A cutting-edge dance company, led by a true experimenter, that tries successfully to work with a managing director who thinks the sun rises and sets on nineteenth-century story ballets will experience a major conflict of cultures and values. The staff will notice these differences in obvious and subtle ways, and the overall enterprise will be hampered by such extreme contrasts.

Heroes The expression of the organization's culture can most often be found in the person filling the key leadership role in the corporation or arts organization. For example, Lee Iaccoca functions as his company's hero-rescuer and uses commercials to project Chrysler's corporate culture as an aggressive, quality-orientated, customer-driven business. On the other hand, Ford uses commercials of its employees articulating their commitment to quality in an effort to stress how much the company values its employees and, in a sense, to make them the heroes.

Arts organizations, especially founder-directed groups, look to the artistic director, music director, and so on, to set the tone for the organization. This person should be a role model who can articulate the mission and values of the organization. A strong leader sets the standards for performance, motivates employees, and helps to carry on the history of the organization. Unfortunately, the true hero-leader is a rare creature. A hero-leader need not necessarily be loved by all of the employees, but he or she certainly needs to be respected.

Arts organizations often experience a great deal of dislocation when a powerful founder-director leaves the organization. The death of Robert Joffery sent the Joffery Ballet reeling for several seasons. George Szell's death forced the Cleveland Orchestra into a long period of readjustment. The departure of Rudolf Bing from the Metropolitan Opera in 1972 created a significant change in the corporate culture of that organization.

Rites and rituals Just as the larger culture in a society has various rites and rituals to assimilate its people, organizations have routines and patterns of behavior that they expect their employees to follow. These may be yearly ceremonies or daily activities, but in either case, the objective is to show employees how to behave. In an organization with a strong culture, the message about what is expected of employees is clearly communicated. The way things get done is usually the first contact an employee has with the culture of an organization. The employee's initial orientation sets out the norms of behavior and performance. If carefully orchestrated, Deal and Kennedy point out, organizations will develop symbolic actions to help reinforce the values and beliefs of the company. They cite a number of social and management rituals that organizations adopt to surround employees with the

Books of Interest

The Addictive Organization by Anne Wilson Schaef and Diane Fassel. "Why we overwork, cover up, pick up the pieces, please the boss and perpetuate sick organizations."

In this chapter, we have been examining the theory and practice of organizational design. Once an organization is in operation, it is important to evaluate constantly its "health." Organizations, like people, can become ill with a variety of diseases. One of the more interesting management and psychology books to be published recently on the subject of organizational illness is *The Addictive Organization*. The authors take the perspective that organizations and individuals exhibit addictive behaviors that lead to their own self-destruction. Schaef and Fassel first identify the addictive system and the terms and characteristic behavior of people in the addictive system. Then they look at the four major forms of addiction in organizations. They wrap up their study with a look at the recovery process and the long-term implications of addictive organizations.

Arts organizations suffer from the same pattern of addictive behavior as large corporations. At the heart of the matter is what is best described as the addictive tendency in humans. Research points to the possibility that humans, by their genetic composition, are susceptible to addictive attachments to a variety of substances. When this research is coupled with a larger culture that stresses and rewards certain types of addictive behavior, you have an unhealthy situation. It is not an exaggeration to say that some people "work themselves to death." Arts organizations are often built on a cultural value system that stresses a willingness to sacrifice yourself for the good of the art form: You will work long hours for very little pay in often hazardous conditions and say you "love it!"

With the understanding that it can sometimes be misleading to quote out of context, here are two quotes from the section on "The Organization as the Addictive Substance" in *The Addictive Organization*.

> *Organizations function as the addictive substance in the lives of many people. We recognize that for many people, the workplace, the job, and the organization were the central foci of their lives. Because the organization was so primary in their lives, because they were totally preoccupied with it, they begin to lose touch with other aspects of their lives and gradually gave up what they knew, felt and believed.* (p. 119)

> *The organization becomes the addictive substance for its employees when the employees become hooked on the promise of the mission and choose not to look at how the system is really operating. The organization becomes an addictive substance when its actions are excused because it has a lofty mission. We have found an inverse correlation between the loftiness of the mission and the congruence between stated and unstated goals.* (p 123)

SOURCE: Anne Wilson Schaef and Diane Fassel, *The Addictive Organization* (New York: Harper and Row, 1988).

---

organization, such as company recognition awards, clinics, and company newsletters.

In the arts, rites and rituals extend throughout various organizations. The mere act of giving a public performance or putting on an exhibition connects the organization to the culture at large. Hundreds of different ways of doing things become established routine within the organization. Rehearsal and production schedules, performer warm-ups, backstage behavior, exhibition installation, and recognition of individual excellence all help to shape the culture of the organization. Each art form has its own sets of rites and rituals in addition to those associated with the individual company. Dancers, actors, singers, musicians, directors, designers, technicians, and craftspeople all have ways of doing things that feel comfortable and make sense to them. In fact, these rites and rituals are often so ingrained that people are not aware of them until conflicting cultures are introduced. A company of actors accustomed to working in a general pattern with a resident director may have trouble adapting to a guest director who brings very different values and approaches to the rehearsal process. Being aware of these rites and rituals and carefully channeling them can help build a strong and positive culture in an arts organization. On the other hand, ignoring cultural clashes of rites and rituals can lead to a breakdown of the spirit of an entire group of artists.

The cultural network The last element that Deal and Kennedy discuss is the internal communication system, which acts as a network to circulate the values of the corporate culture. As noted before, organizations have both formal and informal structures. The "hidden hierarchy" in an organization's culture includes the storytellers, spies, priests, and others who create the overall environment in the organization.[29] The authors accurately identify various characters who play roles in the culture. For example, the storytellers are good at putting incidents in the organization into a reality that others can understand. Storytellers usually add to and embellish the event, and as you would expect, the good ones develop an audience of rapt listeners. Storytellers can help provide a sense of history and perspective for new employees. They can also be conduits for passing along the myths and rituals of the organization. The organization's priests take on the role of confessors, arbitrators of moral dilemmas, and symbols of the mature and serious view of the organization. Spies function as they do in the general society. They look beyond the surface to what is going on behind the scenes. As the authors point out, "Truly effective spies never say a bad word about anybody and are thus much loved as well as much needed. . . . Sharp spies keep their fingers on the pulse of the organization."[30]
Because the cultural network does not appear in the organizational chart or in the memos or published reports, managers must learn to tap into the network to stay on top of what employees are thinking and feeling about the organization. Deal and Kennedy recommend that managers stay in contact with the storytellers and priests, in particular.

In an arts organization, it is particularly important to remain plugged into the cultural network. The ability to manage and shape this network can be especially important when an arts group finds itself in financial trouble, and people begin to fear for their jobs. In addition, as arts organizations attempt to reach out to diversify their audiences and their staffs, entire new cultural perspectives will enter into the mix. For example, employees from different cultural and ethnic backgrounds will bring values and beliefs that may initially conflict with the prevailing culture of the organization. Managers need to be sensitive to the ethical issues of trying to reprogram staff to following the company culture at the expense of their belief system. It is important that everyone be aware of and support the organization's goals and objectives. However, care must be taken not to create an organization with such a rigid corporate culture that it forces people into uncomfortable circumstances.

Corporate Cultures and the Real World

When you start a new job, you are brought into an organization's cultural system. It may be a very strong culture that stresses maximum performance at all times, or it may be a very relaxed culture that stresses slow and steady progress. No matter where the culture fits along this continuum, it will exist. The sooner you recognize it, the sooner you can adapt to it as needed. For example, if you go to work for a marketing director in an arts organization that prides itself on huge leaps each season in its subscriber base, you had better be ready to adopt the attitudes and beliefs that go along with the job, or you will find yourself outside the system and eventually out of a job. On the other hand, once you have established yourself as a manager in the organization and have come to know the culture, you can begin to alter it to better achieve the overall goals and objectives established in your strategic plans. Remember, the culture of an organization is not static. It adapts to changes in the internal and external forces that affect the organization.

Summary

Organizing is the second basic function of management. Organizations are collections of people in a division of labor who work together for a common purpose. Organizing makes clear what everyone is to do, who is in charge, channels of communication, and resource allocation. Managers should organize for results. Organizational structure and charts provide a map of the organization. All organizations also have informal structures, which managers should monitor. Organizations can be designed to use functional, divisional, or matrix structures or a combination of these. Organizations use vertical and horizontal structures and coordination to operate effectively. They may take on organic or mechanistic structures, depending on what they do. All organizations have social systems that define a distinctive culture for that organization. Strong leadership helps define the culture. Organizational design is affected by the organization's culture and social systems.

Key Terms and Concepts

Organization
Organizing
Organizational structure
Organizational chart
Division of work
Informal organizational structure
Agency form of management
Functional department management
Departmentalization
Organization by division
Organization by matrix
Vertical coordination
Chain of command, or the Scalar Principle
Span of control
Delegation
Centralization versus decentralization
Horizontal coordination
Open system model of organizational design
Bureaucracy
Corporate cultures

Questions

1. What are the formal and informal organizational structures of the theater, dance, music, or art department at your college or university? Do these structures effectively support the mission of the department? Explain. How would you reorganize the department to make it more effectively support the mission of the university?

2. Can you cite examples of breakdowns in the vertical or horizontal coordination on a project or production with which you were recently involved? How would you improve the coordination systems to minimize these problems in the future?

3. Based on your own work experience, identify as many of the values, behavior patterns, language, rites, and rituals that formed the corporate culture of the organization.

Case Study

The following excerpt from an article in *the Chronicle of Philanthropy* provides a good example of how the concepts from Chapters 5 and 6, can be put to use.

An Arts Institution's Management Overhaul

Plagued by a persistent deficit, Washington's Kennedy Center turns to techniques used by for-profit companies.

Vince Stehle

WASHINGTON—The John F. Kennedy Center for the Performing Arts here is trying to raise its standards of performance, from backstage to the board room.

After a year-long review, the center is making changes in its fund raising and marketing departments, as well as in other administrative offices, to bring the performance of its management team into line with the quality of its presentations on stage—and to erase a persistent multimillion-dollar deficit.

While many non-profits engage in strategic planning to improve their operations, the Kennedy Center's effort is unusual because of the depth of financial analysis it entailed and because it is based largely in techniques used by for-profit businesses.

The review prompted numerous changes in various departments and has led to more cooperation between divisions to make the center more efficient as a whole. . . .

James D. Wolfensohn, the respected New York investment banker who took over as the Kennedy Center's chairman and chief executive officer 18 months ago, has restructured the senior management team at the institution, delegating day-to-day control to a new official, the chief operating officer.

Many of the changes at the Kennedy Center were suggested by Cannon Devane Associates, a management-consulting firm in Washington. The management-review process began over a year ago, in June, when Mr. Wolfensohn, who works part-time as chairman, accepted an offer from Martin Cannon, the president of the firm, who volunteered to sort out the center's management systems on a *pro bono* basis. The two men had worked together as advisers in the sale of a major hotel chain.

In analyzing the center's management, says Mr. Cannon, he found that many of the same questions facing for-profit business companies applied to non-profit organizations.

"The reassuring message is that there are capabilities that have been largely devoted to the management of the for-profit business sector that are very relevant to the center's business operations," says Mr. Cannon. "And they should not feel embarrassed to call upon those capabilities. Successful businesses do it all the time. . . ."

[The Process]
After several months of analyzing the center's finances and management practices, [Mr. Cannon] says, the management study revealed numerous untapped sources of revenue, including greater potential income from the box office and private fund-raising efforts.

The consultants then began gathering information, including financial data from all the center's departments, data about competing local performing-arts institutions, and demographic information about the Washington metropolitan area. Some of the findings demonstrated that the Kennedy Center had some serious troubles:

- During the late 1980's, more than a third of the center's subscribers had failed to renew their subscriptions, and single-ticket sales had fallen sharply. . . .
- Private donations . . . remained virtually flat from 1985 to 1990, while the costs of raising funds jumped from 13 percent to 18 percent of contributions.

In addition to the vast array of facts and figures, the consultants solicited the opinions of any staff member who wanted to be interviewed. Given the understanding that all interviews would be confidential, over 100 members of the 186-member staff agreed to talk to the consultants. . . .

Taking all the evidence together, Mr. Cannon presented two scenarios. In one, the center could continue business as usual, increasing its annual operating deficit from $3.9 million in 1990 to $6.3 million in 1993. In the other, Mr. Cannon showed how, by cutting costs and increasing revenues, the center could actually realize a $9.2 million surplus by 1993.

Cannon Devane projected that, among other things, administrative overhead costs and production costs could be trimmed by about $1.3 million, in part by establishing a single procedure for negotiating contracts that would cover the Kennedy Center employees and outside producers alike. . . .

[Implementation]
Despite the importance of a healthy balance sheet, most people at the Kennedy Center said they believed that financial considerations must be tailored to the center's mission statement, and not the reverse. So, before committing to the new regime, senior staff members created a new "vision" statement for the center. The statement says the center will "embody, stimulate, and transmit the values of freedom, creativity, expression, and joy inherent in the performing arts" and it will present high-quality, diverse performances and try to attract a wide audience.

Once the new mission statement was adopted, each department began drawing up new plans, setting goals and priorities consistent with the statement. Those plans are still under review by the new chief operating officer, Lawrence J. Wilker, former president and chief operating officer of the Playhouse Square Foundation in Cleveland. Even though Mr. Wilker hasn't made final decisions about the changes that will be made, some new programs and policies have already begun. . . .

New Fund-Raising Efforts Begun
Special festivals The center sponsored a "Texas Festival," which was the first Kennedy Center festival to focus on the arts of a single state. The event, which included performances by the Houston Ballet, the

Texas Boys Choir, and the Dallas Symphony Orchestra, as well as a popular music program known as the Roadhouse Cafe, helped raise $2.4 million in contributions from corporations, foundations, and individuals, which more than paid for the additional costs associated with the festival.

New policy for board members The Center's Board of Trustees agreed to form a fund-raising committee and approved a resolution suggesting, but not requiring, that all trustees either give, or obtain from others, commitments each year to give $100,000, payable over three years.

Giving clubs Two new giving clubs for donors have been set up: the "100 Club," for companies that give at least $100,000 over three years, and "The Trustees' Circle," for individual donors who do the same. The club has attracted 61 members so far.

The center's marketing department has adopted new policies and programs designed to attract new subscribers and, in some cases, to assist fund raisers. For example, last season the marketing department designed a deeply discounted subscription package—three concerts at half price—at the request of the development department. . . .

Another of the management review's findings involved the high cost of attracting new subscribers. . . . The expense of promoting the institution to new customers prompted the management consultants to suggest that subscribing should be made easier, and that complaints be dealt with long before a customer became disillusioned.

A Window Into What Didn't Work

In response, the center has made numerous changes to make it easier for subscribers to exchange tickets and make sure that patrons will have no difficulty finding their way to the six different theaters in the complex or getting something to eat in a hurry before performances.

Still, it's difficult for the administrative staff, most of whom work during the day, to know what's happening in the evening, when most of the center's performances take place.

"We'd get in the office, and the first phone call at 10 o'clock was from an irate patron, and that was your window into what didn't work the night before," says Geraldine Ottremba, director of government relations.

"The obvious solution is to train people to solve problems as they happen."

In addition, she says, when customers now call to complain, the call is not passed around, "in what we affectionately called 'Kennedy Center roulette.'" Instead, all complaints are now routed to a central customer-service office. "Now you have a fairly accurate picture of what your complaint level is, and you have people who are trained to make reparations, whose goal is to recover that patron immediately, and not let them off the phone until they are satisfied. . . ."

SOURCE: Vince Stehle, "An Arts Institution's Management Overhaul," *Chronicle of Philanthropy*, September 24, 1991. Reprinted with permission.

Questions

1. Summarize the main changes made in the Kennedy Center's management structure, as noted in the article.

2. Based on the information in the article, how would you characterize the management structure of the center? For example, is it hierarchical, divisional, or a matrix?

3. What changes in the corporate culture of the center are implied in the reorganization plans?

References

1. John R. Schermerhorn, Jr., *Management for Productivity*, 2d ed. (New York: John Wiley and Sons, 1986), 20.
2. Ibid., 161.
3. Ibid.
4. Ibid., 162.
5. Ibid., 163.
6. Ibid., 190.
7. Ibid., 191.
8. Ibid., 188.
9. James Q. Wilson, *Bureaucracy: What Government Agencies Do and Why They Do It* (New York: Basic Books, 1989), x
10. Schermerhorn, *Management for Productivity*, 163.
11. Ibid., 164.
12. Stephen Langley and James Abruzzo, *Jobs in Arts and Media Management* (New York: American Council for the Arts, 1990).
13. Metropolitan Opera Guild, *Metropolitan Opera 89–90* program book (New York: 1989).
14. Schermerhorn, *Management for Productivity*, 167.
15. Arthur G. Bedeian, *Management* (New York: Dryden Press, 1986), 258.
16. Schermerhorn, *Management for Productivity*, 169.
17. Langley and Abruzzo, *Jobs*, 128.
18. Bedeian, *Management*, 265.
19. Schermerhorn, *Management for Productivity*, 173.
20. Ibid., 174.
21. Ibid.
22. Ibid., 176.
23. Ibid., 177.
24. Arthur Bloch, *The Complete Murphy's Law* (Los Angeles: Price-Stern-Sloan, 1990), 62.
25. Terrence E. Deal and Allen A. Kennedy, *Corporate Cultures: The Rites and Rituals of Corporate Life* (Addison-Wesley, 1982).
26. Ibid., 13, 14.
27. Ibid., 119.
28. Barbara W. Grossman, *Annual Report 1989–90 of the American Repertory Theatre*, 2.
29. Deal and Kennedy, *Corporate Cultures:* 85.
30. Ibid., 94.

7 Staffing the Organization

The last two chapters covered the areas of planning and organizational design. We saw how the mission of an organization becomes the foundation on which the strategic plan is built. Specific goals and objectives are then established. The plan also details how resources are to be used to meet the organization's goals and objectives. The organizing process is designed to move the plan from an idea to a reality. The manager designs the organization to fulfill the plan. The structure of the organization, the lines of communication, and the combinations of vertical, horizontal, and matrix relationships are established among the departments and projects. Departments and other subunits are created to support the plan effectively.

The next stage in the process of creating an organization is staffing it. The human resources required to fulfill the mission and to support the strategic plan of the organization effectively become the key element in the success or failure of the enterprise. The organization's strategic plan must outline its staffing objectives. Descriptions of jobs and the complex working relationships among employees must be carefully factored into the organization.

Arts organizations face numerous challenges when it comes to staffing their organizations. An arts manager must be aware of the laws regulating employment and must be versed in the art of negotiation. Several unions may represent employee groups throughout the organization, and they may have different contract periods. The task of finding the right people for the job, keeping them, and developing them is a never-ending process.

The Staffing Process

Any organization wants to fill its jobs with the best people available. Finding the most talented, qualified, and motivated people to work with you is much harder than it sounds. To make the overall system clear, we will break the staffing process into six basic parts: planning, recruiting, selecting, orienting, training, and replacing.[1] We will also look at how this process varies across the fields of theater, dance, opera, music, and museums. Figure 7–1 provides an overview of the entire human resource management system. As you would expect, there are variations between and within various art forms.

In the business world, the somewhat imposing-sounding phrase *human resources planning* simply translates into analyzing your staffing needs and then identifying the various activities you need to undertake to make the organization function effectively. In many large corporations, a manager identifies specific staffing needs to determine where the staffing resources are required. The manager then works with the human resources or personnel department to find the required

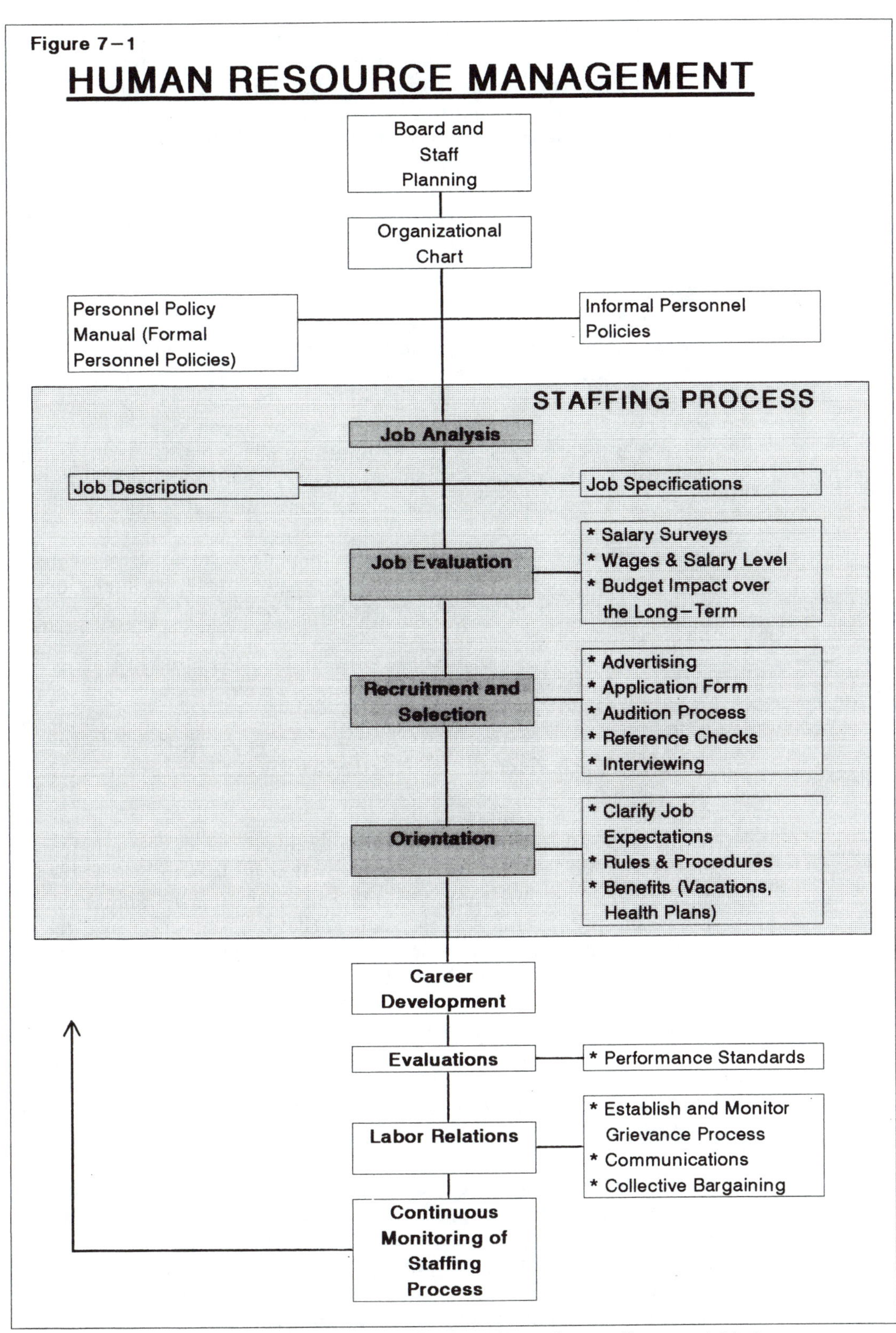

Source: Michael Carrell, Frank Kuzmits and Norbert Elbert, Personnel: Human Resource Management, 3d ed., (Columbus, OH: Merrill, 1989). Copyright (c) 1989 by Merrill Publishing Company. Used by permission of Merrill, an imprint of Macmillian Publishing Company.

staff. Because most arts organizations do not have human resources departments, it becomes all the more important for the manager to have an excellent grasp of the rules and regulations controlling the hiring and firing of employees. Mistakes in these areas can be very costly. More and more organizations face lawsuits because of badly handled personnel decisions. Given the often informal circumstances surrounding the hiring of many people in the arts, it is surprising that there are not more lawsuits. Let's take a look at the steps involved in planning for the people needed to make the performance or exhibition possible.

Planning

The process of finding out who you need must be based on careful thought. What are the strategic objectives of the organization? Do you want to create the best acting company in the world, or do you just want to build your subscriber base? If there is a major economic downturn and you have to cut staff and combine jobs, where do you make the changes? Whatever the circumstances, the planning phase requires that you review five key areas for *all* of the jobs in the organization: work activities, work tools, job context, standards, and personnel qualifications.[2]

Work activities What is to be done? A performing arts organization will obviously need actors, singers, dancers, or musicians. Plans for the season often dictate the range of performer needs the organization will have.

Based on your organization's design (as illustrated in the organizational chart), you should be able to create a clear distribution for the rest of the staff you need. In arts organizations, where staff resources are usually very limited, it is critical that the manager carefully analyzes how to combine the work activities to get the most productivity from each person.

Work tools Some employees will need specific tools to do their jobs. For example, a regional ballet company that decides to set up its own scenery and costume shop, will need thousands of dollars for equipment and space rental. An organization that hires a marketing director, had better be ready to provide the computer equipment and software he or she will need to do their job.

Job context Each employee will have an overall context in which he or she will function. Performers have a set rehearsal and performance schedule, office personnel work within a daily schedule, and all employees work within an overall package that includes contracts, compensation, and benefits.

Standards The manager of an organization must set clear standards for work output and quality. The manager of an arts organization may have various degrees of control, depending on contracts and agreements. The music director of the symphony and possibly a select group from the ensemble may be responsible for maintaining the performance standards at the highest level. On the other hand, complex negotiated agreements may limit the actions that the management may take to dismiss a musician who is not performing up to the standard.

Personnel qualifications The level of education and experience required for each position may vary widely in an organization. The pro-

Sample Job Description

The Theater and Dance Program at Oberlin College invites applications for the position of Ticket Service Manager. This is a full-time, 12-month administrative assistant position supervised by and reporting to the Managing Director.

Responsibilities

The incumbent will have general responsibility for the sales and accounting of individual tickets and subscriptions for productions produced by the Theater and Dance Program, the Conservatory of Music, and other campus organizations. The Ticket Service Manager will perform the following duties:

1. Provide daily operation of a ticket service in a courteous and efficient manner through telephone sales, exchanges, and reservations. In addition, he or she will process mail-order and tape message sales orders.
2. Account for daily income and issue timely accounting and depository reports. In cooperation with other staff members maintain a ledger file of revenue from cash sales, credit card charges, money orders, and intercollege charges.
3. Provide general information in response to inquiries about the college and various performing arts events.
4. Train and supervise a student staff and volunteer assistants in customer relations, subscription and single ticket sales, and general box office procedures.
5. Assist with publicity projects as needed and specifically with the following: (1) manage the ticket service and supervise the concession operations the night of performances; (2) train and supervise a house manager and ushers.
6. Assist with the maintenance and development of mailing lists for arts events.
7. Provide statistical reports on sales, subscriptions, and attendance at arts events during the year.
8. Perform other related duties as required.

Requirements

Three or more years of experience in box office management and sales. Other desired qualifications include a performing arts background, computer skills, and familiarity with office equipment.

Compensation

Salary commensurate with experience. Full benefits.

Applications

Send cover letter, current résumé, and three letters of reference to the Human Resources Dept., 173 W. Lorain St., Oberlin College, Oberlin, OH 44074 EOE/AA.

cess used to select performers requires different personnel qualifications than that used to hire box office salespeople.

Job Description

From this analysis, the manager will be able to create a job description for the new position. The description, which is widely distributed to the labor market, usually covers six areas: general description, responsibilities, specific duties, requirements for employment, compensation, and application method.

The "Sample Job Description" shows how a staff opening might be worded. A shortened version of this description normally appears in a newspaper or job placement listing such as *ArtSEARCH*. Let's look briefly at each section of this description.

General description The opening paragraph describes who is looking for what position and for what length of employment. The opening also clarifies to whom the employee reports and is supervised by. Regardless of the exact wording chosen, these basic ingredients can be used in any job description.

Responsibilities The next section of the description lists the employee's general responsibilities in the job. This example clarifies the scope of the job, the types of shows done, and the departments producing the events.

Specific duties Here you have the opportunity to list the tasks you expect the employee to carry out on a regular basis. In the example, the duties are ranked in order of frequency. Items 1, 2, and 3 are daily and weekly tasks, and items 4 through 7 occur on a monthly or semi-annual basis. You can extrapolate this section to almost any job in the organization. The wording will be different, but performers, shop staff, ushers, and so on, all have lists of specific duties.

Requirements for employment The education, experience, and specific skills required are listed in this section. You can also list additional skills you would like the employee to have. This does not mean that the employee must have these skills. In fact, he or she may bring unanticipated skills to the job that will benefit your organization.

Compensation There are differing opinions about what to list for salary information. In some cases, unions may require that the salary be stated in the advertisement. In other cases, you may use a range to suggest some latitude in the salary to be offered. For example, "in the mid-20s" may mean as little as $22,500 to as much as $28,500. Other listings will simply say "compensation commensurate with experience." This could create budget problems for a manager when it comes time to negotiate a salary offer. For example, you would waste time by taking an applicant who is making $40,000 through the entire application process for a $30,000 job. One way to cover this contingency is to be sure that you clearly state the experience level required for the job in the requirements section. If you require a MFA (Master of Fine Arts) or equivalent experience, the applicant may take that to mean at least three years of work. This is based on the fact that most MFA programs require three years to earn the degree. In the long run, you may save yourself and the applicant a great deal of confusion if you are clear about the compensation levels at the beginning.

The term "full benefits" means the following:

- Health insurance
- Life insurance
- Disability insurance
- Retirement benefits

The type of benefits package an arts organization can offer is a budget decision in many cases. The costs of benefits can run as much as 30 percent of an employee's salary. Therefore, the total compensation of salary and benefits must be carefully designed within the limits of the organization's resources.

Application method The final section of the job description explains what is required for the application, when the application is due, and to whom it should be sent. This is the appropriate place to state your concerns about seeking applications from special constituencies. You will also want to include your participation in equal-opportunity employment.

The Overall Matrix of Jobs

In the initial stages of forming an organization, a great deal of time must be spent analyzing the minimum number of people required to operate the enterprise. This task is best done by placing the organizational chart and the various key staff and support positions onto a large writing surface. As you begin to get a clearer picture of the overall scope of the jobs required to make the organization function, your job descriptions and posting will have the potential to reflect how you are going to achieve your objectives. The number of full-time and part-time employees and how they all relate to the season or exhibition schedule becomes the foundation of your operating budget. When salary and benefits often account for up to 80 percent of an organization's budget, it is critical that the manager keep the overall picture of the staffing as clear as possible. This overview should make the planning process easier. For example, when it comes time to initiate a new program or project, being able to look at the overall organizational structure can prove helpful. Being able to see where you can make adjustments and predicting the possible impact of staffing changes should make the planning process less difficult.

Constraints on Staffing

The costs related to staffing a position play a key part in the constraints placed on managers. For example, if you hire an administrative assistant for $20,000, your first-year costs must also include another $8000 for benefits and taxes. If you consider a $28,000 staff position over a five-year period, assuming an inflation rate of 5 percent, your salary and benefits cost for this position will be in excess of $34,000 by the fifth year. Over five years, you will have paid out more than $154,682! Assuming that you are able to justify the expense for the staff addition, you must still face the ever-increasing complexity of laws and regulations that control hiring in the United States.

Government regulations Arts organizations are not exempted from the rules and regulations on hiring. The Equal Employment Opportunity Commission (EEOC) rules generally apply to all private and public organizations that employ 15 or more people. In addition, some state laws supplement various federal laws. In Chapter 4, we discussed the impact of the political/legal environment on arts organizations. Here are examples of eight major pieces of federal legislation that affect the hiring process:

- The Title VII Civil Rights Act (1964) and the Equal Employment Opportunity Act (1972) prohibit employment discrimination based on race, color, religion, sex, or national origin.
- The Equal Pay Act (1963) prohibits wage discrimination based on sex. The law requires equal pay for equal work regardless of sex.
- The Age Discrimination Act (1967; amended 1973) protects people from 40 to 70 years old against discriminatory hiring.
- The Rehabilitation Act (1973) provides for affirmative action programs for hiring, placing, and advancing people with disabilities.
- The Mandatory Retirement Act; Employment Retirement Income Security Act (1974) was designed to prohibit mandatory retirement before age 70. The law also provides for some pension rights for employees.
- The Privacy Act (1974) gives employees the right to examine letters of reference in their personnel files.
- The Pregnancy Discrimination Act (1978) requires pregnancy/maternity to be treated as a legal disability.

- The Americans with Disabilities Act (1990) requires businesses and public services to open up jobs and facilities to disabled people.

Recent efforts to pass a law allowing for maternity and adoption leaves has faced hard lobbying by the business community. Laws in this area will eventually be passed that will have an impact on arts organizations.

Organized labor An organization required to employ people working under a union contract must adopt rules and policies for employment that fulfill specific legal procedures agreed to by the union and the management. Although union membership has been dropping steadily since 1956,[3] a variety of unions represent various employment groups in the arts and entertainment industries. Unions represent artists and craftspeople, such as actors, singers, writers, directors, choreographers, and technicians. In addition, specific contracts with unions like the Teamsters (they may control the loading and unloading of trucks at the performance space) might be required in larger metropolitan areas.

The human resource function of an arts organization must take into account the myriad of rules and regulations that typically accompany a union contract. We will discuss this area later in this chapter under the heading "Unions and the Arts."

Recruitment

Depending on the situation, you may have a great deal of choice in filling a staff position and very little choice in filling a union position. The range involves the limits placed on the overall contractual relationship between the employees and the employer. If you need an extra electrician for a lighting setup in a union theater, the local sends over whoever they have. If, on the other hand, you need to fill the position of head electrician for the theater complex, you take steps similar to filling a salaried staff opening. The three steps in recruiting a position are advertising the opening, screening possible candidates, and critically evaluating possible candidates for a list of finalists.

Internal As you may know, many jobs are filled before anyone hears about the openings. The internal recruitment process is common in companies that have a promote-from-within policy. For example, if you are hired as the assistant to the marketing director for the museum and you are aware that the promotion policy favors internal candidates, you may have an additional motivation to do your best work.

External Many arts organizations use a dual policy of internal and external recruitment. As organizations seek to create more multicultural staffs, active external recruitment has become a regular procedure. When doing external recruiting, there are various avenues to pursue, depending on what you seek. Some organizations may arrange auditions just to build files of possible performers for the future. Specialized publications like *ArtSEARCH*[4] can be used to seek out executive staff, designers, technicians, and craftspeople. Trade newspapers covering the arts may be found in some of the larger cities. General publication newspapers may also be used. Mailings to schools and colleagues may also generate applications. Finally, a professional recruitment service can be hired to find top candidates. Although "head-hunter" services may be an extra expense, they may also generate the most likely candidates for higher level positions, such as museum director, or artistic director. Searches for executive-level personnel within an arts organization may be carried out by a designated subcommittee of the board.

Recruitment philosophy There are two fairly common philosophies about recruiting.[5] One approach assumes a traditional "selling" of the organization, and the other takes the "realistic," or real-time, approach. In the selling approach, you stress the most positive features of the job and the work environment. Essentially, you try to present an up-beat picture of the organization. The real-time approach tries to depict accurately what day-to-day life is like in the organization. You try to present an objective view of the work situation and are not shy about answering questions about the less positive aspects of working in the organization. Organizations often blend these two approaches in their recruitment campaigns, but the tendency is to sell the organization. After all, who wants to paint a picture that might scare off the applicants? People involved in the recruitment process have been known to use the realistic approach to discourage candidates that they perceive as not being "right" for the company.

Recruitment difficulties Arts and other nonprofit organizations have complex personnel needs like any other business. Arts organizations hire people for vastly different jobs in meeting their objectives. Recruiting a soprano for an opera, hiring a marketing director, filling an administrative assistant position, or negotiating a union contract with the musicians or stagehands requires different priorities and strategies. Staff recruiting for salaried positions is often difficult because many organizations do not offer competitive pay rates. Artists, on the other hand, often negotiate contracts through their agents that exceed union scale minimums. Employees who work for wages that are negotiated by union representation often receive pay at rates comparable to those of private industry. This may lead to pay disparities that have a negative impact on recruitment strategies.

Trying to attract candidates to staff positions with no advancement possibilities, low salary, minimal—if any—benefits, and an overwhelming workload is an impossible task. It is not surprising that there is a high turnover rate in lower-level staff positions. Ironically, the people in these positions often do the basic work that keeps the organization going, such as payroll taxes. Competent, skilled administrative staff members are needed to ensure that the organization complies with all of the federal, state, and local laws pertaining to collecting and reporting taxes. Some of the lowest paid positions in an arts organization probably fall into this category.

Selection Process

Auditions Different selection processes are used for different employee groups. The performers may all be selected by audition, depending on how the schedule is organized. For example, most resident theater companies now hold auditions in New York and a few other selected cities at various times during the year. Very few theater companies have actors in residence for a full season. Ballet companies, on the other hand, have yearly auditions for a resident group of dancers.

Smaller dance companies try to provide at least 26-week contracts for their dancers. Lead dancers for special performances in the repertory may be contracted as the need arises. The larger regional opera companies, which have lengthy seasons, tend to audition resident choruses, dancers, and musicians and to hire the principal singers and conductors on a show-by-show basis. Smaller regional opera companies often work with a minimal staff and contract for all singers, dancers, and musicians on a show-by-show basis.

The audition process requires a great deal of data management. Records, including photos, lists of special skills, and so on, must be organized so that they can be easily retrieved. An audition space with a piano, tape player, dressing room, and warm-up areas must be secured.

Ultimately, the organization's survival depends on its ability to select the right talent for the roles available. Because the performers are part of the product that the audience purchases, it is critical that the artistic management establish criteria that match the company's desired quality level.

Traditional application process The typical pattern for filling many staff positions in an arts organization follows these seven steps: formal application, screening, interviewing, testing, reference check, physical exam, and hiring.[6] Let's take a brief look at each stage.

Formal application Standard forms are used to take applications for job openings. Arts organizations often invite applications and request a cover letter, a résumé, and three or four references. In some cases, organizations actively seek out specific individuals and ask them to apply for the opening. The advantage of the application form becomes clear when you begin reviewing the candidates for the job. By using a similar format for gathering data, it may be easier to see which candidates have the qualifications you seek. When no formal application form exists, the person responsible for reviewing the files must develop a check list of qualifications and requirements for the job.

Screening The next step requires narrowing the list of applicants by eliminating those who do not match your search criteria. Further fine-tuning of the applicant pool usually reveals a short list of qualified candidates for the job. This work may be done by an individual or by a committee. For example, if the board is trying to hire a new artistic director, a search committee will be established and chaired by one of the board members. The screening process may become a very difficult stage in the hiring process because of strife within the organization. The search committee may be divided about the kind of person being sought, or the committee may arrive at a set of finalists that the rest of the board finds unacceptable. Needless to say, a significant number of variables can add to the complexity of the screening process.

Interviewing The interview is equivalent to the audition for a staff position. If the search is being conducted by a staff member (not a search committee), the usual procedure is to interview several of the top candidates and to schedule second interviews for the best candidates. The committee approach may involve interviews with the top two or three candidates over a day or two by a wide range of people. Because the costs of conducting staff searches can run into the thousands of dollars, it is important to take the time to narrow the list of finalists before beginning the interview process. When pressed, some of the finalists may withdraw their candidacy, leaving you to fall back on other candidates on the list.

During the scheduled interviews, care must be taken to deal fairly with the candidates. Questions about age, marital status, family, national origin, handicaps, or religion are not legal. The same questions should be asked of each candidate, and they may be structured in such a way that you can legally discover whether the prospective employee can meet your work schedule, can communicate and write effectively, and can perform the specific tasks you require.

Prospective employees should not be permitted to leave the interview process with the impression that you were engaged in discriminatory activities. Your organization may have to answer to a lawsuit if you do not manage your interview process carefully.

Testing, reference check, physical exams, and hiring The decision to hire the candidate who is best suited for the job often rests on intangible emotional responses. After you evaluate and compare the composite skills of the people most likely to fill the staffing needs of the organization, you may still be left with a high degree of uncertainty. Further background checks and additional on-site interviews may help. Ultimately, a degree of chance is always involved in hiring.

Many corporations use detailed screening tests for applicants in an attempt to narrow down the variables involved in hiring. Psychological, medical, or specific skills tests are often administered in large companies. Because of the very high cost of hiring the wrong employee, corporations are not shy about spending several thousand dollars to hire the right person. On the other hand, the company's needs must be balanced against the individual's right to privacy.

Arts organizations must work with very limited recruitment budgets. High staff turnover creates an even greater burden on the budget because too much time is spent seeking replacements for people who quit. Few arts organizations have the time to conduct an extensive background check on a prospective employee. As a result, hiring practices are often fairly casual.

Legal issues may enter into the hiring process. If you narrow your candidates down to two or three finalists, the potential for sending misleading signals to the candidates is very high. You will need sufficient documentation to stand up in court about why you did not hire candidate B or C—should your decision be challenged. When you make any hiring decision, you face some risk that the rejected finalists will take legal action. This may sound pessimistic, but more organizations face lawsuits over hiring practices each year.

Hiring Tip

A saying is often heard in the personnel field: "Hire the person who best fits the job, not necessarily the best qualified." This statement implies that you need not force yourself to hire the person with the most qualifications. Remember, the key factor in attaining the maximum productivity from a staff member is finding someone who matches the overall job environment. The person you hire must be compatible with the corporate culture of the organization. In some cases, hiring the most-qualified (or overqualified) person can lead to problems.

Orientation The fourth stage in the process begins once the hiring has been completed. There are countless variations on employee orientation. Some organizations have fixed probationary periods (three months) in which very specific activities are planned to acquaint the employee with the organization. In other cases, a new person is hired and told to direct any questions to a specific individual or mentor.

The most important element in the orientation process is the socialization of the new employee into the organization. New employees are often anxious about their new jobs. They want questions to be answered and clarifications made about where they fit into the entire operation. The informal interactions that new employees encounter will help shape their perceptions of the entire organization. To avoid problems later, it is best to schedule a review of the overall policies with new employees near the end of their probationary period. It is also important to note the date that you reviewed all of the relevant material with the new employee. You may later incur a problem with an employee who claims to have never been told about a specific policy. It may prove helpful to document when you reviewed a specific problem area with the employee.

Training and Development

In corporate America, millions of dollars are spent each year on employee training and development. Many employers need to train their workers in such basic areas as reading, writing, and simple mathematics. Due to their limited resources, arts organizations do not have such training programs. Training usually takes place only after costly errors have been made by a new employee. Although employee training is recognized as a real cost to organizations, few take into account the financial impact of not having a training program.

On-the-job training Most arts organizations use some variation of on-the-job training (OJT). The formal approach includes very specific work abilities that are tested by the supervisor at specific time intervals. The less formal approach usually includes quick demonstration sessions, and then the new employee is expected to get on with the task at hand. More rigorous OJT structures include some combination of job rotation, coaching, apprenticeships, and modeling.

Job rotation In this system, employees move around to different areas to receive training in specific activities.

Coaching As the term implies, a new employee receives very specific help with a skill related to the job. For example, an experienced stagehand may guide a new crew member through the operation of a follow spot. The stagehand watches the employee's performance and offers suggestions on improving his technique as he runs the spot.

Apprenticeships The apprentice system is used extensively in the arts for training. Ideally, the apprentice works alongside a more experienced employee. If the system is really to function effectively, apprentices should be given specific tasks that allow them to assume substantial responsibility.

Modeling In modeling, a new employee watches the performance of the supervisor or trainer. Personal demonstrations of what is expected help form a consistent presentation of the organization. This is especially important for employees who come into contact with the public. For example, people who sell subscriptions or museum memberships are involved in performance-related skills. They must be able to act out the script they have been provided with to make the correct sales presentation. Watching and listening to a more experienced staff member go through the "scene" is a useful way to train a person.

Replacement and Firing

There may be a number of reasons to replace an employee. You may have made a selection error that resulted in a poor match between the individual and the organization, or the person you hired may have outgrown the job. You may move someone to a new job and create a vacancy due to reorganization. The person you hired may have violated the rules and procedures of the organization, leaving you no choice but to fire him or her. You may experience a slowdown or budget cuts that require you to lay someone off, or, an employee may develop an illness and be unable to work for an extended period. You may need to replace someone due to a retirement or death or the employee may quit.

Firing No part of the staffing process is more troubling than firing an employee. Obviously, care must be taken when firing someone because of the legal ramifications. A poorly handled employee termination can cost an organization millions of dollars. *Wrongful-discharge suits*, as they are called, are becoming commonplace. Assessing the risks of firing an employee has led to better evaluation and documentation procedures. Verbal and written warnings must usually precede a termination. Although a union contract may stipulate very precise steps that must be taken, it should not be assumed that it is impossible to fire a union employee. Clauses that provide for the rights of management usually address this issue. On the other hand, many staff people work with no contracts whatsoever. Others work under what are called *employment-at-will* agreements. Simply stated, they can be fired at any time with limited notice, but they may also quit with the same limited notice. Typically, the notification period is two weeks to 30 days. Some upper-level management staff have very detailed contracts developed by their lawyers and the organization's legal advisers. Precise language is often used to cover all aspects of compensation, evaluation, retirement, and termination.

In the News—Personnel Actions Make News

The board member quoted in this *New York Times* article is quite a bit less direct than the headline. As you read through this article, try to ascertain the reasons for the staffing change.

Whitney Museum Cans Director

NEW YORK—Ending a long struggle between the Whitney Museum's trustees and its director, Thomas N. Armstrong III, the board voted to dismiss Armstrong from the position he held for 15 years.

The outcome was announced after a board meeting Monday afternoon by William S. Woodside, president of the museum, whom Armstrong had been instrumental in recruiting as a trustee in 1979.

In making the announcement, Woodside praised Armstrong for his dedication and contributions to the museum.

But "as the board has contemplated the changing climate in which private cultural institutions will need to operate in the years ahead," he said, "it perceived the need for leadership appropriate to the demands of the new environment."

Among the reported reasons for the board's decision were dissatisfaction over the failure of the museum's proposed expansion, unhappiness with Armstrong's leadership of the curatorial staff, and what was seen as his lack of serious engagement with the museum's direction.

SOURCE: "Whitney Museum Cans Director," *New York Times*, March 7, 1990.

Unions and the Arts

Some of the key unions involved in the arts are Actors Equity Association (AEA) for actors and stage managers, American Federation of Musicians (AFM), American Guild of Musical Artists (AGMA), American Guild of Variety Artists (AGVA), American Federation of Television and Radio Artists (AFTRA), United Scenic Artists (USA) for scenery, costume, and lighting designers, and International Alliance of Theatrical Stage Employees and Motion Picture Machine Operators of the United States and Canada (IATSE). Large organizations like the Metropolitan Opera in New York City must deal with multiple unions. The management must negotiate contracts with everyone from the musicians to the people who hang the posters in the display cases. Museums located in the larger cities must also work with unions who represent employees from many different groups, such as security guards.

Definition and Purpose

The classic definition of a *trade union* is a "continuous association of wage-earners for the purpose of maintaining or improving the conditions of their working lives."[7] Unions arose to fight the exploitation of employees, which was, more often than not, the norm. Although it

In the News—Management and Union Conflict

The following article illustrates what happens when new management enters an ongoing situation in which there is long-standing union representation.

Multimedia Focuses on Concessions from WKYC Channel 3 Staff

Sandra Livingston

Multimedia Inc., the new majority owner of WKYC Channel 3, is seeking major concessions from the station's news staff and technical crews, according to union officials.

The unions say the company has asked for cutbacks in wages and health and retirement programs; elimination of all 10 holidays in the base contract covering the news staff; and elimination of certain restrictions that prevent management from performing union jobs.

John Llewellyn, the station's general manager, declined to comment on contract proposals.

"Multimedia is negotiating a local agreement," he said, "an agreement that reflects this market place."

Since buying a 51% stake in the station from NBC in December, Multimedia has seemed determined to create a purely local pact that would lower Channel 3's labor costs.

As a NBC station for 26 years, WKYC had labor contracts that were part of the network's master pact with technicians or that shared the same negotiators as contracts covering news staffs at other stations.

"They stated when they came into town that they wanted totally new agreements with both unions," said Joan Kalhorn, executive director for the Cleveland local of the American Federation of Television and Radio Artists (AFTRA). "What they are asking for is so serious . . . that on a long-range basis, it would be very difficult for the union to function there."

About 45 workers—including anchors and reporters—are covered under the agreement with AFTRA. A separate contract covers 58 full-time technicians—such as camera operators—who are members of Local 42 of the National Association of Broadcast Employees and Technicians (NABET).

Both unions are operating under the terms of expired contracts. Contract talks continue without a deadline.

While AFTRA is concerned about a broad range of proposed cutbacks, NABET's key issue is wages.

"They're claiming that . . . we're making New York pay scales in a Cleveland environment," said Don Huston, president of NABET Local 42. "Our people have lived under this contract for (26 years). . . . To make Cleveland wages would be very difficult for our people to accommodate."

Huston said his members make weekly base wages ranging from $895 to just more than $1,000, not including overtime. He said at WJW Channel 8, the only other local station with NABET members, they make weekly base pay of about $800.

Huston said Multimedia offered a base weekly range of $765 to $790. He said the company also proposed that workers pay at least 25% of their health care premiums.

Huston said the issue of going on strike is "very touchy" with his members, who struck for 18 weeks three years ago and would like to continue negotiations. "But if hard pressed, we would take whatever means we felt were necessary," he said.

SOURCE: Sandra Livingston, "Multimedia Focuses on Concessions from WKYC Channel 3 Staff," *Cleveland Plain Dealer* (March 21, 1991).

might be argued that the unionization of the arts has created a division between salaried artists and employees who are paid a wage, the reality is that unions are here to stay. The union's primary responsibility to the workers is to derive benefits from the working relationship with the employer through a written contract. This contract is carefully negotiated by individuals elected by union members to represent them and a designated group representing the management. Although there are thousands of variations on the terms in a contract, the six key areas are compensation and benefits, job specifications, grievance procedures, work rules, seniority rules, and working conditions. The life of the contract is generally limited to two or three years.

Disputes

The agency most often involved in labor and management disputes is the National Labor Relations Board (NLRB). This organization investigates unfair labor practices by employers and unions. A NLRB representative listens to both sides of a dispute and renders a decision aimed at resolving the conflict. If either party is unhappy with the ruling, the court system is the next step. The high cost of litigation motivates both sides to try to reach an out-of-court agreement.

Horror stories abound in the arts about union abuses. The most common complaint is *featherbedding*, or creating jobs that are not really essential to the project. The unions are often blamed for creating a very high overhead for professional productions. However, the union mandate is to achieve the best wages and working conditions for its membership.

The 80s saw a major shift in the way in which the business community dealt with unions. Led by President Ronald Reagan's dissolution of the air traffic controller's union, company after company simply let unions call strikes and then went out and hired replacement workers. Management became much bolder in demanding concessions from the unions. Another strategy followed by many companies was to work in a more cooperative association with labor. Some companies actually began listening to the workers' ideas about how to improve quality and productivity. These trends trickled down into negotiations between arts organizations and some of their unions.

Unfortunately, the "us" versus "them" attitude is still very much a part of the day-to-day relationship of labor and management. To a large extent, the corporate culture of the arts organization can play a big part in forming the overall attitude about employees and the perceived value of their contribution to the organization's goals and objectives. If the organization's values express the attitude "We are here to do quality work as creatively and efficiently as possible, and we appreciate and reward people who have these work ethics," the odds are that the relations with the union will be fairly positive. However, if management's attitude is, "You can't trust them, they always goof off, and they are slow to get the show done," a work environment filled with suspicion and mistrust is reinforced.

In the News—The Negation Process

This excerpt illustrates the often-adversarial relationship that exists between unions and management.

Filmmakers, Unions in Tentative Agreement

New York Times News Service

NEW YORK—Negotiators for a consortium of Hollywood producers and two New York unions reached a tentative agreement settlement Tuesday that would end a bitter 19-week stalemate that all but shut down movie-making in New York City, a vital part of the city's economy.

The unions, representing production crafts workers and cinematographers, agreed to studios' demands for money-saving work-rule changes that would reduce overtime and weekend pay.

In return, the unions won substantial increases in health and pension benefits. But some rank-and-file members said the loss of overtime made ratification by the union's members less than certain.

SOURCE: New York Times News Service, "Filmmakers, Unions in Tentative Agreement," *New York Times*, March 21, 1991.

Maintaining and Developing the Staff

If an organization is to be successful over the long run, it must have a dedicated and experienced staff. The only way to build such a staff is to monitor the work environment constantly. A good manager should be aware of the staff's changing needs. The degree of intervention exercised by the manager depends on whether problems have arisen that require correction.

The psychological atmosphere of the work place changes every day. One of the most important parts of the manager's job is staying attuned to the mood of the work place. You can employ several strategies to help you stay in touch with your employees. Organizations must develop ongoing systems to assess regularly employee concerns in the work place. Annual or ongoing evaluations, scheduled project assessments, production meetings, informal lunch or dinner meetings, and awards for outstanding performance or achievement all form a menu of choices that an organization must have available. (See Chapter 9, Performance Appraisal System).

Career Management

If an organization places a high value on employee retention, a career management system must be established. Employees need to believe that they are learning and growing in their jobs. Some of the ways to help employees develop a long-term commitment to the organization is to offer support for additional training, provide leaves of absence for outside study, and solicit employee input about job and work expectations. Obviously, there are limits on the amount of career enrichment available for every level within an organization, but the creative application of these ideas can help promote an organizational culture that places a high value on people. For example, it would be a mistake to assume that someone functioning as a receptionist is only capable of answering the phone and directing inquiries. It is true that this job is not a staff position with a great deal of potential for career development. However, by carefully designing the job to provide additional duties, such as assisting with gala event planning or conducting donor research, you may be able to make the job more challenging for an employee.

The "Right Staff"

The importance of staffing the organization cannot be stressed enough. All of the neat and tidy organizational charts, beautifully detailed strategic plans, forceful mission statements, and carefully designed marketing and financial campaigns will be of no use without the people to make it all happen. To function effectively as an organization, you must have the personnel with the skills and dedication suited to the mission. As you will see in the next chapter, the success or failure of an organization is directly related to the effectiveness of its leadership. Finding the right people for the jobs you have and building a team of productive staff members is one of the most difficult tasks a manager faces. In situation after situation, the failure to assemble the right combination of people on the work force leads to the failure of organizations to achieve their aims. A symphony with a brilliant conductor is only as good as the musicians in the orchestra. The finest collection in a museum will fail to live up to its potential without an effective curatorial staff. A dynamic choreographer or director needs equally dynamic dancers, actors, or singers to grab the audience's interest and support.

Personnel Odds and Ends

As this story demonstrates, employee problems can be manifested in various ways.

News of the Weird
Chuck Shepard

In May, Los Angeles Philharmonic bassist Barry Lieberman was suspended without pay for assaulting colleague Jack Cousin as they were leaving the stage after a performance. Lieberman alleged that, because of an ongoing dispute, he was justified in shoving his bass into the back of Cousin's legs to trip him as they were filing off the stage.

SOURCE: Chuck Shepard, "News of the Weird," *Cleveland Plain Dealer*, January 6, 1991.

Summary

The staffing process can be broken down into six parts: planning, recruiting, selecting, orienting, training, and replacing. The planning process assumes that you are staffing the organization to realize strategic objectives. Job design helps integrate the staffing plan with specific job responsibilities and duties. Organizations must function within the laws that affect hiring personnel. Union contracts and stipulations are a fact of life in the arts. Arts managers must be well-versed in negotiating contracts and structuring their organizations to work effectively with unions. The two major recruitment methods are internal and external recruitment. Recruitment options include auditions and traditional application and screening processes. When interviewing candidates for jobs, managers must carefully follow legal guidelines. The hiring and orientation of new staff can be assisted through formal procedures to ensure that consistent information is presented. Job training and long-term staff development are a key component in building an experienced and productive staff. Firing and replacing staff can be legally risky if handled improperly.

Key Terms and Concepts

Human resources planning
Job posting
Job description
Job matrix
Equal Eemployment Opportunity Commission (EEOC)
On-the-job training (OJT)
Wrongful discharge
Employment-at-will
Unions
National Labor Relations Board (NLRB)

Questions

1. Based on your own employment experiences, give an example of how the job requirements differed from the official job description. Relate the problems or benefits of the situation.

2. Discuss the pros and cons of unions in the arts. Make a case for each side of the argument.

3. Do you agree with the union position described in "Multimedia Focuses on Concessions from WKYC Channel 3 Staff"?

4. Have you ever gone through a formal job orientation? Was it an effective tool for bringing you into the work environment?

Case Study

This article from the *New York Times* highlights a very important staffing issue facing arts organizations. The casting of a performance obviously says a great deal about the values and beliefs of the arts organization. As you read this article, try to keep in mind the perspective of other art forms besides theater.

Should Equal Opportunity Apply on the Stage?

Hal Gelb

James Earl Jones as Big Daddy, the patriarch of a powerful white Southern family?

"I'm not interested in seeing it, nor do I think it has any value other than as an actor's exercise," said Douglas Turner Ward, artistic director of the Negro Ensemble Company, when Mr. Jones was being considered for the role in a New York revival of "Cat on a Hot Tin Roof" to be mounted by Fran and Barry Weissler during the next year.

"Cat on a Hot Tin Roof," Mr. Ward continued, "came out of a specific historical context which we are all a part of, a Mississippi in which the existence of a Big Daddy was founded on a racist world, and blacks weren't allowed access to any sort of power. I don't doubt that Jimmy could play the character wonderfully. But what interpretive weight are we giving to this role except to say, 'Oh, we're using this to overcome the social disadvantage of black actors'?"

Yet that is precisely what a current campaign to cast minority-group, female and handicapped actors in parts traditionally played by nonhandicapped white males is attempting to do. And in fact, though Mr. Jones will not play Big Daddy in the forthcoming revival, he has already done so — during a symposium sponsored by Actors' Equity and designed to educate producers, directors and others in the merits of nontraditional casting. He has also played white characters in "The Iceman Cometh" and "The Cherry Orchard."

The outspoken opposition of Mr. Ward and other black playwrights to such affirmative-action programs is significant, for the particular concern of Actors' Equity is the under-representation of blacks and other members of racial minorities. The organization has compiled statistics demonstrating that between 1982 and 1987, when such minorities accounted for some 17 percent of the population nationwide, fewer than 12 percent of the acting jobs in regional theaters, where most of the stage work is these days, were filled by actors of color. For Broadway plays, the figure was a mere 6 percent.

To remedy this anomaly, Equity is working closely with the Non-Traditional Casting Project, a nonprofit organization based in New York. The project claims some progress in its two years of existence, but still finds considerable resistance. As noted, some of the strongest resistance comes from black playwrights, who, while supporting the project's general economic goals of increased employment, object to nontraditional casting on historical, esthetic and even social grounds.

Mr. Ward, for one, finds great danger in, in effect, rewriting history. To cast, for example, a black actor as Big Daddy in "Cat on a Hot Tin Roof" is, in his view, to diminish what he sees as the play's depiction of a racist society and a family whose wealth and power are based on the exploitation of blacks.

Lonne Elder, whose "Mummer's Play" was recently produced at the American Place Theater in New York, is concerned more generally that a writer's vision be realized in all its particularity. "A writer has a certain feel about the character, in terms of what his background is, what type of person he is, what he does," says Mr. Elder. "Neil Simon's people are indigenously Neil Simon's people. It would be ludicrous to say, 'Well maybe I'll go with a black in this part.' It might be fine for employment parity, but it would be deadly for the theater."

Moreover, Mr. Elder feels, to assert that white and minority actors are interchangeable is to deny their individual identities, the very subject of art. "Everything that can affect you—pain, joy, etc.—will also affect me, but when you get into this whole area about, 'Well, we can just be anybody, just be a human being,' you're saying you don't have any definition, you didn't come from some place. Then you become an invisible person."

August Wilson, author of "Fences," agrees that cultural difference are crucial in art. He, too, rejects the idea of casting blacks in parts normally played by whites, "because it denies them their basic humanity, their right to stand on a stage as who they are: African-Americans." To stage an all-black "Death of a Salesman" or "Agnes of God," he contends, would be a mistake. "The important thing is that blacks would not act or talk like white characters in a white play. They have their own culture and their own sensibility, which informs even the way they think. So the whole approach would be different."

For Mr. Wilson, nontraditional casting is simply another instance of blacks being asked to deny their blackness in order to participate in society. "Whenever they say,'O.K., we'll restrict this color-blind casting thing to plays that are universal,' the 'universal' plays they come up with are always white plays, in which they're going to put a black person on stage and have him deny his humanity in order to participate in the universal. I'm saying the universal exists in black life."

While the Non-Traditional Casting Project would probably not wish to deny it, it finds the situation more complicated. The co-chairman of its board, Clinton Turner Davis, himself a director, complained that while a white actor is perceived as a human being, the third-world actor is seen "as a Hispanic human being, an African American human being." Consciously or unconsciously, he argued, people who make casting decisions—playwrights included—assume minority actors are limited to embodying the thoughts and aspirations of their own races.

Moreover, the project receives strong support from other prominent producers and directors, both minority and white. Joseph Papp, head of the New York Shakespeare Festival, has been producing multiracial theater for three decades. Nontraditional casting, he feels, works particularly well in such productions as A.J. Antoon's version of "A Midsummer Night's Dream" of last January, in which the Festival's casting added new thematic sense to a classic play. With black actors as inhabitants of the forest in a Brazilian locale dominated by white colonial rule, nontraditional casting was "integral to the whole approach of the play, and had a historic basis."

More problematic, according to Mr. Papp, are "color-blind" productions of modern realistic plays, with actors cast regardless of race, simply because they are the best qualified by personality and craft. Given America's race-consciousness, he says, choosing an actor of color to play a white character without incorporating some recognition of the character's race and its effect on relationships will result in "a fairy tale."

Charles Gordone, who takes issue with the Non-Traditional Casting Project, nevertheless believes that minority actors cast in realistic American plays can illuminate both the script and minority experience, but it has to be done without ignoring either the actor's ethnicity or historical context. At the American Stage in Berkeley, Calif., Mr. Gordone cast Hispanic performers as the migrant laborers in "Of Mice and Men," and a Creole as Stanley in "A Streetcar Named Desire."

Lloyd Richards, artistic director of the Yale Repertory Theater and dean of the Yale School of Drama, champions nontraditional casting insofar as it adds a stimulating new dimension to a play. He thinks audiences should be "able to accept and utilize their own imaginations to focus on the essence of what is being presented rather than stumbling over the fact of an actor's size, weight, pigment, whatever."

The greatest impetus for nontraditional casting, however, has come from the ground up—from actors for whom it has become a bread-and-butter issue. There is a crush of minority actors entering the profession, and few roles available to them in Western classics and the modern American realistic genre. According to Charles Fuller, a playwright who generally supports nontraditional casting, actors have to "turn to what is already there and say, 'Let us try this,'" simply to survive.

The playwright Amiri Baraka, who helped create the community-based black arts movement in the mid-60's, takes a more jaundiced view, seeing the current embrace of nontraditional casting by actors as one more example of 80's careerism. He feels that the push for nontraditional casting hides the fact that third-world theaters "have insufficient self-determination and resources" to do the plays and roles *they'd* like to do, and that there are few minority administrators in regional theaters, where the repertory remains unchanged. "It's still the great books," he says, "with colored covers."

Other playwrights, too, believe that multiracial casting is often a substitute for productions of plays by minority dramatists. Mr. Ward finds this "a period of intense difficulty" for independent black theaters. Ironically, just when Mr. Wilson's plays emerged—not from a black company, but from the new breeding grounds of the Yale Repertory Theater and regional theaters—Mr. Ward's Negro Ensemble Company had to cancel its season.

Color-blind casting and the production of plays by minority playwrights are not mutually exclusive goals, says Mr. Ward, but "to place the main emphasis on nontraditional casting is out of balance." In his view new plays about the black experience will do more to provide minority employment than racially mixed productions of the classics.

Ultimately, however, the actors can only perform what's out there, and Charles Gordone threw it back on the playwrights, black and white, to create a multiracial theater—by writing for multiracial casts and by dealing with the conflicts that arise within a multiracial society. "Once there are writers who write about an American experience that is all-inclusive, you'll have a vital and vibrant American theater."

SOURCE: Hal Gelb, "Should Equal Opportunity Apply on the Stage?" *New York Times*, August 28, 1988.

Questions

1. Do you agree with the goals of the Non-Traditional Casting Project? Discuss.
2. Do you agree with the idea that nontraditional casting is directing energies away from the problems of Third World theaters?
3. How do the issues raised in this article apply to opera singers and ballet and modern dancers? Do casting choices related to ethnicity have the same impact on these art forms? Explain.
4. What should the role of museums be in providing "equal opportunity"?

References

1. John R. Schermerhorn, Jr., *Management for Productivity*, 2d ed. (New York: John Wiley and Sons, 1986), 241.
2. Ibid., 243.
3. Howard M. Wachtel, *Labor and the Economy*, 2d ed. (New York: Harcourt Brace Jovanovich, 1988), 373.
4. *ArtSEARCH* is published by the Theater Development Fund, New York, NY. A yearly subscription costs $48.
5. Schermerhorn, *Management for Productivity*, 249.
6. Ibid., 250–53.
7. Sidney Webb and Beatrice Webb, *The History of Trade Unionism* (London: Longmans, Green and Co., 1894), 1.

Additional Resources

Many texts cover the field of personnel management in depth. A quick check of business books in a university bookstore should turn up at least an undergraduate-level book in this area. Three excellent sources for more information about personnel issues may be found in the following books.

Carrell, Michael, Frank Kuzmits, and Norbert Elbert. *Personnel: Human Resource Management*, 3d ed. Columbus, OH: Merrill, 1989. This text was used in developing this chapter.

Langley, Stephen. *Theatre Management and Production in America.* New York: Drama Book Publishers, 1990. Chapter 4 provides much information on personnel for all levels of theater.

Wolf, Thomas. *Managing a Nonprofit Organization.* Englewood Cliffs, NJ: Prentice-Hall, 1990. Chapters 3 and 4 provide clear information about putting together a work force and establishing personnel policies.

8

Fundamentals of Leadership and Group Dynamics

We are now ready to examine the complex areas of leadership, the management communication process, and group dynamics in arts organizations. This chapter is designed to make you aware of the many theories that exist in the field of management psychology. I strongly urge you to explore the list of books at the end of this chapter.

Up to this point, we have created an organization, given it an overall structural framework, established strategies and plans to realize its mission, and begun staffing the enterprise with the best people we can find. Before we move into the specific operational areas of finance, budgeting, scheduling, marketing, and fund raising, we need to finish building the organization's interpersonal structure. Every day, arts organizations face the changing dynamics of people working together. With sensitive and adaptive leadership, the organization will go far. As you will see in this chapter, developing an organization with effective leadership is a continually challenging process.

Leadership Fundamentals

The subject of leadership is explored in numerous books each year. Stop by your local bookstore, and go over to the section on business books. There you will find dozens of titles on the topic. The search for the best way to develop leadership skills and to use those skills to create an organization that flourishes is a popular topic in today's business literature.

What is the essence of leadership? Simply put, *leadership* is the manager's use of power to influence the behavior of others.[1] Power, as we will use the term, is defined as the ability to get someone else to do something you want done.

In all of our discussions of leadership, keep in mind that leadership success is a necessary, but not the sole, condition for managerial success. It is also important to remember that although a good manager should be a good leader, a good leader is not necessarily a good manager. People have ranges of skills, some of which are more developed than others. For example, someone identified as an excellent leader may not be particularly good at planning and organizing. Some managers may be wonderfully organized with detailed plans but lacking in leadership abilities. Let's look at the two basic leadership modes: formal and informal.

Formal and Informal Leadership Modes

Formal leadership is leadership by a manager who has been granted the formal authority or command right.[2] The director of the play, the conductor of the orchestra, and the chair of the board of directors have been given formal authority by the organization to act in behalf of the organization. *Informal leadership* exists when a person without authority is able to influence the behavior of others.[3] Often informal leadership grows out of specific situations where an individual steps in and takes over. For example, suppose that an inexperienced student stage manager is unable to control the cast. A cast member with some stage managing experience steps in and starts giving orders. Because other students have respect for her, they listen to this informal leader and ignore the formal leader.

Theory X and Theory Y Approaches to People

Before examining the details of various leadership theories, consider again a topic we touched on in Chapter 3. Douglas McGregor's classic book, *The Human Side of Enterprise,* was noted in the summary of management evolution. His Theory X and Theory Y contrasted the fundamental beliefs that managers have about the people who work for and with them. The Theory X manager assumes that people dislike work, lack ambition, are irresponsible, resist change, and prefer to be led rather than to lead. The Theory Y manager works with people from the opposite perspective. He or she assumes that people like to work, are willing to accept responsibility, are capable of self-direction and self-control, and can be imaginative, ingenious, and creative.[4] McGregor's theory raises the issue of the self-fulfilling prophecy. This psychological term simply translates into a way of viewing how people will perform in their jobs. In other words, if you treat your staff, cast, and crew like idiots, they will tend to fulfill your expectations.

This topic is fundamental to the underlying attitude a manager has about the people he or she works with and supervises. The psychology of the work place is usually very complex. If you assume a leadership position without having developed an overview about the people you work with, you will probably run into a series of personnel problems that will limit your effectiveness.

Before you can develop your leadership skills, you must seriously evaluate your attitudes about work and people. In many arts organizations, both X and Y attitudes operate. Needless to say, these conflicting approaches usually lead to varying degrees of employee satisfaction. In addition, some arts organizations function with leadership that borders on tyrannical, while other organizations appear to be without leaders. The leadership sets the tone for the entire organization. The corporate culture of an arts organization is established and reinforced by its leadership.

Power: A Leadership Resource

The word *power* often has negative connotations. Yet without power, it would be impossible to operate most organizations. As we defined it, power is the ability to get someone to do what you want. However, in most arts organizations (or any organization), you have only as much power as your co-workers are willing to give you.

Let's begin our investigation of leadership power by posing three questions:

1. What sources of power are available to the manager?
2. What are the limits to the manager's power?
3. What guidelines exist for acquiring and using power?

Sources of Power

Two sources of power are available to the manager: position power and personal power.[5] The first comes with the job you occupy, and the second is directly attributable to you.

Position power As we have seen in Chapter 6, the organizational design process should establish the working relationship between and among employees in the organization. No matter how little vertical or hierarchical structure exists in the organization, managers are given power by their designated positions. For example, the production manager has more power than the technical director, the technical director has more power than the master carpenter, and so forth. Management texts identify three types of position power: reward, coercive, and legitimate power.

Reward power is the capability to offer something of value as a means of controlling others.[6] For example, a position may carry with it the power to grant raises, promotions, special assignments, or special recognition.

Coercive power is defined as the ability to punish or withhold positive outcomes as a way of controlling others.[7] For example, if you have ever received a verbal reprimand or a demotion or if you have ever been fired from a job, you have been subjected to coercive power. Control over the work schedule may be used as a form of coercive power. (Scheduling could also be used as a reward power in some cases.)

Legitimate power is the ability to control others by virtue of the rights of the office.[8] It is asserted in the phrase, "I am the boss, and therefore you must do what I ask."

Personal power Along with the position you hold, you bring your unique attributes and talents to the situation.[9] The two types of personal power are expert power and reference power.

Expert power is simply the ability to control others because of your specialized knowledge.[10] This could include special technical information or experience that others in the organization do not possess. For example, a stage manager with production experience may have expert power when it comes to planning a scenery shift on stage. This would allow the stage manager to exercise more power in a production meeting when alternative ways of doing a set shift are being discussed.

Reference power is derived from a more personal level of interaction with employees. Reference power is the ability to control others because of *their* desire to identify personally with the power source.[11] This use of power is often found among strong founder-directors of arts organizations. Their charismatic personality and forceful approach to managing the organization are used as a way of controlling others.

It is fairly easy to apply the five types of power—reward, coercive, legitimate, personal, and reference—to a conductor, a director, or a choreographer. Each of these leadership positions requires the use of some combination of these powers. You have probably realized that some individuals are better than others at using the power they have been given.

Limits of Power

Now that we have looked at the sources of power, let's look at some of the limits of power. In the organizational setting of the arts, the power to control others is a potential more often than an absolute. Although history provides many sad examples of individuals abusing the power they had over others, there are limits to power. In arts organizations, as we have seen, several different groups of employees work to support the organization's stated goals and objectives. Within each employee group, differing degrees of power are exercised. The union stage crew has a different relationship to the power structure of the organization than the senior staff has. However, whatever the differences may be within each work group, there are limits to how effectively power can be used to control work output.

Acceptance theory A theory put forth by Chester Barnard about power limitations is called *acceptance theory*. Simply stated, power is only realized when others respond as desired—that is, when they accept the directive.[12] Acceptance theory states that people are most likely to accept orders or requests when one or more of these four conditions are met:

1. They truly understand the directive.
2. They feel capable of carrying out the directive.
3. They believe that the directive is in the best interests of the organization.
4. They believe that the directive is consistent with their personal values.

Zone of indifference Another theory that applies to the use of power is called the *zone of indifference*. This theory states that power in organizations is limited to the range of requests and directives that people consider appropriate to their basic employment or the psychological contracts they make with the organization.[13] A directive that falls within the zone of indifference tends to be accepted and followed automatically. For example, if a marketing research assistant in a museum is asked to check the membership list for zip code distribution in comparison to census data reports, he would not react negatively. However, if his supervisor asks him to pick up her dry cleaning on the way to work, the odds are good that the supervisor has crossed the zone of indifference. The assistant will react by saying, "That is an inappropriate request to make."

Both of these theories can be easily applied in an arts setting. Trying to ignore these theories may make the leadership role very difficult. Put yourself in the position of a cast member, intern, or crew member, and think about how the acceptance theory may affect your interaction with your supervisor.

Guidelines for Using Power

Consider using the guidelines that follow when you find yourself exercising formal or informal power. They are from an article by John R. Kotter in the *Harvard Business Review*.[14]

1. Don't deny your formal authority. It is acceptable to act like the boss if you keep your perspective and remember that you are dependent on the good will and cooperation of the people who work for you.

2. Don't be afraid to create a sense of obligation. Doing a few fa-

vors or clearing the path so that employees can get their jobs done will help establish their obligation to follow your direction.

3. Create a feeling of dependence. Although care must be taken not to create a negative dependence (employees can't make a decision without you), it can be helpful to establish a situation in which people depend on your help to make their job easier. This will make it easier to gain their cooperation later.

4. Build and believe in expertise. A few solid examples of your having accomplished something will help build a belief by others in your expertise. No one likes working for know-nothing bosses who do not seem qualified to hold the positions they do.

5. Allow others the opportunity to identify with you as a person. When you create an environment in which the people you work with know and respect you as a person, they are more likely to follow your direction and supervision.

Approaches to the Study of Leadership

Management researchers have developed several theories that attempt to predict why some people are better leaders than others. The first studies of leadership examined the personal traits and psychological characteristics of people in leadership roles.

Trait Approaches to Leadership

The earliest research was based on the assumption that a person with particular traits has leadership potential. The idea was to establish an inventory of traits and to match them to people. This early research focused on physical and psychological attributes. However, there proved to be little correlation between these attributes and leadership.[15]

Behavioral Approaches to Leadership

Other researchers have tried to formulate a leadership model by studying recurring patterns of behavior by people in leadership positions. The research focused on the leader's orientation toward tasks and people. Leaders who were highly concerned about the tasks to be done exhibited certain behaviors: planning and defining the work to be done, making clear assignments of task responsibility, setting work standards, and following up on task completion, and monitoring. The people-oriented leaders tended to emphasize other behaviors: developing social rapport with employees, respecting the feelings of others, and developing a work environment of mutual trust. These styles of leadership are diagrammed in Figure 8–1. A practical application of this matrix is the relationship between a stage director and the cast. Consider your own experience, and try to place an arts leader you have worked with somewhere within this matrix.

Contingency Approaches to Leadership

Circumstances in the work place change, and these changes may require different leadership approaches. Some management researchers questioned the idea that any particular leadership style is effective in all situations. Out of their studies came the theories called *contingency* or *situational* leadership. Here are two examples of different leadership styles required in a typical arts setting.

A membership manager for a museum must coordinate a renewal and new members drive each year. Once it is planned, the entire opera-

Figure 8–1

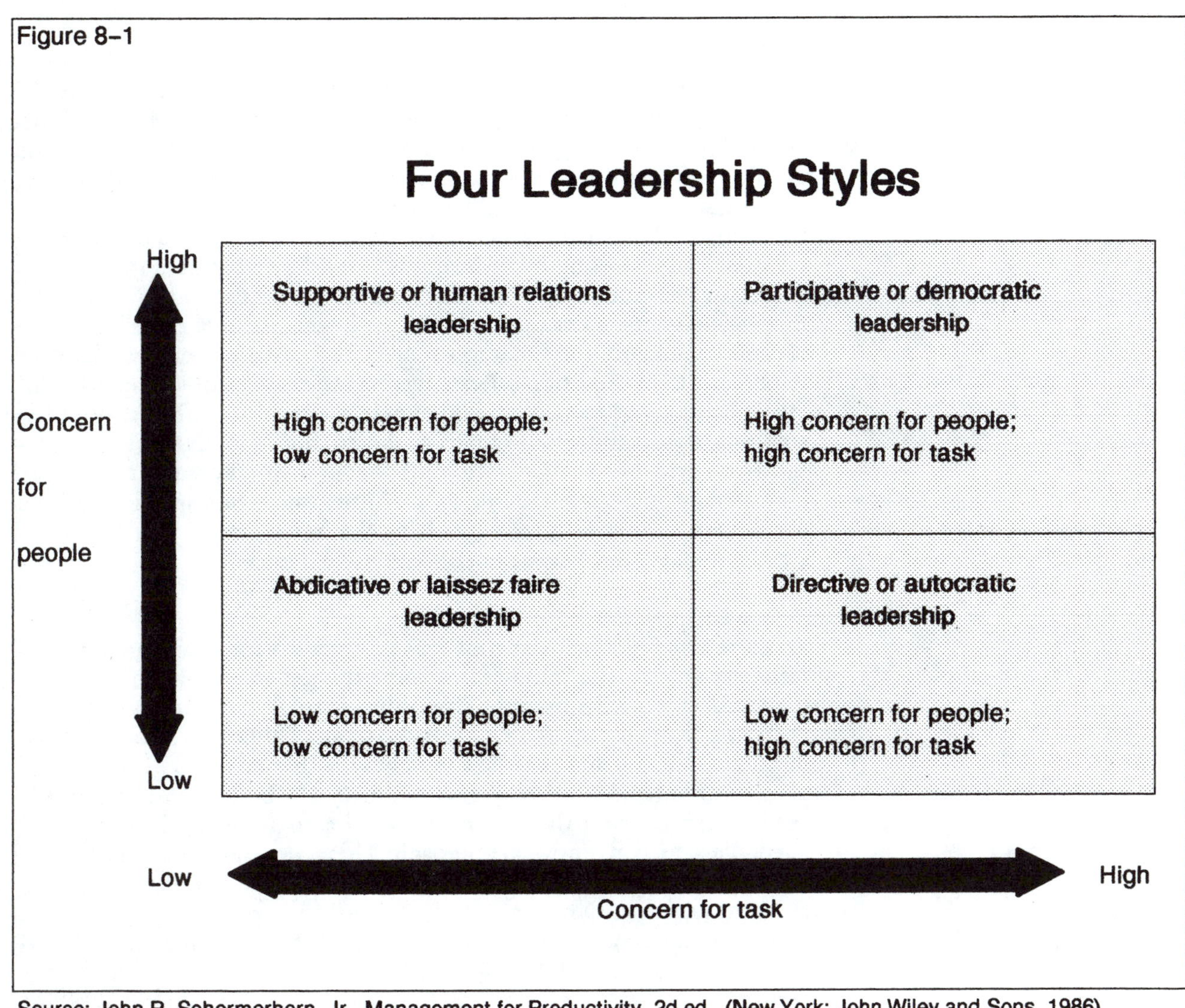

Source: John R. Schermerhorn, Jr., Management for Productivity, 2d ed., (New York: John Wiley and Sons, 1986). Copyright (c) 1986 by John Wiley and Sons, Inc. Used with permission.

tion is very task-oriented and specific. The leadership requirements in this situation would be directed toward ensuring that employees were being accurate completing routine tasks. At the same time, because this project requires repetitive work on the part of the staff, the manager's challenge is to keep people motivated and productive. Therefore, the manager might structure the day with frequent breaks or some other form of stress relief for the staff.

After finishing the big renewal campaign, the membership manager has been appointed to chair an ad hoc committee to study and improve management and employee relations. Different leadership skills will be required. The goal is defined, but the specific tasks are not indicated. Other committee members will be volunteers, and the manager will have little direct control over them. This situation requires strong group leadership skills. The manager must also develop clear objectives for all phases of the study. Deadlines will have to be set, objectives defined, and committee procedures established.

Applications to the Arts

Management theory tries to be scientific about creating "experiments" and "controls" in an attempt to "test" the theories. In reading the literature, it becomes apparent that no one theory can explain why some people are dynamic, productive leaders and others are not.

In an arts organization, keeping the creative spirit alive is a full-time job for a manager. Leadership directed at trying to run an arts organization "just like any other business" could be a formula for disaster. Although it is true that an arts organization must function in a business-like way, arts managers must address the larger issues of the relationship of their organization to the larger society and culture. Authors like Warren Bennis (see "Additional Resources" at the end of this chapter) point out that the trend to two- and three-part leadership—distributed leadership—of organizations only tends to dilute their effectiveness in pursuing their overall goals. Instead of leadership, the result is more bureaucracy. Instead of looking to the future, managers spend their time dealing with the routine of supporting the bureaucracy. (See "Kennedy Center Adopts New Management Plan" in Chapter 7.)

Future Leadership?

As more arts organizations move away from the founder-director leadership structure, the trend seems to be toward adopting the multiple-manager leadership model. The older, intuitive leaders with great charismatic appeal are fading from the scene. Corporate structures and distributed leadership may be the only way that arts organizations can gain the required fund raising credibility in the community, but it seems doubtful that this is a formula for artistic leadership that goes beyond a safe, conventional approach. There are many examples of upstart arts organizations of 1970 that are now cornerstones in regional arts consortiums. Obviously, artistic leadership need not succumb to conventionality just because it is accepted in the community. However, since so much funding for the operation of arts organizations comes from ticket sales and local fund raising, there is a point at which controversial leadership becomes a detriment to the organization.

Arts managers will eventually find themselves solving critical problems and making big decisions through groups. It might be considered a type of "art by committee." The trend toward the committee-style management of arts organizations may prove irreversible. If this is the case, a leader with great skill as a negotiator will be required.

Motivation and the Arts Work Setting

No discussion of leadership would be complete without examining the area of individual and group motivation and the communication process. To lead effectively, you must understand some basic concepts about what makes individuals want to work and create. At the same time, most of the activity that occurs in an arts organization revolves around groups. Organizations are divided into work groups that should match the operational and planning objectives that the organization has established. Creating, maintaining, and keeping these work groups productive is one of the manager's leadership responsibilities. To work with individuals and groups, you must communicate your expectations and objectives, and the people you work with must effectively communicate their progress and problems to you.

Let's look at the area of motivation first. The people who make up a large portion of an arts organization are usually highly self-motivated. The discipline and motivation required to become a singer, dancer, actor, designer, or musician are not universally found in society. Ideally, professional performers need not be told to learn their lines. However, we do not live in an ideal world. People respond in often unpredictable ways to various challenges. Therefore, even the most gifted and motivated may benefit from carefully structured communications that help them achieve their goals.

Theories of Motivation

Management texts usually devote large sections to the subject of motivation. Motivation theories and applications arise from research in psychology. These applications are directed toward the work place and making workers more productive. The research includes content theories and process theories. The *content theories* of motivation "are concerned with identifying what is within an individual or the work environment that energizes and sustains behavior. That is, what specific things motivate people."[17] The *process theories* "try to explain and describe the process of how behavior is energized, directed, sustained, and finally stopped."[18]

Theories based on the content approach include Abraham Maslow's hierarchy of needs and Frederick Herzberg's two-factor theory. Popular process theories include Victor Vroom's expectancy theory and the reinforcement theory of B.F. Skinner and others. In keeping with the scope of this text, the major theories are summarized, and sources for further readings are suggested at the end of the chapter.

Content Theories

Maslow's hierarchy of needs Abraham H. Maslow's 1954 book *Motivation and Personality*[19] created a foundation on which many business psychologists have built. Maslow proposed that human beings have five levels of needs arranged in a hierachy of importance. These fell into lower-order needs (physiology, safety, and society) and higher-order needs (esteem and self-actualization). The system is based on the assumption that only unmet needs act as motivators. The other key principle is that these needs are arranged in a strict hierarchy. The implication is that an individual can move to the next level only after satisfying the needs in the next lower level.

Maslow's theory has been embraced by much of the business world, but that doesn't mean that it can explain all facets of human behavior. Cultural differences, the reality that strict hierarchy doesn't always describe how people behave in a work environment, and the fact that people's needs change over time were not easily accommodated in Maslow's theory.

In an arts organization, providing a job in a comfortable work environment (physiological need) that does not endanger health (safety need), has a degree of group stability (social needs), recognizes good performance (esteem need), and provides opportunities for creativity (self-actualization need) shouldn't be an impossible task. However, if employees fear for their lives whenever they go out on the stage because the lighting system is dangerous, for example, their safety and physiological needs are not being met.

Two-factor theory Frederick Herzberg and B. Syndermann's 1959 book *The Motivation to Work*[20] became the cornerstone for another content theory of motivation. Their study focused on what they called the *two-factor theory* of motivation. The *maintenance factors* refer to those elements of work that are not motivators as long as they are adequately met. These factors include salary; good relations with subordinates, peers, and superiors; status; and work conditions. The assumption is that if one or more of these needs are not met, they can become motivators. For example, if you are unhappy with your salary, it becomes a motivator, and you will be prompted to do something to correct the unmet need.

The second set of factors are the *motivational factors.* They consist of such things as achievement, recognition, advancement, personal growth, and responsibility. Job enrichment is one practical management application of Herzberg and Syndermann's theory. They argue that just increasing someone's salary or making a change in the work environment will not immediately translate into harder work.

The limited scope of the study on which the two-factor theory is based (limited to about two hundred engineers and accountants) invalidates the theory to some critics. For example, by only studying professionals, Herzberg and Syndermann do not address the fact that maintenance and motivational factors might differ for hourly employees. In other words, the staff members in the marketing department of a performing arts center will probably have different perceptions about motivators than the stagehands who unload the trucks at the loading dock.

However, this should not invalidate the two-factor concept. Managers may obviously adjust the range of maintenance and motivational factors for particular work groups. For example, most employees like recognition. Prominently displayed "employee-of-the-month" plaques with photographs and accompanying praise for some accomplishment can boost morale. Unfortunately, the two-factor theory doesn't provide much guidance on how the motivational factors can be translated into measurable increases in productivity.

Process Theories

We will take a quick look at two of the process theories. Essentially, these theories look at how people think about their jobs and their work. It is assumed that people find their own sources of motivation and dissatisfaction in the work place.

Equity theory The equity theory of motivation is based on the work of J. Adams Stacy in the 1960s.[21] The theory states that a perceived inequity functions as a motivator. When employees believe that they are not being treated equitably, they will try to change the source of the perceived inequity. Employees perceive inequities whenever they feel that they are not rewarded for their work at the same level as someone else who works equally hard. To resolve the equity conflicts, Stacy predicts that employees will change how much work they do, try to get their salary increased, rationalize the inequity, or quit.[22]

Equity issues most often arise with highly separated work groups. For example, an arts organization might have union stage employees who receive $15 to $20 per hour for their labor and a marketing research assistant with a master's degree who receives the equivalent of $7.50 per hour. It will not take long for the research assistant to pick up on these wage differences, thanks to the informal communication system in most organizations. According to Stacy's equity theory, the research assistant will probably create some rationalization to minimize the inequity, or she will quit. If she approaches the head of the marketing area for a pay increase and is told that the budget is too tight, she may go back to work, but she will probably reduce her work output. The inequity may not go away. In fact, the problem may grow as the unhappy employee lets other employees know that they are receiving a lot less money for their hard work than others in the organization. The employees may also say to themselves, "If that's all you think I'm worth, then that's all the work you are going to get from me." The net result is dissatisfied employees and less work output.

This issue of equity is a hot topic in the business world. Lobbying groups for the business community are hard at work in Washington, trying to control the growth of the concept of comparable worth. If the marketing assistant is doing work of similar value to the organization as the stagehand, why isn't she paid the same rate? Should the organization pay equally for equal work?

Arts managers would do well to read up on the equity theory. They must anticipate these issues and formulate some strategies to help employees. The employee's perspective is important here. For example, when it is time for contract negotiation with union employees, nonunion employees will start talking about wage inequities. Managers might explain that the stagehands don't work year-round, so they do not make as much as imagined. In fact, they may make less on average per year than salaried employees.

Expectancy theory Another motivation theory often cited in management textbooks is the *expectancy theory*. Simply stated, Victor Vroom's theory postulates that people will be motivated to work if they expect that they will be adequately rewarded for their effort.[23] The expectancy theory has four components.

Choice is a key element in the process. *Choice* means that an individual has the freedom to select from various alternatives. He or she can exert a great deal of effort or only the minimal amount necessary to get the job done.

Expectancy is defined as the belief that the work selected can or cannot be accomplished. This is a totally subjective issue. If you are given a task—such as updating a 20,000-name mailing list in two days or building an entire set of stage platforms without any assistance in three days—your expectancy might be zero. Using these examples, your expectancy will probably be higher if you were given six months for the mailing list and four weeks for the construction project.

The third component of the theory is *preference*, which refers to the values people attach to the possible results of their efforts. If you are told that you will receive an extra vacation day if you finish the mailing list or the construction project early, you must decide whether the value of the reward is worth the extra effort.

The last element of the model is *instrumentality*, which is Vroom's somewhat obtuse title for the probability people subjectively assign in linking their level of performance to their potential reward.

Does this model offer any help to arts managers? Yes. For example, you can influence expectancy by establishing a general attitude in your work group that the work is important and does make a difference. You can also hire and train people in the work group who are willing to accept the attitude you desire. You can influence preferences by developing an ongoing system of listening to employees' needs and guiding them toward results.[24]

Never underestimate the power of perception, and never assume that the people who work for you have the same values and assign the same priority to the work to be done. Your effectiveness in the leadership role is dependent on your ability to motivate the people with whom you work. Understanding what motives *them*, what *they* perceive as a reward, and what *they* value in the work place is a key element to your success.

Reinforcement theory The last area of motivation falls under the broad heading of reinforcement theory. The motivational theories described previously approach behavior from the perspective of how people perceive the value of work, how they satisfy needs, or how they try to resolve inequities. Reinforcement theory focuses on the external environment in which people function. The use of positive and negative reinforcement is the motivating force that managers use in their leadership roles. Let's take a quick look at this topic.

Organizational behavior modification (OBM) is an approach that uses the principles of B.F. Skinner's research on human behavior.[25] *Operant conditioning*, a key element in the research, assumes that you can control behavior by manipulating its consequences. By using positive and negative reinforcement, you can increase desired behaviors or eliminate undesired behaviors. Some other key concepts in the system include the *law of contingent reinforcement*, which states that for a reward to have maximum impact, it must be delivered only if the desired behavior is exhibited, and the *law of immediate reinforcement*, which states that the quicker the delivery of the reward after the desired behavior, the greater the reinforcement value.

Does behaviorist theory have a place in an arts organization's leadership system? Yes, if carefully applied. Let's consider a few examples. Something as subtle as nodding occasionally during a meeting as your assistant makes a presentation about a new marketing plan can have a positive reinforcing effect. As an example of behavior modification through negative reinforcement, suppose that you always make a point of saying, "I thought we had a no-smoking rule on stage in this theater" whenever you find the crew head smoking in the theater. Then one day you see that the employee isn't smoking, and you walk by without saying anything. You stop nagging the employee when he stops the undesired behavior. Negative reinforcement, by the way, is not necessarily the best way to approach behavior modification. Unfortunately, for most of their lives, people hear only about the behaviors that they are not supposed to engage in. In some organizations, negative reinforcement is the main operating principle.

As an approach to organizational leadership, behavior modification has been criticized because it focuses solely on extrinsic reinforcers. The complex reasons behind a particular behavior pattern are of little interest to the leader who relies on behavior modification. Critics argue that self-motivated artists, who are independent and creative, will laugh at attempts to influence their behavior through simplistic, positive-reinforcement techniques. However, praise is a powerful leadership tool, and as a positive reinforcer, most people do not seem upset when it is used sincerely.

An aware arts manager who carefully and thoughtfully uses some components of operant conditioning usually can't go wrong. A director, choreographer, or manager of any type will usually get better results with positive reinforcement than with negative reinforcement. Berating and belittling people usually instills hostility and resentment among employees. Managers who believe that the only way to get top performance from their employees is through terror tactics are sadly out of touch with reality.

Theory Integration

The manager's objective is to be as effective as possible in getting people in the organization to achieve the results that support the organization's goals and objective. It isn't by chance or through the efforts of one person that an organization reaches and exceeds its goals. The motivational theories are tools to be used by the manager. Within an arts organization, some employee groups are motivated by extrinsic rewards, others by how they perceive their role and status, and still others by the need to achieve some degree of self-actualization. It will take time and experimentation to find the best mix of motivators in any work situation.

Group Dynamics

A fact of organizational life is that leaders must work effectively with many different groups. Whether the group is formal or informal, when you put several people together, a collective behavior pattern emerges that is usually different than an individual acting singly. Arts organizations are made up of several groups: a cast, corps de ballet, ensemble, crew, board, committee, sub-committee, task force, and so on. The effective leadership of all of these various groups can result in a dynamic, creative organization that has a positive impact on the community. By the same token, ineffective group leadership can result in low-quality events and productions, poor use of resources, high turnover of staff and board members, labor problems, and marginal community support. Let's look at some of the basic terms and concepts of group management and leadership.

Group Management Terms

A *group* is a collection of people who regularly interact with one another in the pursuit of one or more common objectives.[26] A *formal group* is created by the formal authority structure within an organization to transform inputs into product or service outputs.[27] For example, a theater company sets up a formal group (e.g., a cast) by deciding to do a play and present it to the public. A board of directors creates a formal group when it selects a personnel search committee to find a new museum director. Organizations may establish *permanent work groups* to carry on specific operational activities. For example, the production staff in an opera company or the curatorial staff in a museum may meet regularly as a group to make plans, assign work, and evaluate progress. *Temporary groups*, such as a personnel search committee, may be established to accomplish a particular task. The group is disbanded after it completes the job.

An *informal group* is "one that emerges in an organization without any designated purpose".[28] These informal groups can satisfy employee needs for socialization, security, and identification.[29] Informal groups can also help people get their jobs done by establishing a network within an organization. For example, a production manager in an arts organization may establish a formal working relationship with the crew heads through regular staff meetings. However, within the crews, various informal groups will form that can help or hinder the overall operation. An informal group may form within the crew (usually with an informal leader), centering around the belief that the production manager is incompetent. This informal group may try to influence others in the formal work group about the manager's incompetency. Soon, the production manager finds that things are not getting done or are being done in the way that the informal group decides is best. Direct intervention by the formal leader of the group may be the only way to disrupt the influence of the informal group.

Types of groups Various types of groups are formed in organizations, including command, task, interest, and committee groups.[30]

Command groups are established in the organizational chart in the working relationship between supervisors and subordinates. Figure 6–2 depicts the command group relationship between the managing director and the directors of marketing, finance, fund raising, and facilities.

Task groups are groups of employees who work together to complete a project or job. A major portion of the activity in arts organizations is accomplished by task groups.

Interest groups form when employees unite around a particular issue. The members of this group could be from different work groups who are brought together to resolve a short-term problem. For example, when a theater company announces that, due to a shortfall in fund raising, all medical benefits for the regular staff will cease, an interest group forms to deal specifically with this issue.

Finally, we have the committee, which has been rudely defined as "the only life form with twelve stomachs and no brain."[31] The operations of a committee are summarized under Old and Kahn's law: "The efficiency of a committee meeting is inversely proportional to the number of participants and the time spent on deliberations."[32] A more formal definition of a "*committee* is a group of two or more people created to perform a specific task."[33] Organizations establish *standing committees* to fulfill ongoing needs (for example, the Finance Committee) or *ad hoc committees* to fulfill specific needs (for example, the Search Committee). Numerous books offer suggestions for making committees function effectively in organizations. Such issues as committee composition, size, clarity of purpose, and ability to bring resources to bear on a problem are covered in a variety of texts and business books. The disadvantages of compromised decisions, long deliberation periods, and the expense are often cited in the literature. However, committees do tend to proliferate in organizations. Care must be taken to avoid using the committee approach to avoid taking individual responsibility for decisions.

Stages of group development The study of groups shows that when a new group is formed, it typically undergoes four stages: forming, storming, initial integration, and total integration.[34]

In the *forming stage*, the group tries to establish its purpose, define its operational rules, establish the identity of members of the group and what they have to offer, and define how people will interact with each other.

The *storming stage* may be very emotional or relatively calm, depending on the personalities of the group members. For example, an ad hoc committee to examine employee benefits that is made up of staff and hourly workers could experience substantial personal style differences that take some time to work out.

During the stage of *initial integration*, the group reaches an agreement about how it will work together on the issue at hand. Members may agree to disagree about some topics and focus instead on the things that they agree on.

Assuming that the committee chair or group leaders have done their job, the group should reach *total integration* and become a functioning and productive unit. Constructive ways of handling disagreement will be found, and group discussions will allow differences to be expressed. Group members will feel more confident about their specific responsibilities and will help keep the group focused on the problems that must be solved.

A group such as a cast of a play will go through some variation of this process as it moves from auditions to rehearsals. Other ensemble efforts share similar patterns of development. When a committee is formed by the board of directors, patterns similar to those noted take place. An arts manager must watch vigilantly for committees that become dysfunctional. For example, some committees never achieve integration. The committee output is often slow in coming or is marked by minority reports by differing subgroups that form within the larger committee.

Group norms and cohesiveness *Group norms* is a familiar phrase related to leading and managing groups. *Norms* are the rules that guide group behavior.[35] The leader of a group must establish *behavior norms* ("One person talking at a time, please") as well as *performance norms* ("We must finish deliberations and report to the board by March 1").

At the same time, a leader must develop a cohesiveness among the group if it is to be effective. *Cohesiveness*, in this case, refers to the degree of motivation of members to stay in the group. For example, a running crew for a production is a task-specific group that often requires a high degree of cohesiveness. You can use specific circumstances such as having to do a complex scene change in a limited time, as a way of building cohesiveness among a group. For example, if the performance norm is to complete the scene change in one minute, the group may be challenged to beat that norm and do the shift in 45 seconds. When this new norm is established, the group usually feels some sense of collective accomplishment, which is a way of building cohesiveness.

Successfully managing groups in an arts organization requires careful thought about establishing norms, performance expectations, and building cohesiveness. In arts organizations, group performance extends from the board through the construction shops. Let's look now at some of the problems that can arise with groups.

Dysfunctional group activities and some cures One of the well-noted problems with groups that are too cohesive has been termed *groupthink*. In an article in 1971, Irving Janis defined the groupthink phenomena as "a tendency for highly cohesive groups to lose their critical evaluative abilities."[36] Unless a member of the committee or workgroup is designated as the devil's advocate, there is the danger that groupthink will establish itself in an organization. The peer pressure to appear to agree is enormous. A group leader should make it a point to have conflicting points of view aired before the group.

Some of the symptoms of groupthink are rationalizing data that contradict the expectation, self-censorship by group members, and creating an illusion of unanimity by stopping the discussion of a topic prematurely. As an example of groupthink, imagine a design-development discussion that includes a director, the designers, and key technical staff. The production manager, who is running the meeting, knows that the proposed set design is too big and expensive to produce, but the designer and director do not want to hear that. In fact, the director has said on several occasions, "Don't tell me what you can't do, tell me what you can do." The technical director has tried to tell everyone that this design is more than the shop can handle. Every time the technical director tries to bring up the subject of time, money, and personnel constraints, the production manager cuts off the conversation. The schedule dictates that construction start immediately. The group "decision" is really nothing more than a groupthink trap. The technical director knows it can't be done, but goes along with the group decision anyway. The shop proceeds to construct the set as designed. Later, when the show is overbudget and behind schedule, the technical director is asked, "Why didn't you say something before we started building the set?"

Strategies for Making Groups More Effective

Figure 8–2 illustrates many of the common problems that occur when people get together to function as a group. An aware leader must act immediately to stop these dysfunctional activities from disrupting the group.

It would be wise to build in some simple group behavior patterns as norms at the beginning. The first set of behavior patterns fall under the heading of *task activities*. The second group of patterns are called *maintenance activities*, and they support a set of healthy group interactions.

Edgar H. Schein, *Organizational Psychology*, 2d ed. (Englewood Cliffs, NJ: Prentice Hall, 1970) p. 81 lists these task activities:

1. Initiating: Setting agendas, giving ideas, defining problems, and suggesting solutions.
2. Giving and seeking information: Offering information directly related to the problem, asking others for ideas, and seeking facts.
3. Summarizing: Restating the highlights of the discussion can help keep everyone on track.
4. Elaborating: Clarifying ideas by citing relevant examples can help keep the group working effectively.

The maintenance activities include the following:

1. Gatekeeping: Allowing various members of the group to talk. Sometimes one person will try to dominate the discussion and direct the group to his or her opinion by monopolizing the discussion.
2. Following: Going along with the group and agreeing to try out an idea.

Figure 8–2

Dysfunctional Group Activities

1. ***Being Aggressive***
   Working for status by criticizing or blaming others, showing hostility against the group or some individual, deflating the ego or status of others.

2. ***Blocking***
   Interfering with the progress of the group by going off on a tangent, citing personal experiences unrelated to the problem, arguing too much on a point, rejecting ideas without consideration.

3. ***Self confessing***
   Using group as a sounding board, expressing personal, nongroup–oriented feelings or points of view.

4. ***Competing***
   Vying with others to produce the best idea, talk the most, play the most roles, gain favor with the leader.

5. ***Seeking sympathy***
   Trying to induce other group members to be sympathetic to one's problems or misfortunes, deploring one's own situation, or disparaging one's own ideas to gain support.

6. ***Special pleading***
   Introducing or supporting suggestions related to one's own pet concerns or philosophies, lobbying.

7. ***Horsing around***
   Clowning, joking, mimicking, disrupting the work of the group.

8. ***Seeking recognition***
   Attempting to call attention to self by loud or excessive talking, extreme ideas, unusual behavior.

9. ***Withdrawing***
   Acting indifferent or passive, resorting to excessive formality, daydreaming, doodling, whispering to others, wandering from the subject.

Source: J. William Pfeiffer and John E. Jones (Eds.), 1976 Annual Handbook for Group Facilitators (San Diego, CA: Pfeiffer and Company, 1976). Used with permission.

3. Harmonizing: When appropriate, reconciling differences and promoting compromise can help keep the group going.
4. Reducing tensions: Using humor as an antidote when the situation becomes emotional. This can help shift the energy of the group long enough to put the conflict in perspective.[37]

Distributed Leadership

For any organization to function effectively as a group, there must be a healthy interchange among its members. Arts organizations, especially performing arts organizations, spend a lot of time engaged in group activities. The management of these various group efforts calls for the recognition of the concept of *distributed leadership*. Simply put, it means that the leadership responsibility is shared by the group members. As a member of a committee, a work group, or a cast, you share a responsibility to keep the group from becoming dysfunctional. Leaders who point out effective strategies and dangerous behaviors have the best chance of bringing distributed leadership to life for the group. As noted earlier, distributed leadership can create another layer of management and can add more bureaucracy in an arts organization. However, if all members of the group adopt the attitude that being a leader means making decisions, the organization does not have to become mired in inaction.

Communication Basics and Effective Leadership

Underlying the entire area of leadership is the assumption that good communication and listening skills are used daily. Success as a leader directly relates to your ability to send, receive, interpret, monitor, and disseminate information. However, because the process of communication is so simple and, at the same time, so complex and subtle, we often overlook the obvious when we hunt for the source of a problem. The consequences of miscommunication—ranging from the simple "go" on a cue by the stage manager to the complex report by the director of finance to the board—can be devastating. A missed special effects cue may be life threatening to a performer, and a misunderstood financial report may lead to bankruptcy for the enterprise. Almost all organizations say, "We have a communication problem around here." Whether this is true or not is irrelevant. If the phrase is repeated often enough, the perception that a communication problem exists will be created.

Let's look at some of the basic terms and definitions in communications and then explore some strategies to minimize the problems.

The Communication Process

We will use the following definitions as a starting point. "*Communication* is the creation of meaning through the use of signals and symbols. Furthermore, *meaning* is defined as the perception that takes place when we formulate the relationship between two statements or images. Lastly, signals and symbols are key components in a message. *Signals* mean the messages which a communicator feels are beaming from a source, and they suggest very limited but concise meaning. *Symbols* suggest broader and more complex meanings assigned to the verbal and nonverbal language of the communicators."[38]

Suppose that a museum director walks into the museum on Monday morning, scowls at everyone, goes into the office, and slams the door shut. This nonverbal symbolic behavior communicates a wealth of information to the office staff. Imagine that the director of a play watches a scene and says to the cast in a monotone, "Very good." The message is mixed. The verbal tone communicating a half-hearted endorsement contradicts the meaning of the words." "Very good" might mean, "You did fine, but I really was not impressed."

As you can see from these examples, the communication process carries many nuances that have different meanings to people. Figure 8–3 depicts a simplified overview of the communication process. Let's briefly review what takes place in a typical interchange between two people.

The communication process includes a sender, who encodes and delivers a message through a communication channel, and a receiver, who decodes the message and perceives a meaning. The sender receives some feedback or an acknowledgment that the message has been received. At the same time, the communication channel is directly affected by noise that interferes with the message. Noise, in this case, means anything that disrupts the message or the feedback.

Perception

For the communication process to be effective, both the sender and the receiver should be aware of four key elements that modify the perception of the communication by each party. These four elements are stereotypes, the halo effect, selective perception, and projection.[39]

Stereotypes When you speak of "dumb dancers," "techie types," or "musicians!" you are communicating in stereotypes. When you refer to the board of directors as if it were simply of one mind, you are stereotyping the individual members as if they all acted and thought the same way. If you are to become a credible leader, you must drop the stereotypical thinking from your perceptional system.

Halo effect The halo effect is the perception of an individual based on one strong attribute. For example, a person who shows up late for a rehearsal or a meeting more than once will suddenly be known throughout the organization for "always being late." The halo effect can also be used positively. For example, a recent report on the long-term funding prospects for the organization may make a staff member a star just in time for the annual board meeting when, in fact, this individual has been coasting all year and does not deserve the praise.

Selective perception Selective perception refers to noticing only those incidents or behaviors that reinforce what you already strongly believe about a situation or a person. You may choose not to see problems that particular employees are having because it is inconsistent with your perception of them.

Projection When you project, you assign your personal attributes to someone else. The classic example of projection is when you assume that everyone who works for you shares your attitudes and beliefs about their job and the organization.

Formal and Informal Communication

Within an arts organization, formal and informal networks exist to communicate with and among employees. Managers must give constant attention to how well both systems are serving the organization's communication needs. Memos, small group meetings, forums, newsletters, and annual meetings make up a part of the organization's formal communication system. The informal communication system

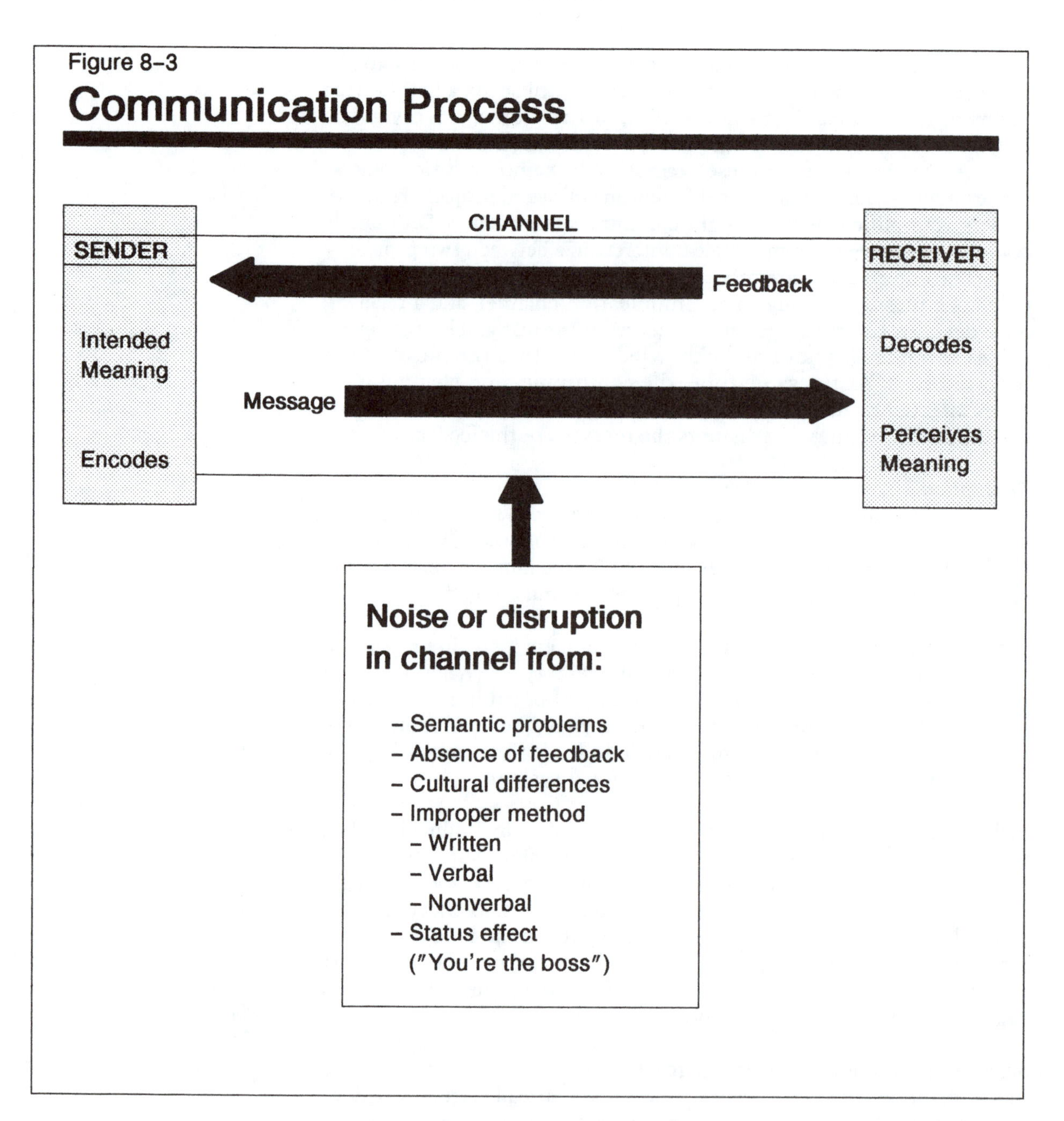

exists at every level in the organization. Phone conversations, waiting in line to use the copy machine, coffee breaks, and rehearsal breaks may all be touch points for informal communication. The informal system may seem impossible to manage, but by simply recognizing its existence and monitoring the information (or misinformation) being communicated, the manager can creatively intervene when required.

Conclusion

This chapter provided background on one of the most important areas in operating an arts organization. One question remains: What makes a good leader? As you have seen, it takes a great deal of hard work to be an effective leader. Having a vision of where you want to go, and being skilled in such areas as communication, interpersonal relations, observation, and situation analysis are equally important.

In fact, being a leader means playing a role to some degree. Some people are very comfortable performing on a stage, making a presentation in front of a group, arguing a point, or carrying on an intensive negotiation. People you work with perceive your performance as a leader in much the same way that an audience perceives a performance and develops an impression of a character during the course of a play. A complex combination of body language, tone of voice, and, ultimately, the conviction with which you deliver your lines, form your co-workers' overall opinion about your leadership. If you are unsure and do not act committed to the idea or project, it will be hard to convince your "audience" that you have the leadership needed to see something through to the end.

Is there a lack of good leadership in many organizations? Sadly, yes. Coping with ineffective leadership is the topic of Muriel Solomon's *Working with Difficult People.* (Englewood, NJ: Prentice Hall, 1990). Chapters such as "When Your Boss Is Belligerent," "When Your Boss Is Arrogant," and "When Your Boss Is Exploitative" paint a grim picture of the work place.

Good management and good leadership do exist. Although there are few studies of arts organization, some people like Tom Peters (*In Search of Excellence*) are reporting on companies that meet their goals and keep their people happy and productive. For arts organizations, with their never-ending struggle against limited resources, it is especially critical for the leadership to recognize and reward the hard work and sacrifices of its employees.

Summary

Leadership is the use of power to influence the behavior of others. Power means getting others to do what you want. Formal leadership is granted to a manager by the organization. Informal leadership arises from special situations. A manager can draw on position power and personal power. Power is limited by acceptance theory and the zone of indifference.

Leadership theories have developed from trait studies that tried to identify leadership qualities by evaluating personal attributes. Behavioral theories are based on the study of the leader's attitudes about tasks and people. Contingency theories work from the concept that leadership approaches must be adjusted based on the particular situation.

An effective leader must understand motivation theories and how they apply in the work place. Content theories work from the perspective of specific things that motivate people. Process theories examine how behavior is initiated, directed, sustained, and halted. Maslow's theory states that people are motivated by unmet needs. Herzberg and Syndermann's two-factor theory states that maintenance and motivational factors affect worker performance. Maintenance factors that are not met act as motivators. Achievement, recognition, advancement, and responsibility are examples of motivating factors.

The process theories include equity, expectancy, and reinforcement theories. Equity theory states that perceived inequities act as motivators. Expectancy theory states that people will be motivated to work if they believe that they will be rewarded for their efforts. Reinforcement theory, which is based on operant conditioning, deals with rewards or punishment for desirable or undesirable behavior.

Real-life situations require that managers recognize that different work groups are motivated by different things. Theory integration is a possible model.

Managers must lead and effectively work with groups. Both formal and informal groups are a part of every organization. Group dynamics include understanding what happens when people are brought together to achieve certain objectives. Norms of behavior and cohesiveness are key elements in group development. Like people, groups can become dysfunctional over time. Groupthink is one symptom of ineffective group management.

An effective leader understands and uses the communication process between people and among groups. Elements of the process include the sender, the receiver, the channel, and the effects of noise on communication.

Key Terms and Concepts

Leadership and power
Formal and informal leadership
Position power
- reward power
- coercive power
- legitimate power

Personal power
- expert power
- reference power

Acceptance theory
Zone of indifference
Trait approaches
Behavioral approaches
Contingency approaches
Maslow's hierarchy of needs
Herzberg and Syndermann's two-factor theory
Equity theory
Expectancy theory
Reinforcement theory
Organizational behavior modification
Group dynamics
Formal and informal groups
Command, task, interest, and committee groups
Group development stages
Group norms and cohesiveness
Groupthink
Distributed leadership
Communication process
Stereotypes
Halo effect
Selective perception
Projection

Questions

1. The use of power is a key component in leadership. Discuss examples from your work experience in which power was used effectively or ineffectively.

2. What are some additional examples of acceptance theory and the zone of indifference in the psychological contract people have with an organization?

3. Can trait theory be effectively applied in evaluating arts leadership? Explain.

4. A directive or autocratic leadership style is often exhibited in an arts setting. Is it possible to have a strong artistic vision for a project and a participative leadership style? Explain.

5. Cite examples in which situational leadership (contingency approaches) worked or failed.

6. Analyze a recent motivation problem you encountered in your work or educational setting. What steps would you have taken to motivate the individual or group involved?

7. Can you cite a recent example of a dysfunctional group? How would you have solved the problem knowing what you now do about group behavior?

References

1. John R. Schermerhorn, Jr., *Management for Productivity*, 2d ed. (New York: John Wiley and Sons, 1986), 275.
2. Ibid., 276.
3. Ibid.
4. Ibid., 46.
5. Ibid., 279.
6. Ibid.
7. Ibid.
8. Ibid.
9. Ibid., 280.
10. Ibid.
11. Ibid.
12. Chester Barnard, *The Functions of the Executive* (Cambridge, MA: Harvard University Press, 1938), 165–66.
13. Schermerhorn, *Management for Productivity*, 280–81.
14. John R. Kotter, "Acquiring and Using Power," *Harvard Business Review*, 55 (July–August 1977): 130–32. Adapted from Schermerhorn, *Management for Productivity*, 282–83.
15. Schermerhorn, *Management for Productivity*, 284.
16. Warren Bennis, *Why Leaders Can't Lead* (San Francisco: Jossey-Bass, 1989).
17. Donnelly, Gibson and Ivancevich, *Fundamentals of Management*, 304–5. This quote cites a reference to a text by John P. Campbell, et al., *Managerial Behavior, Performance and Effectiveness* (New York: McGraw-Hill, 1970), 341.
18. Donnelly, Gibson and Ivancevich, *Fundamentals of Management*, 305.
19. Abraham H. Maslow, *Motivation and Personality* (New York: Harper and Row, 1954).
20. Frederick Herzberg, and B. Syndermann, *The Motivation to Work* (New York: John Wiley and Sons, 1959).
21. J. Adams Stacy, "Toward an Understanding of Inequity," *Journal of Abnormal Psychology*, 67 (1963), 422–36.
22. Schermerhorn, *Management for Productivity*, 338–40.
23. Victor Vroom, *Work and Motivation* (New York: John Wiley and Sons, 1964).

24. Donnelly, Gibson, and Ivancevich, *Fundamentals of Management,* 313–16.
25. B.F. Skinner, *Science and Human Behavior* (New York: Macmillan, 1953); B.F. Skinner, *Contingencies of Reinforcement* (New York: Appleton-Century-Crofts, 1969).
26. Schermerhorn, *Management for Productivity,* 359.
27. Ibid.
28. Ibid., 361.
29. Ibid.
30. Donnelly, Gibson, and Ivancevich, *Fundamentals of Management,* 346–47.
31. Arthur Bloch, *The Complete Murphy's Law* (Los Angeles: Price-Stern-Sloan, 1990) 48.
32. Ibid., 71.
33. Arthur G. Bedeian, *Management* (New York, Dryden Press, 1986) 508.
34. Schermerhorn, *Management for Productivity,* 366–67.
35. Ibid., 370–71.
36. Ibid., 374.
37. Ibid., 375–76.
38. John J. Makay, and Ronald C. Fetzer, *Business Communication Skills: Principles and Practice,* 2d ed. (Englewood Cliffs, NJ: Prentice-Hall, 1984), 5–6.
39. Schermerhorn, *Management for Productivity,* 310–15.

Additional Resources

Bennis, Warren. *Why Leaders Can't Lead.* San Francisco, Jossey-Bass Publisher, 1989.

Loden, Marilyn. *Feminine Leadership.* New York: Times Books, 1985.

Mintzberg, Henry." The Manager's Job: Folklore and Fact." *Harvard Business Review,* July–August 1975.

Tichy, Noel M., and Mary Anne Devanna. *The Transformational Leader.* New York: John Wiley and Sons, 1986.

Zaleznik, Abraham. *The Managerial Mystique: Restoring Leadership in Business.* New York: Harper and Row, 1989.

Additional Titles

Bardwick, Judith M., *Danger in the Comfort Zone.* (New York: American Management Association, 1991).

Hunsaker, Phillip L., and Alessandra, Anthony J., *The Art of Managing People.* (New York: Touchstone Book published by Simon and Schuster, Inc. 1986).

Control: Management Information Systems and Budgeting

9

Before we move on to the topics of finance, marketing, and fund raising, we need to examine one more area in our look at the overall management process: control. We saw how planning helps us set the organization's direction and allocate its resources. We studied the organizing process to see how best to bring together people and resources. The leadership part of the process focused on directing people in the utilization of resources. We will now look at control, the part of the management process that ensures that the right things happen, in the right way, and at the right time.[1]

In an arts organization, the word *control* carries connotations that often make people uncomfortable. People generally do not like to think of themselves as being controlled by others. At the same time, however, they are not comfortable in situations that could be described as being "out of control." If an arts manager is to lead an organization successfully, systems of control must be in place and must function effectively. Far too often we hear of the results of a lack of a control process in an arts organization. When you read an article about a dance company that ran up an unanticipated deficit of $800,000 in one season, you have to ask yourself how this could happen. The assumption is that the budgetary and financial control systems must have broken down completely. After all, $800,000 does not just appear in a budget report one day.

Control as a Management Function

We will use the term *control* to mean "a process of monitoring performance and taking action when needed to ensure the desired results are achieved."[2] The elements that enter into this process and affect how well the system works are uncertainty, complexity, human limitations, and the degree of centralization in the organization.

Uncertainty exists in all planning. Without the ability to see into the future, every organization must plan for uncertainty, and the control system must take this element into account. One useful way of accommodating uncertainty is to evaluate regularly the progress being made in meeting the defined objectives. This becomes a control point at which you may make adjustments in the activities being performed.

Issues related to *complexity* are sometimes more elusive. Over time, organizations grow and become more diverse.The controls required to monitor activity in an organization often lag behind growth.

For example, if you shift from processing all of your ticket and subscription revenue through your own box office to a new performing arts center, your old control for tracking revenue will probably be inadequate for the new system.

All control systems must take into account *human limitations*. Errors will be made. An incorrect amount will be entered in the computer, an order form will be misplaced, a costume or set piece will be constructed incorrectly, or a purchase order will be lost. The control system must recognize that these things will happen and must have in place adequate monitoring capabilities. For example, ticket sales might be counted by two different people before being deposited in the bank.

The basic design of your organization may require different control processes because of the degree of *centralization* or *decentralization*. If you operate a decentralized organization, authority will normally be delegated to more people in middle and lower level management positions. Control systems that ensure accountability will be required. For example, if the scenery construction shop is five miles from the administrative offices, you do not want to make a staff member drive over to the office every time a purchase order is needed. Instead, you will delegate the authority to approve purchases to a staff member at the shop, who will submit the purchase orders and invoices for the week to a member of the accounting staff for review. The control points now include the staff member who authorizes the purchase and the staff member who reviews the purchase before entering it into the accounting records.

Elements of the Control Process

There are four elements of the control process: establishing performance objectives, measuring results, comparing the actual outcome with the objectives, and implementing corrective procedures.[3] Figure 9–1 shows this process.

The first area to examine in the control system is the *output standards*. What were the expectations about how much, how good, how expensive, or how timely the work being performed was? How many membership or subscription orders were processed in a day, in a week, in a month? Was the set built on time and within budget? Did the fund raising campaign reach its target figure? Was the work done accurately?

The control system extends into areas that may not be as easily quantified. For example, what is the appropriate output standard for the rehearsal process of a play, opera, dance, or concert? Assuming that all of these events have a deadline for opening night, the person in the leadership role (director, choreographer, and so on) must make it clear through the rehearsal schedule what will be expected during the preparation stage for the event. However, if no one monitors the process, the control system breaks down. For example, if the artistic director is directing the play and spends the first five weeks of a six-week rehearsal period on the first act, who is left in a control position to take corrective action? The stage manager may point out that the play is far behind, but if the person in charge of the whole operation does not stick to the schedule as written, there isn't much to be done.

For many arts organizations, there is no solution. The artistic director may hold others to the established output standard while personally ignoring it. The net result is an organization in a constant state of panic about getting the production ready for the opening.

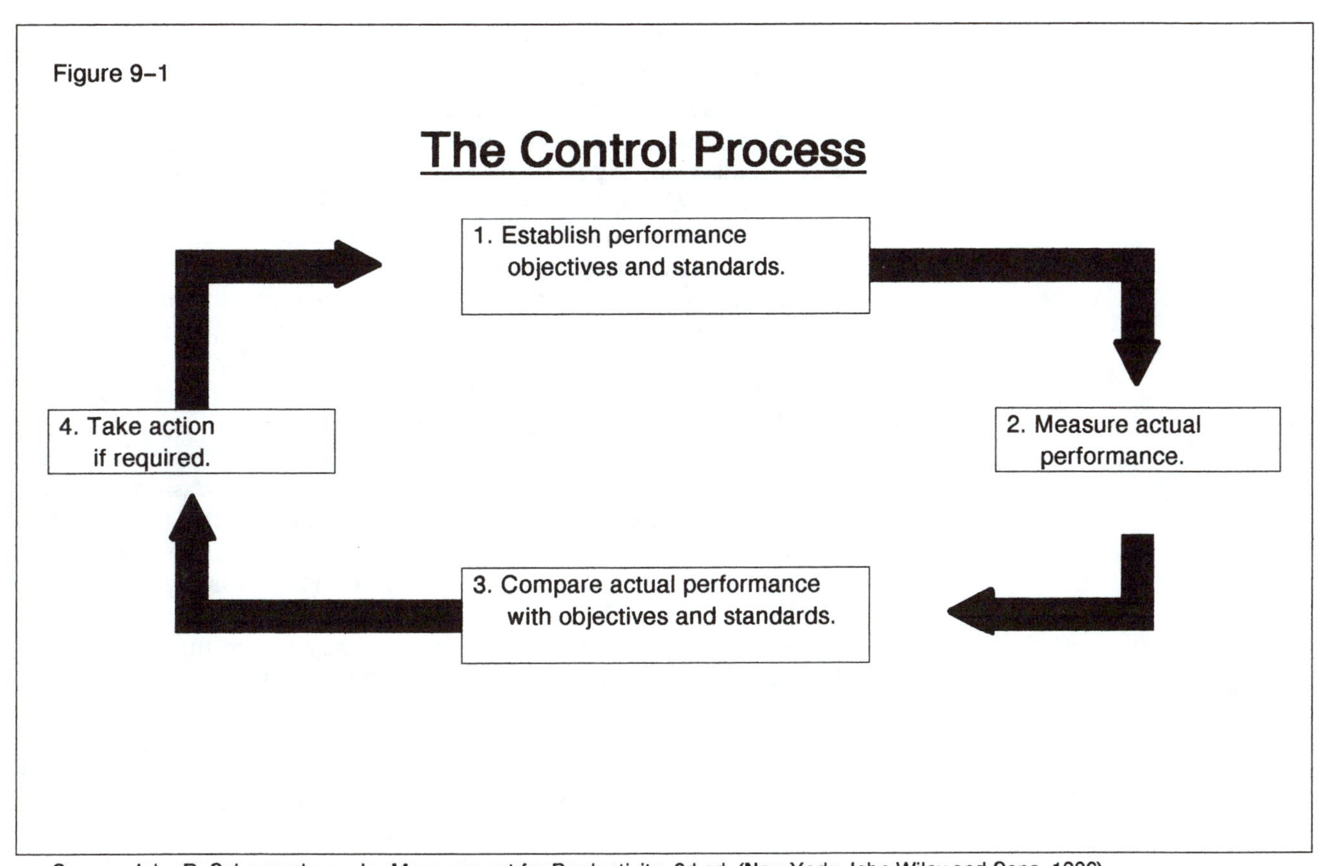

Figure 9–1

The Control Process

Source: John R. Schermerhorn, Jr., Management for Productivity, 2d ed. (New York: John Wiley and Sons, 1986).

One solution might come from an effective partnership between an artistic director and a managing director. They would need an agreement that makes it clear that the managing director has some real authority to monitor the schedule and take corrective action when required. For example, the managing director would point out at the appropriate times that the published rehearsal schedule is not being followed. Adjustments in the schedule would be made, and it is hoped, the show gets back on schedule.

In the real world, of course, it is much harder to get people to accept intervention in their projects. It is often the case in arts organizations that people occupy multiple positions of control. The artistic director may be very effective at setting the output standards for a guest director and others in the organization while reacting violently to being criticized for falling behind schedule.

Another component of the control process is to set *input standards*. The process involves evaluating the effort that goes into a task. One way to evaluate the work is to look at how well the person used the available resources. For example, a staff member might ask for two extra helpers to complete the subscription orders within the six weeks allotted for the task. If the orders actually take nine weeks to complete and halfway through the schedule, two more people had to be hired, the supervisor might wonder about the staff member's ability to estimate the resources needed to complete a project.

Once you establish objectives for output, you face the issue of *measuring actual performance*. In some cases, a manager can define clearly the quantitative measures and communicate them to the employees. For example, a manager expects at least 45 subscription orders to be processed each day. The actual number of orders filled in one day give the manager a specific piece of information. A lower output would lead to an investigation of the work process, and it may be found that by changing the order in which the work is completed, the average number of orders filled daily exceeds 45.

Since many areas in an arts organization deal with specialized craft and custom construction techniques, it is much harder to make accurate projections about the performance level of a staff member. Suppose that eight chairs must be built for a dining room scene in a play. The shop supervisor asks the props master how long it will take to build the chairs, and they agree on five days as the output standard. At the end of that period, only three chairs have been completed. The shop supervisor notes that the expected output level was not met, and he intercedes and changes the input standard. Two extra people are assigned to assist the props master complete the remaining chairs.

The previous examples demonstrate how a manager will compare the actual performance with the standards and make adjustments to correct any problems. The success of any of the projects cited in these examples depends on active and involved management of the control process.

In an arts setting, the critical work of all of the creative artists must also undergo a similarly active interaction with management. For performers, the roles they act or sing and the music they play represent a complex mix of talent, ego, and ensemble interaction. How do you set standards, evaluate performance, and take corrective action when a performer does not measure up to the expectations? You can call on your communication skills to tactfully present the problem, suggest alternatives, and ultimately, if the work does not meet the expected standards, replace the performer. Circumstances may prevent taking such direct action, however. For example, a union contract may prevent or hinder abrupt changes in casting.

In some situations, you may have no recourse. For example, suppose that the scene designer you hired to do the sets for your opening production misses the deadline for submitting the plans. Your shop staff cannot start building the set, and the entire construction process begins behind schedule. Your only recourse would be to refuse to hire that designer for your next show. By the time you confront the problem of failing to meet an output standard, it is too late to take much corrective action.

Management by Exception

One way of creating a control system is to establish a *management by exception* (MBE) process within the organization.[4] Essentially, the MBE process (shown in Figure 9–2) works as a part of the comparative element outlined in the control system. Once you establish clear performance standards and communicate them throughout the organization, you focus your energies on the exceptions to the norm. In this approach, you spend time on the less-than-standard performance. However, you can boost morale and productivity if your management team recognizes and rewards people who exceed the standards.

For management by exception to work, internal control must be at a high level in the organization. High standards for performance

Figure 9–2

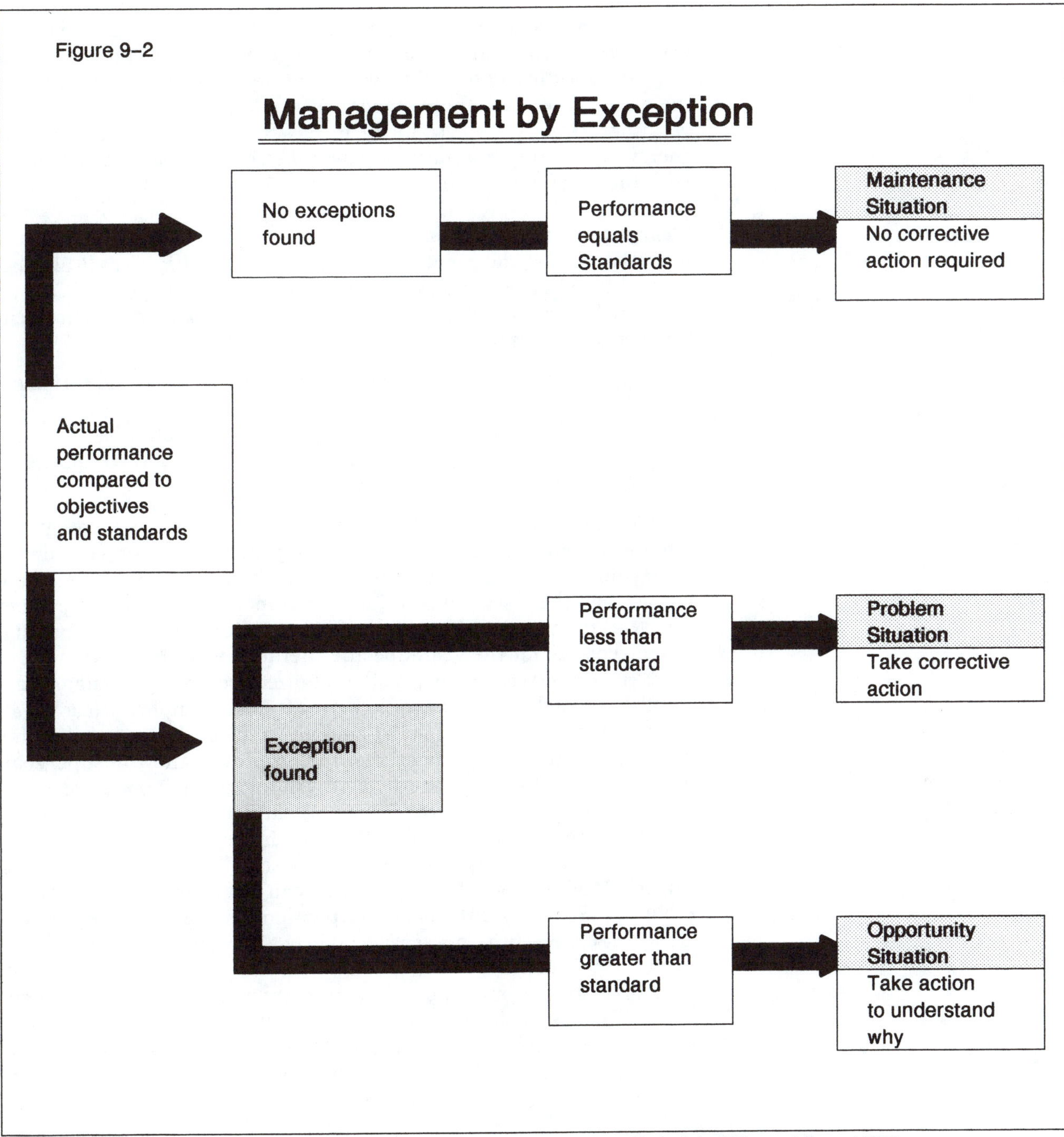

Source: John R. Schermerhorn, Jr., Management for Productivity, 2d, ed. (New York: John Wiley and Sons, 1986.)

must be central to the culture and value system of the organization. From this strong culture should come attitudes among your staff that support them in setting high goals for their own work output. An arts manager who approaches the staff from McGregor's Theory Y perspective, as you may remember from Chapter 8, will assume that staff members want to do a good job. Of course, if clear standards are not communicated to employees, you cannot expect even the most highly self-motivated people to meet your expectations.

Another element in the MBE process is a system of external controls. Every organization needs some ongoing policies and procedures to guide work behavior and to state expected standards clearly. When the organization sets these standards (smoking policy, break periods, vacations, sick days, and so forth), it frees the manager from expending energy on routine expectations. The manager need only be concerned with the exceptions to the external controls.

Management by Objectives

In the late 1960s, the concept of *management by objectives* (MBO) began to be applied widely in the business world. Simply put, MBO is[5] an integrated planning and control system that involves a formal agreement between a supervisor and subordinate concerning:

1. the employee's performance objectives for a specific period of time;
2. the plan(s) to be used to accomplish the objectives;
3. agreed upon standards for measuring the work accomplished;
4. and, procedures for reviewing results.

When properly applied, management by objectives is integrated into the overall strategic plan for the organization. For example, if one of an organization's goals is to increase the level of donations from corporations, and the objective is to increase corporate giving by 10 percent this year, then the development staff can specifically develop quantitative objectives for the year. Specific methods would be developed to meet the objectives (phone, mail, and direct contact campaigns), and standards of achievement would be set for each employee (each staff member is given a specific dollar amount as the goal for a specific time period). During regular meetings, the employee and the supervisor would evaluate the employee's progress in meeting the objective.

As you can see, the MBO process can be very time consuming. When you begin to account for all of the time spent drawing up objectives, meeting regularly to review and revise objectives, and documenting the MBO of each employee, you can begin to see some of the problems. You may also encounter problems if the objectives you set are too easily reached. In a sense, you begin to establish lowered performance expectations.

In an arts organization, different employee groups have different time frames in their work. As a system, MBO does not make much sense for performers. Elements of MBO may make sense in administrative areas in the organization, provided that the time and commitment to support the extensive demands of MBO really exist.

Performance Appraisal Systems

When an arts organization grows big enough to keep a staff employed on a year-round basis, a performance appraisal system will have to be established. *Performance appraisal* simply means formally evaluating work performance and providing feedback so that performance adjustments can be made.[6] Performance appraisal is part of the overall control system for the organization. If the system is working well, it should provide employees with constructive feedback about their strengths and weaknesses and concrete suggestions for improving and developing their careers. The objective of the appraisal system should be to benefit the employee and the supervisors and to help the organization reach its goals.

Appraisal methods The business of evaluating people should be tailored to the organization's overall design and structure. Arts organizations would probably be put off by the various *numerical rating scales* devised by business specialists. For example, there is dubious value in giving an employee a rating of 7 on a scale of 1 to 10. However, a rating of "unsatisfactory" on a *behavioral scale* in a specific job area ("relates well to others") might draw more attention.

Arts organizations find the *critical incident appraisal* technique to be more acceptable. The supervisor keeps a running log of positive and negative work performance over a given time period and reviews it with the employee at specific intervals. Arts organizations also could use a *free-form narrative* to evaluate employees. This essay format usually notes overall job performance, specific accomplishments, strengths, and weaknesses.

Timely feedback The annual evaluation is a key part of an organization's overall control system, but does not provide feedback on work performance on a day-to-day basis. An effective control system must give employees regular feedback about their work output. Some organizations tend not to comment about the good work someone is doing until the annual evaluation. The net result is that the employee spends a year working in a vacuum. Even worse, a serious problem in an employee's work habits will be left unattended for a year.

A good manager realizes that each employee has a different need for feedback. Some people need constant monitoring, and others are happy when left alone. An appraisal system must be flexible enough to accommodate a range of employee needs.

Summary of Control Systems

The control process extends into all areas of an organization. The planning, organizing, and leading functions of management interact with the control systems in a way that should provide an effectively managed organization.

At the beginning of this chapter, we noted that the word *control* is a source of discomfort to many people. How can you have a dynamic, creative arts organization and still have effective control systems? The two elements do not need to cancel each other out.

In an arts organization, there will ideally be an element of creative chaos. The creation of an evening of theater, dance, opera, or music or the installation of an exhibit will develop a life of its own. It may be messy. It may be filled with conflict and passion. It has been my experience that no two events will ever come together in exactly the same way.

What, then, is an effective way to establish a control system in this environment? One simple approach is to recognize that producing art is not the same as producing large quantities of a product or service. The unique event being presented needs a set of mutually agreed on rules, regulations, and guidelines specific to that project. Recognizing that the event must interface with an ongoing organizational structure, the challenge to the arts manager is to find a way to create bridges between the two approaches.

For example, the financial and accounting aspects of the organization require a great deal of control. Rules, regulations, and laws must be obeyed. You can increase compliance with the rules by making them clear and simple to follow. If, to purchase three yards of fabric for a costume, a staff member must fill out six forms in triplicate and have them signed by two different people, the odds are good that people will do whatever they can to avoid using the "correct" procedures. In this example, the organization would benefit from a different control system for its purchasing procedures.

Management Information Systems

Let's now look at a key supporting system that makes the organizational control work effectively: the management information system. A *management information system* (MIS) is formally defined as "a mechanism designed to collect, combine, compare, analyze, and disseminate data in the form of information."[7] For an arts organization, a well-designed MIS should serve as an almost invisible element. The design, implementation, and maintenance of the MIS may not be particularly exciting to people working in the arts. In fact, many organizations never establish a formal MIS; one evolves. The evolution of the MIS often comes from the crisis management style exercised by many organizations.

For example, it is the middle of summer before you find out that subscription sales revenue is down 15 percent and your cash flow has all been spent to meet creditors' bills and last month's payroll. This has never been as big a problem before. What has happened? In this case, maybe the MIS, as it existed, simply did not get financial information about sales and accounts to you quickly enough.

The MIS may be too informal. Suppose that you are planning a major tour in which your ballet company will perform in five large cities. Two days before the tour starts, you are informed by the management in the first city that only 35 percent of the house has been sold. Whenever you asked about how sales were going, you were told, "Orders are coming in at a steady pace." Because you were led to believe that the sponsor would easily be able to sell 80 percent of capacity, based on previous dance company performances, you signed a contract based on a percentage of the house, not a flat fee. This example of a lack of a hard data delivered in a timely manner could very well mean bankruptcy for the dance company.

Both of these examples demonstrate the importance of a good MIS. A key function of the MIS is to help managers make decisions. To make a decision implies a choice. To exercise choice means that you select from alternative plans of action. The choices you have may become increasingly limited as time passes. In the case of a subscription campaign, you have to make the sale before the season opens. If you learn early enough that sales are down, you can implement planned courses of action to increase sales. If you learn too late about the shortfall in revenue, all you can do is plan for an operating deficit.

Let's look at how to establish an effective MIS so that many of these problems can be avoided.

Data and Information

When we defined a MIS, we used the terms *data* and *information*. Each of these terms implies a great deal about the MIS. With the advent of the personal computer, the term *data* has found its way into our daily vocabulary. Data comes to us in the form of facts and figures, which we then process to form a meaningful conclusion. We disseminate this conclusion to others in a regular pattern of information within the organization. The actual number of subscriptions sold and the revenue collected each day at the box office represent raw data that we process and disseminate to those who need the information.

Data and *information* are not neutral terms. Because people process data, certain biases may affect this part of the process. For example, 25 subscriptions sold in one day may seem like a basic piece of data. You might ask how this number compares to the number sold at this time last year or how the number compares to projections of expected sales to date. If, on the other hand, the box office manager only took in revenue from 18 sales and the other seven sales were phone calls from people who said they would be renewing, the data collected implies something quite different. The box office manager may have been telling you what she thought you wanted to hear because she wanted to present as optimistic a sales picture as possible. The point is that the MIS you have in place is meaningless if individuals manipulate the data to present misleading information.

Management Information Systems in the Arts

As shown in Figure 9–3, a MIS extends into all areas of the organization. A small arts organization does not require a supercomputer to function as its central databank. Although computerization would certainly assist the management decision process, simply walking from department to department to gather information is a far less costly alternative. For example, daily reports from the stage manager and technical director to the production manager support the design and production information system. Reports made in weekly staff meetings by the production manager to the managing director complete the cycle of data gathering and distribution.

Most arts organizations start with a small staff of two or three people. The MIS exists as an informal communication among people who are often part of a well-established social unit. One person may deal with accounting, finances, and logistics, while someone covers marketing and fund raising. A personnel system isn't even needed. As the organization grows, more staff members are added, and specialization and departmentalization occur.

Organizational design has a direct impact on the MIS. For example, the MIS of a regional theater company with three theaters in different locations in the city must take into account the potential problems of decentralization. How will these remote locations operate in relation to the accounting department? How will accounting know about purchases unless the MIS includes the accounting department in the ordering stage? If the information system is required to keep track of and control funds expended, it cannot record purchases based on invoices that may come 15 to 30 days after an item has been purchased.

Computers and the Management Information System

Computers and management information systems seem to have been made for each other. The computer's ability to store large amounts of data and distribute it through networks within an organization has had a major impact on the business world. The ability to gather, store, and manipulate data is now very cost effective. The smallest arts organization usually has at least one computer to do the bookkeeping or to manage a mailing list. Large businesses spend more on a computerized

Figure 9–3

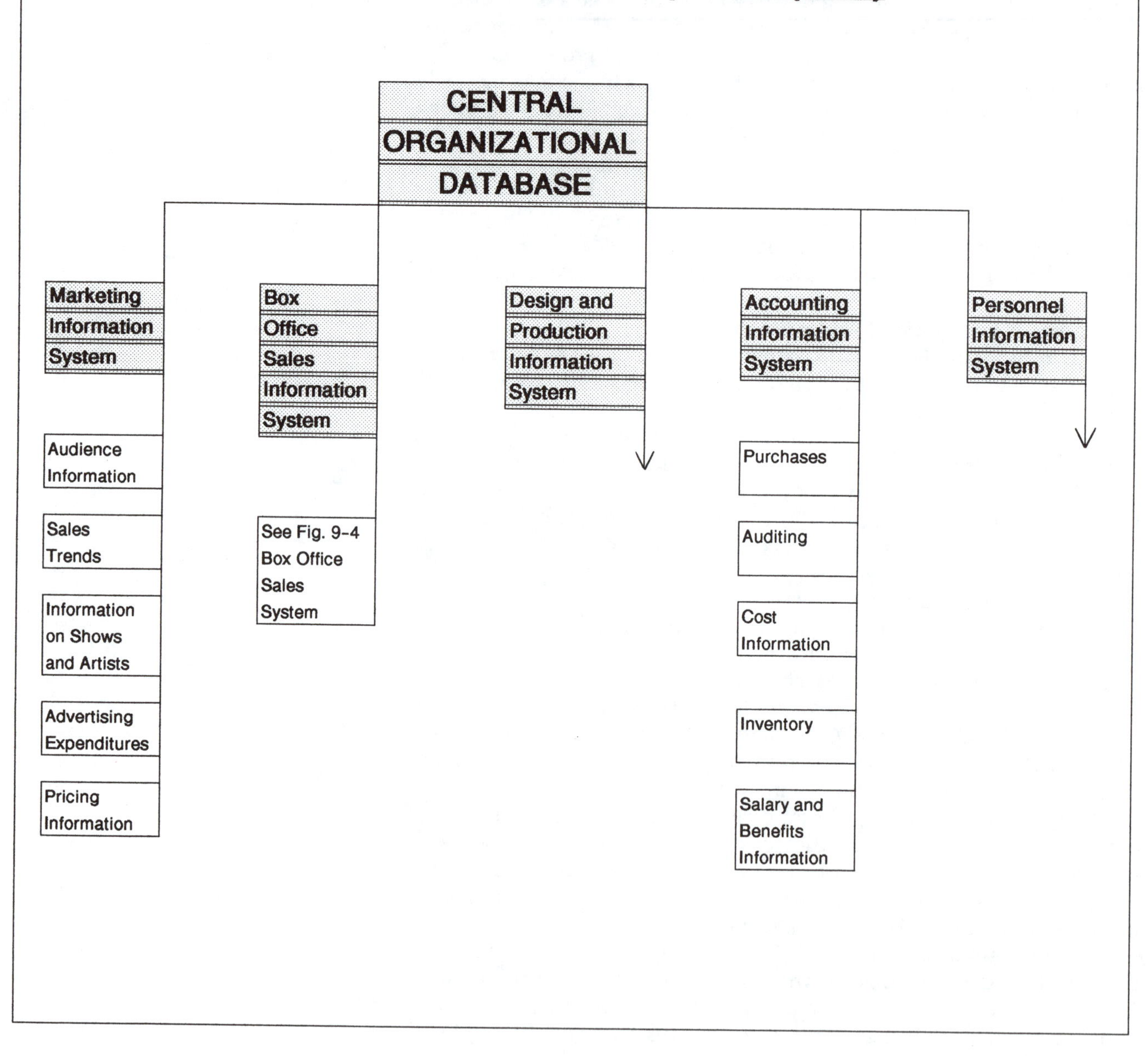

MIS than ten regional theater companies spend for all their needs. Whatever the scale of operation, careful planning is required if the maximum benefit of computerization is to be realized.

Figure 9–4 shows a simple network connecting a box office with the marketing, fund raising, box office, and facilities departments. Word processing and facsimile capabilities enhance the effectiveness of the overall system. Typically, such a system provides daily reports of

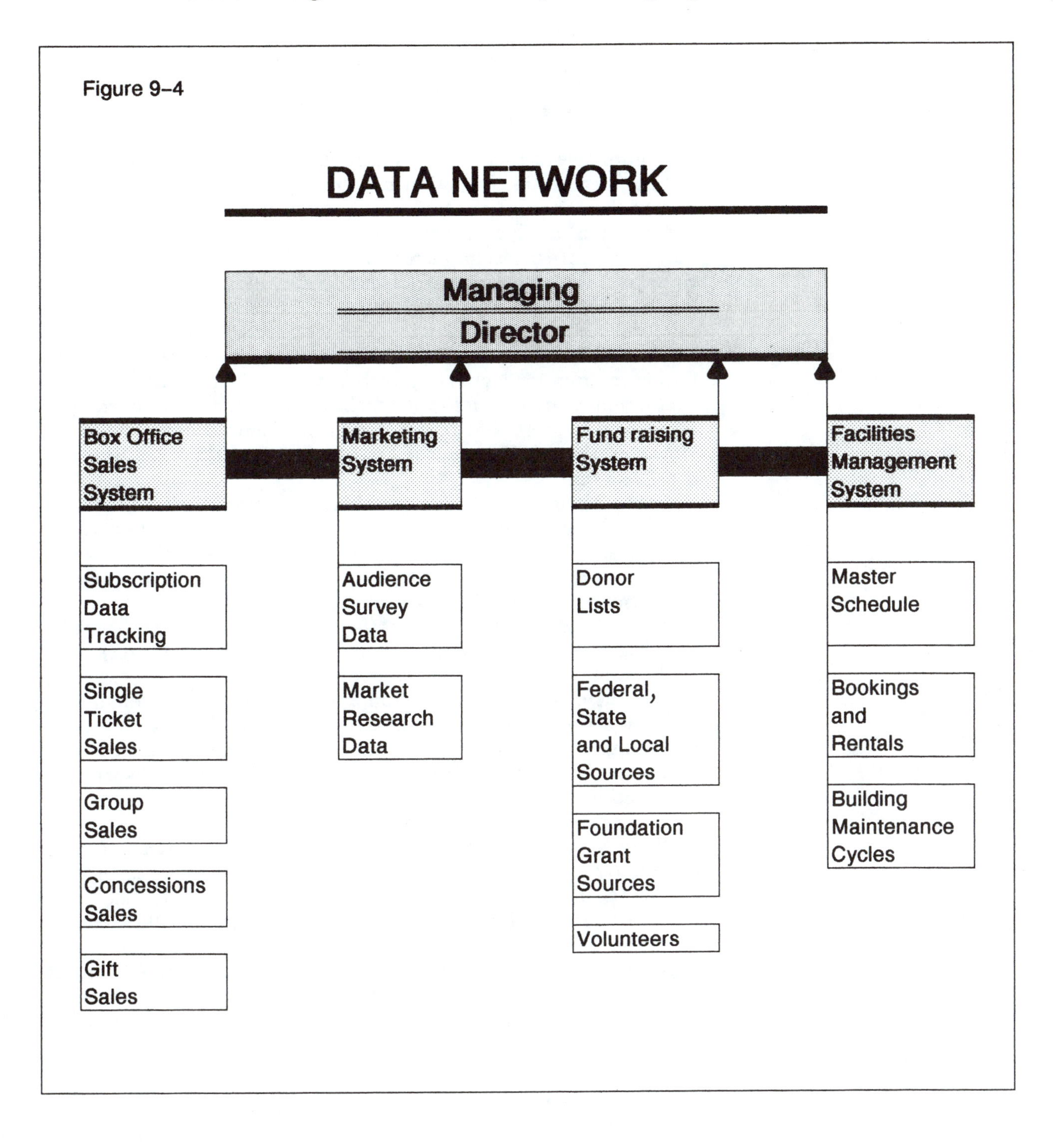

ticket and subscription income, as well as year-to-date and previous-year comparisons.

An Effective Management Information System

The purpose of any MIS is to facilitate the accomplishment of the organization's objectives through improved problem solving and decision making. In shaping and revising a MIS, three factors must be taken into account. There are uncontrollable, partially controllable, and fully controllable factors that determine how effective the MIS will be.[8] Let's look at each type.

Uncontrollable factors Some factors, such as organizational structure and the organization's relationship to its external environments, are beyond the control of the MIS. For example, in a highly decentralized organization in which subgroups have a great deal of autonomy, it may prove difficult to implement a MIS effectively. A regional opera company may have administrative offices in one location, rehearse in two or three different spaces, build sets and props in yet another locale, and perform in two different venues during the year. This structure would make it difficult to set up a MIS linked by computers. The flow of information would be possible by using various data transmission formats, such as facsimile machines and modems hooked up to remote computers that report back to the administrative offices. However, the resources needed to pay for all of this may limit the overall system.

An unstable internal or external business environment also has an effect on the MIS. For example, suppose that you are trying to track audience response by collecting data from your mailings. Between the first and second mailing, a fiscal crisis forces the organization to drop two shows from the season. How will you collect meaningful data about the effectiveness of a second mailing if the season package keeps changing?

Many arts organizations simply do not have the resources to install computer networks. The uncontrollable factor in this case may be that there are ten personal computers in the organization, all use different software, and all are isolated from each other. The fund raising staff uses one type of software to track the donors, and the box office uses a different software to collect sales and subscription information.

Partially controllable factors It is possible to gain some short-range control over a poorly structured MIS by bringing available resources to bear on the problem. In the previous example, for instance, a managing director might be able to intercede and make sure that the different computer systems use a common software. Data files could then be copied from one machine to another, and information could be shared over what is called a *shoe leather network*. For this to succeed, people in the organization must understand the importance of creating a data gathering and information dissemination system.

Fully controllable factors A MIS that has fully controllable factors is supported and encouraged by the organization and its management. In this ideal world, a staff member would be designated to oversee the data and information system. This would be a senior-staff position with support staff to assist with system maintenance. The MIS would be fully integrated into the overall organizational operation.

Integrated computer systems and ongoing operating procedures would support regular data gathering and storage. For example, an effective MIS would allow a marketing staff member to track subscription sales by type of purchaser over the previous five years by accessing a database of subscriber files. When a staff member in the press office needs to look up information about a singer who was in an opera produced by the company three years ago, the information would be available in a database of artist biographies.

Common Mistakes

Care must be taken to avoid some common mistakes in establishing a MIS. As noted earlier, many arts organizations evolve and grow without paying much attention to any MIS. Much time is often wasted hunting down information that should be readily available. However, trying to force a MIS on an organization that is not yet ready for it can damage the credibility of the system. Even when the MIS is accepted, there are still pitfalls to avoid. Here are four problems with bringing a management information system into operation.

1. More information is not always better. The issue here is quality, not quantity. Data that are translated into too much information may turn out to be more of a hindrance than a help. It does not take a great deal of effort to overwhelm people with too much information.

2. Do not assume that people need all the information they want. When designing a MIS for an organization, it is important to review with the various staff members exactly what information they need to be more productive. People tend to request more information than they will possibly have time to process and synthesize.

3. Despite receiving more information, decision making might not improve. Information does not translate into more effective management. In some cases, too much information may result in decision paralysis for some managers.

4. Don't assume that computers can solve all of your information management problems. Arts organizations, which are resource poor, have benefited greatly from the continuing decrease in cost and increase in power of computer systems. The greatest benefits come when a well-designed software system is carefully integrated with a clear vision of the organization's information management needs. However, organizations tend to forget that time is needed to train people to use a computer-based MIS effectively. A poorly designed and managed MIS will probably be abandoned by the users, and everyone will return to the old procedures.

Management Information System Summary

If you ask a staff member in an arts organization how well the management information system is working, you might get a puzzled look in reply. If, on the other hand, you ask them whether the monthly account statements that detail the expenses of the department are informative, the staff person will probably say yes. In this case, the MIS appears to be working.

The effectiveness of a MIS in an arts organization often boils down to integrating the existing systems that produce data and information. The accounting reports of expenses, bills, and payroll, the box office reports of sales, and the fund raising reports of donor amounts may simply need to be pulled together in an overview for the staff and board. Probably the most effective way to quickly pull together a MIS is to ask some very basic questions about what information the organization and the staff need on a daily, weekly, monthly, quarterly, and annual basis. A chart that lists the data, reports, and so on, needed for each time frame can become the foundation for a very simple and effective MIS.

The Future

Arts organizations as well as many other businesses have only begun to tap into the potential of computers as an organizational support and a MIS tool. The current application of computers in arts organizations has focused on automating manual procedures. This has resulted in productivity gains in isolated parts of the organization. It may now take minutes to gather the information that used to require days.

As computer networks become more common and computer processing speed increases, the arts manager of the future will need to be creative in extending the effectiveness of the automated MIS beyond record keeping and data management. The new technology is leading systems that will allow the multimedia use of computers. For example, designers, directors, and data managers will be able to update designs, visualize a staging, or manipulate data many times faster than ever imagined.

Budgets and the Control System

In this final section on control systems, we will look at the area of budgeting. Budget control can very quickly become the major focus of an arts manager's job. Being able to project revenue accurately and to monitor and control expenses for a small organization with a limited budget is an extremely valuable set of skills to possess. The very survival of an arts organization often depends on being able to keep current with income and expenses. As a control center, budgeting is a key element in the overall MIS of the organization.

What exactly is a budget? One common definition of a *budget* is "a quantitative and financial expression of a plan."[10] A budget represents an allocation of resources to activities. If the control process is to be effective, the person supervising the use of the funds must be held responsible for the budget. Depending on the organization's culture, the budget development and implementation process may range from highly structured and formal to informal. Formal budgeting implies a proposal and review process before budget changes are approved. Depending on the structure of the organization, proposing and making changes in the budget could involve very precise procedures. See Figure 9–5 for an overview of the process.

Budgetary Centers

Before an arts organization begins to fulfill its mission, a budget must be prepared. Even the smallest organization will need to look at two key centers: revenue and expenses. The *revenue centers* include income from the sale of tickets or memberships, donations, grants, concessions, program advertising, rentals, and consumer goods. *Expense centers* generally follow the organizational structure that has been established. These two components form the overall *operating budget* of the organization.

Budgets as Preliminary Controls

A budget is a preliminary control because it establishes the resources allocated to achieve the objectives outlined in the operational plans for the organization. An opera company that allocates one-third of its production budget to one show of a five-show season is making a statement of artistic priority. The monthly budget reports should give the production manager a clear sense of whether the resources allocated

Figure 9–5

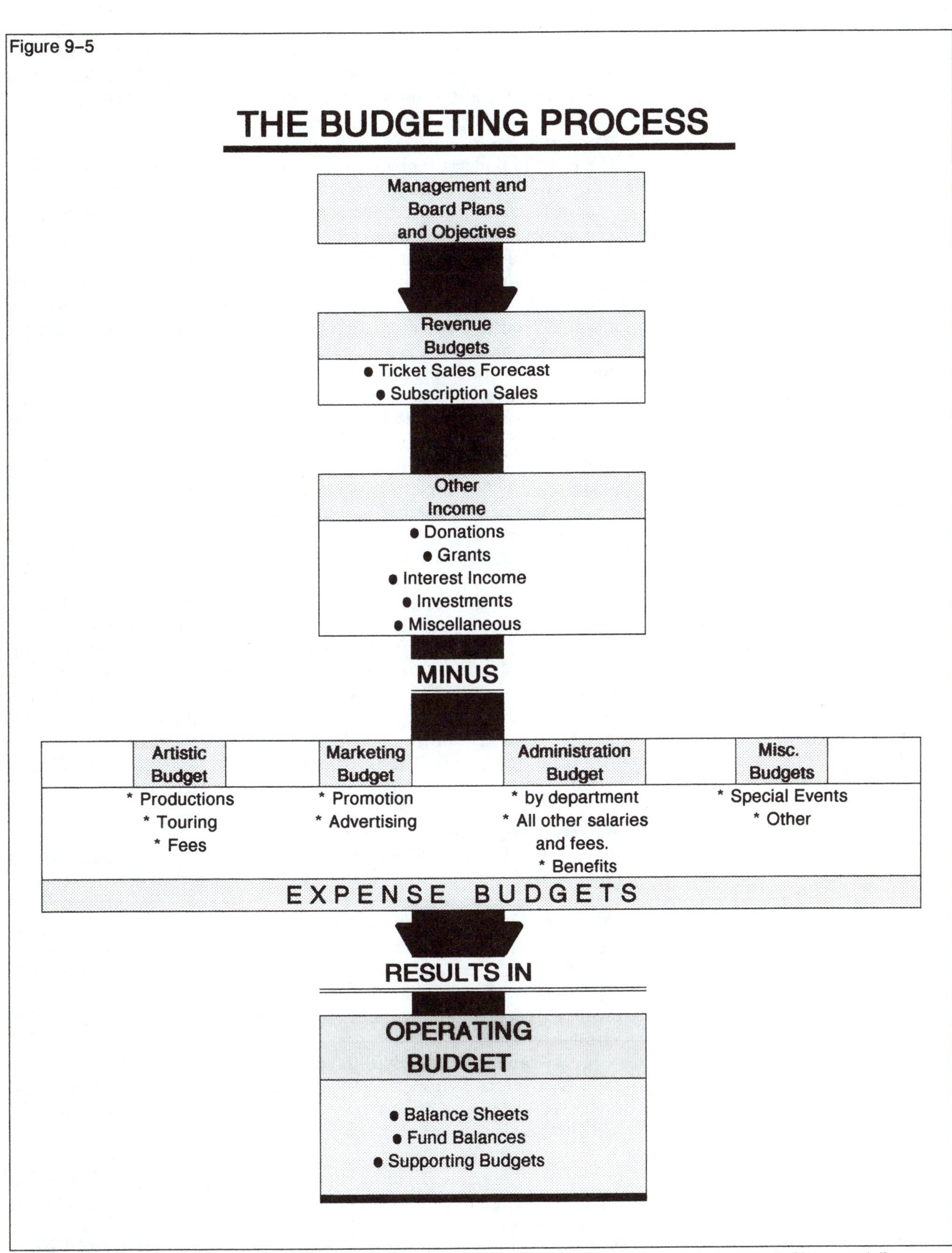

Source: James H. Donnelly, Jr., James L. Gibson, and John M. Ivancevich, Fundamentals of Management, 7th ed. (Homewood, IL: BPI/Irwin, 1990). Copyright (c) 1990 by Richard D. Irwin, Inc. Used with permission.

for each show are being used as projected. A budget functions as a preliminary control when the production manager tells the designers that they have a specific amount for sets, costumes, and so forth.

Types of Budgets

The types of budgets that an organization could use include fixed, flexible, zero-based, operating, short-term, and long-term budgets.[11]

In a *fixed budget*, allocations are based on the estimated costs from a fixed base of resources. For example, the salary budget is set for the year, and resources are allocated to cover that expense.

A *flexible budget* assumes that activity levels will influence resource use. For example, a museum's payroll budget may respond to increased traffic flows due to programming. If the museum has a big exhibit that brings in a larger than average number of paid admissions (increased revenue), it may also have to hire two additional part-time guards to control the crowds (increased expense).

A *zero-based budget* is a planning and objective-setting tool that is used by some arts organizations. Managers cannot assume that the funds they had the previous year will be there again. Revenue and expense must be justified in relation to the plans and objectives for the whole organization. Compared to the fixed budget with its yearly incremental increases, the zero-based budget can be an effective tool to help keep an organization from creating budgets filled with underutilized line items.

An *operating budget*, as already noted, is normally a yearly budget created to carry out the organization's operational plans.

A *short-term budget* describes expense and revenue activity during a period of one year or less. An individual project or production may have a short-term budget that covers its activities and is situated within the overall operating budget of the organization. A *long-term budget* is used for projects or programs of more than one year, such as a five-year fund raising campaign. The objective of a long-term budget is to allocate resources that are not subject to year-to-year budget cutting.

The Budgetary Process

As noted in Figure 9–5, the budgeting process usually begins with a projection of the organization's various sources of revenue. In arts organizations, revenue comes from a variety of sources. The organization's MIS should be able to provide detailed reports of all revenue from the previous year or years. The evaluation of the previous revenue distributions and the comparison with the budget projections for the coming year are the next step. Care must be taken when projecting revenue. A few too many optimistic revenue projections could lead to a midseason budget crisis.

The same type of activity takes place in the organization's various expense centers. Comparing the recent year's expense patterns and evaluating the project and programming plans for the next year, must be done in the context of the overall revenue for the organization.

The next stage in the process compares and adjusts budgets based on expected revenue. If, after subtracting revenue, expenses still exceed the available resources, a series of revisions are made in the expense budgets.

Arts organizations sometimes find themselves adjusting revenue projections to match expenses. This is a dangerous game that usually leads to chronic financial trouble for the organization. For example, unrealistic projections of fund raising or subscription sales revenue may result in a balanced budget to present to the finance committee of the board, but such budget practices are nothing short of fraud.

Budget Reality

A manager in an organizational setting soon discovers that the control process in budgeting extends into anticipating behavior patterns by staff members. Staff members become very territorial about budgets, for example. Some people overestimate needs, while others try to make the budget allocation process a competition for resources within the organization. (One tool for controlling this problem is to work with zero-based budgets.) Very few staff members will loudly proclaim, "I don't need this much. Here, take back some of my budget."

Budget Controls

Trying to control a budget can be a full-time activity. Theoretically, the organizational MIS will provide the manager with the required information about revenue and expenses. For budget control to be effective, the manager must have this information as quickly as possible each month. Without computers, this work becomes more time consuming, but it is not impossible. Simple year-to-date percentage expectations for revenue and expenses can give a manager some element of control over a budget. For example, at the end of a specific period of time, a specific percentage of the budget should have been expended. The manager's efforts can then be focused on variances from the expected distributions.

Another key element in the budget control process involves the authorization procedures for expending funds. A system of review and approval must exist if the organization is to control its budget effectively.

Finally, the manager must recognize and develop strategies for dealing with the political nature of budgets. Intraorganizational politics play a role in the budget control process. Staff members try to obtain as many resources as possible for their work areas. The information a manager receives may have been filtered by department heads to distort the true budgetary condition of a subunit within the organization.

Chapter 11 examines in detail the actual workings of a budget. Sample budgets are shown within an overall financial management information system.

Summary

Control is the process of monitoring performance and making adjustments as required to meet planned objectives. Uncertainty affects all planning. The degree of complexity, human limitations, and the degree of centralization also influence how effective an organization's control system will be.

Output and input standards must be clearly established. Measurement standards must be in place so that a manager can compare what was done with what was expected. A control system requires that there be mechanisms in place for correcting work that does not meet the standard. Management by exception (MBE) allows a manager to focus attention on variances to expected performance. Management by objectives (MBO) encourages the integration of planning objectives and work objectives.

Performance appraisal systems are formal methods for providing feedback to employees on a regular basis. Numerical rating scales, behavioral rating scales, critical incident method, and free-form narratives are techniques used to appraise work output.

Effective control systems depend on data and information gathered from the organization's management information system (MIS). The MIS extends into all areas of the organization and is influenced by factors related to the controllability of the information flow through the organization. The organizational design might promote or hinder the effectiveness of the MIS. Some of the possible shortcomings in a MIS include providing too much data, providing irrelevant data, or assuming that computerizing operations will improve the MIS.

Budget control systems are a critical component of an organization. Controlling the distribution of resources and monitoring how effectively the resources are used is a full-time job. Organizations must identify revenue and expense centers and project monetary activity accordingly. A budget can function as a preliminary control on a project by defining the limits on the available resources. Budgets can be fixed, flexible, or zero-based and cover a short or long term. Budget controls concentrate on the timely monitoring of revenue and expenses for all areas of an organization.

Key Terms and Concepts

Organizational control system
Output and input standards
Internal and external controls
Management by exception (MBE)
Management by objectives (MBO)
Performance appraisal system
Management information system (MIS)
Budgets
Fixed and flexible budgets
Zero-based budgets
Short- and long-term budgets
Budgetary process
Budget control system

Questions

1. What is the relationship of control to the manager's other functions?

2. What are the four steps in the control process? Give a specific example of the control process in an arts setting.

3. Describe the typical steps involved in applying the management by exception process. As a system, how does MBE affect an organization's planning process?

4. From your personal experience, describe a situation in which the control system for an organization did or did not work well. Offer suggestions for appropriate improvements, if applicable.

5. What are the four main appraisal methods used in a control system? Briefly evaluate some of the things you have accomplished in the last year.

6. Define the term *management information system.*

7. What will be some of the future applications of MIS computers in the arts?

8. How is a budget part of the control system?
9. What are the five types of budgets? Which type or types would be most effective in supporting an arts organization?
10. Outline the budget process.

Case Study

The following article points out problems with an organization's control system. You should be able to find numerous flaws in the system used to keep staff activities in check.

Radio Station Officials Got Free Cruises

Officials with a college's public radio and television station have received tens of thousands of dollars worth of free ocean cruises through fund raising promotions of questionable value to the stations.

Last year, the station manager and his wife were sent on an all-expenses-paid cruise to China, Korea, and Japan by a travel agency that had a promotional contract with the stations—a contract that the station manager signed. The travel agency also paid $2700 toward a cruise package for one of the radio station's program directors, records show.

A glossy, full-color promotional brochure sent to 33,000 of the station's donors described the 18-day excursion as a once-in-a-lifetime opportunity and boasted of the ship's excellent chefs and fine dining. Ports of call included Shanghai, Hong Kong, Pusan, and Nagasaki, with three nights in Peking.

Although the cruise was heavily promoted as a fund raiser to benefit the area's only public broadcast outlets— with $200 from every cruise package sold being donated to the stations—the chief benefit appears to have been to station officials.

The cruise packages given to the station manager, his wife, and the program director were worth a total of $11,000 to $15,000. The cruise line sold more than $125,000 worth of cruise tickets through the station and received thousands of dollars worth of free advertising.

The stations, meanwhile, got a $4800 "contribution" from the cruise line and an additional $7250 from auctioning off two cruise packages donated by the cruise line for the station's annual membership drive. The 24 people who bought the cruises received a $200 tax write-off per person.

The station manager said that he and his wife were asked by the travel agency to host the trip because the agency thought that having the couple host the trip would sell more cruises.

But the station manager, his wife, and the program director, who has been on most of the cruises that the station has sponsored, were listed in the station's promotional brochures as the hosts of the cruise.

A special assistant to the president of the college and legal counsel to the college said that he approved the idea of the station manager taking his wife along and said that the manager asked his permission before doing so.

As manager of a college-owned station, the station manager is a public employee, the legal counsel said. State law prohibits public employees from using their official position "to secure anything of value" for themselves that they would not ordinarily receive in the performance of their official duties.

The station manager said that the official duties that required him to go on the cruise were "being the head of the station and having the expertise in the East."

The station's marketing director said that the station had done only one cost–benefit analysis of the cruise promotions since they began in the early 1980s, and those records showed that *the promotion cost more than it raised.*

According to documents provided by the station, a 1986 cruise promotion brought in a $6800 "contribution" from the cruise line and $9500 from a donated cruise that was auctioned off.

But to get that money, it cost the station $20,941 for such things as providing 71 free 30-second ads on the FM radio station, numerous television spots for the cruise, additional air fare for the auction winners, staff salaries, and full-page ads in the station magazine.

The travel agent, who began arranging the cruises for the station in 1984, said that the station manager complained during the negotiations for last year's China trip that the station wasn't getting enough out of these promotions.

During the negotiations, records show, the cruise line initially offered to pay most of the costs of the station manager's wife's cruise, which was described in a letter the travel agent wrote to the station as "a favor" from the cruise line, but that created "an additional expense (for the station) that is difficult to explain."

In the end, the station paid for part of the program director's cruise, and all of the station manager's wife's cruise.

The travel agent said it was routine for cruise lines to provide free trips in an effort to entice groups to sell cruise packages.

"It's the carrot they hold out. Everyone wants a free cruise," she said.

The marketing director said that the station is no longer receiving free cruises for station personnel because the college's Alumni Association is now cosponsoring the promotion. She said that the cruise line now handling the arrangements simply donates one cruise for the annual fund raising auction.

According to the marketing director, the station decided last year that the cruise promotions were taking up too much of station personnel's time and were not producing enough revenue.

SOURCE: The information in this case study was based on an article by Gary Webb and published in the *Cleveland Plain Dealer* in 1988.

Questions

1. Was there a conflict of interest when the station manager signed a contract that gave him free cruises?

2. The case study alleges that the station manager and his wife used station funds to cover their personal expenses. If you were on the board of directors for the station, how would you handle this issue?

3. The legal counsel said that the station manager sought permission before accepting the cruise, and the station manager claimed that his travel was part of his job responsibilities. Was there a violation of the control mechanism implied in the regulations prohibiting state employees from accepting gifts?

4. If the cost–benefit analysis indicated that the station was taking a loss on the cruise promotions, why do you think they continued to offer them for four years?

5. What kind of controls should have been in place to prevent the station from entering into such a costly promotional campaign?

References

1. John R. Schermerhorn, Jr., *Management for Productivity*, 2d ed. (New York: John Wiley and Sons, 1986), 397.
2. Ibid.
3. Ibid., 398–99.
4. Ibid., 400.
5. Ibid., 414–16.
6. Ibid., 404–05.
7. Arthur G. Bedeian, *Management* (New York: Dryden Press, CBS College Publishing, 1986), 588.
8. Schermerhorn, *Management for Productivity*, 447.
9. Ibid., 449–50.
10. Ibid., 428.
11. Ibid., 429–30.

10 Economics and the Arts

Arts organizations, as we saw in Chapter 4, function in multiple environments. Economic, political, cultural, demographic, technological, and educational environments shape and change an arts organization over time. The economic environment is the focus of this chapter. The objective here is to gain an understanding of many of the economic forces with which an arts organization interacts and the impact of the economy on the organization. This chapter and the next one on financial management are not intended as a substitute for courses in economics and finance. If you are currently an undergraduate interested in a career in arts management, you should take courses in these areas before you graduate.

The economy has a direct effect on artists and arts organizations every day in the United States. Staying aware of the economic environment allows an arts manager the opportunity to prepare plans of action designed to ensure the survival of the organization. An arts manager might consider the following questions related to the economy:

1. Will there be a downturn or an upturn in the economy?
2. Will people have more or less disposable income?
3. How will inflation affect operating costs?
4. How will changes in interest rates affect the budget and the organization's other investments?

Ultimately, the answers to these and many similar questions remain uncertain. The economic information that a manager needs to make decisions is often out of date or contradictory. One report says that the economy is coming out of a recession, and another says that the recession will continue for six more months. An abundance of common sense and skepticism are essential. Rather than plotting out abstracted data on charts and graphs, an arts manager's time is better spent analyzing the threats and opportunities that exist in the economic environment. The objective is to translate the potential impact of these conditions into practical courses of action to help the organization survive.

Introduction to Economics

This section will highlight some of the economic principles that have an impact on arts organizations. It is assumed that the reader has not yet taken any courses in economics. If you have a background in this subject, you may want to skip ahead to "Microeconomics."

Let's begin by defining some basic terms. An *economy* is "a system of organization for the production, distribution, and consumption of all things people use to achieve a certain standard of living."[1] *Eco-*

nomics, on the other hand, is "the study of how people and society choose to employ resources to produce goods and services and distribute them among various persons and groups in society."[2] The study of economics can be further broken down into macroeconomics and microeconomics. *Macroeconomics* is concerned with the entire economy or large sectors of it. *Microeconomics* focuses on the individual units (a household or business or arts organization) that make up the entire economy.[3] Microeconomics also studies how individual markets (a place where buyers and sellers come together) are organized, grow, develop, and change over time.[4]

Inherent in the discussion of the economy is the assumption that *goods* (tangible things such as a car) and *services* (intangible things such as a live concert) are scarce. That is, there is not enough of either to satisfy everyone's needs. Prices for goods or services are based on two factors: (1) the demand for them, and (2) how scarce or plentiful they are. At the most basic level, a dance concert performance scheduled for one night is a scarce entertainment service. The "product," in this case, is the dancers' performance. The price that audience members will have to pay to see the performance depends in part on the overall market for one-night dance concerts in that area, and may be very high. On the other hand, the level of demand for dance events in the area might be so low that the tickets must be inexpensive to attract an audience.

People desire goods and services because these things provide a form of satisfaction to the purchaser. The complex interaction of individual tastes and preferences, advertising, the socialization process, and the educational system influences what people perceive as desirable. Millions of dollars are spent each year on advertising designed to convince people that they will find satisfaction if they consume this product or that service. Producers of goods and services attempt to increase the demand through advertising and ultimately make a profit. Nonprofit arts organizations seldom, if ever, have the money budgeted to advertise their product with the same frequency as many consumer goods. These limited resources result in a fairly low level of awareness among the general public about the fine arts and the live performing arts in U.S. society.

The Economic Problem

Economists argue that the underlying *economic problem* is "the combination of scarce resources and unlimited wants."[5] Scarcity forces us to make choices about which goods and services we are willing to pay for with the resources that we have available. These choices lead to what is referred to as an *opportunity cost*, or "the cost that equals the amount of the good that must be given up in order to have more of another."[6] For example, if we have a limited natural resource that can be processed to manufacture both weapons and hospital instruments, we would say that the opportunity cost of making more weapons will be that we will have fewer hospital instruments. These tradeoffs take place at all levels in an economy.

For an arts organization or an artist, the concept of opportunity cost could be applied along similar lines. For example, if you decide to become a concert violinist, you will have to forgo other opportunities because you will only have a limited amount of time to train, rehearse, and perform. The opportunity cost of becoming a soloist might therefore be a social life or a part-time job.

Organizations can have the same problem. For example, arts groups often believe that they can do more than they can with the resources that they have available. There is, for example, an opportunity cost if the organization decides to undertake a tour to all of the schools in a two-hundred-mile radius. The organization will not be able to do other things if it expends the required human, financial, and time resources on this new project. Unfortunately, many arts organizations start up new programs without evaluating the opportunity costs. As a result, staff time and budget resources are overextended.

Three Problems for an Economy

When we talk about an entire economic system, we need to remember three basic questions:[7]

1. What and how much should be produced?
2. How should the means of production be organized?
3. For whom should the goods and services be produced?

In the U.S. economy, the question of what and how much to produce is determined by a complex mixture of the market system, government control, and politics. Recent trends to deregulate or remove some elements of government control from segments of the economy (agriculture, airlines, banking) have met with varying degrees of success. On the other hand, thousands of laws and policies exist at the federal and state levels that affect what and how much should be produced in the economic system. For example, the existence of a system of tax exemptions and grants fosters a system of government support for arts organizations that could not depend solely on the marketplace to provide enough income to operate. By freeing some organizations from paying income tax, the government in effect subsidizes services of benefit to the whole society.

Organizing the means of production has proved to be very difficult. There is little agreement in the United States about the best way to organize the economy. In an ideal world, some economists see the market system as the key to solving the issues related to distributing all resources to maximize productivity of the work force and to produce the optimum mixture of goods and services to further everyone's standard of living. Until recently, the planned or command economy, such as the centrally controlled systems of what was formerly the USSR, was seen as the major alternative to the market system. In fact, this system had limited success. Much of the turmoil in the USSR and in Eastern Europe continues to revolve around the changes brought about by shifting to a more market-driven economy. Combinations of market and command systems are in place in most contemporary economies in Europe.

The third question, for whom to produce the goods and services is no easier to answer. The distribution of the total output of the economic system is uneven. Should only those who contribute to the system get something back? The U.S. economic system answers that question by providing social support for those members of society who do not directly contribute to the total output of goods and services. However, the political issues related to supporting all members of society have yet to be settled.

Government funding, for example, is used for everything from AIDS research to grants for new composers. Recent battles over NEA funding further reflect the political battles over who should get a share of the wealth. Looking at the world around us, it is easy to see that neither the free-market economies nor the centrally planned systems are providing for all of their people.

Arts organizations have been directly affected by the shifting political attitudes about all three of these questions. Since the early 1980s, much time and energy have been spent in reducing the government's role in the day-to-day operation of the economy. For example, the competition for funds has increased as policies have shifted from government support to private support.

Macroeconomics

An arts manager operates an organization in a macroeconomic system that is always undergoing change. The macro level in any economy is concerned with such issues as the overall price level or the unemployment rate, not the cost of an arts subscription or a new car. Because an arts organization is a part of the entire system, economic conditions will have an impact on such areas as the organization's strategic planning or financial management.

The Role of Government

In the U.S. economy, the government provides the following:

- the legal and institutional structure in which the markets operate
- intervention into the allocation of resources in areas of the economy in which public policy deems it beneficial to intervene
- redistribution of income through taxes and payments of entitlements
- stability in prices, economic growth, and economic conditions in general

In the world of macroeconomics, the scale of operation and the amount of financial activity defy comprehension. When you read that the *gross domestic product* (GDP)—the market value of all final goods and services produced by the economy in a year—has increased by 1.2 percent, it may not mean much to you. When the numbers being tossed around are in the trillions of dollars, it is difficult to relate to them at the micro level of an organization. When you read that the Federal Reserve Bank has dropped the prime interest rate by 1 percent or has increased the money supply, you may wonder if this has any relationship to your impoverished arts organization. The answer is that it does. For example, an increase in the GDP may translate into a healthier economy in which people have more money to spend or to give. If the cost of borrowing money declines as a result of actions taken by the Federal Reserve Bank, an organization will not spend as much on loans, which in turn will free up resources that can be used for other purposes.

Although an arts organization may not immediately feel the impact of changes in the macroeconomic environment, adjustments in strategy and operational objectives will eventually become necessary. For example, if the GDP is falling, and the entire economy appears to be headed into a downturn, an organization should activate a series of operational adjustments to cope with these external changes.

Microeconomics

The arts organization is a small business. It functions in the world of microeconomics and, as such, is subject to a constantly changing market. The term *market* refers to all of the sales and purchase activity that affects a product or service. In the case of an arts organization, the market activity that takes place occurs in the larger market of the entertainment industry. This industry includes film, television, theme parks, personal recreation, and spectator sports. Let's take a brief look at some economic principles from the perspective of the arts.

Law of Demand

Trying to identify the demand for a particular show or exhibition requires that a manager weigh numerous factors that influence the behavior of consumers. The well-known factors of supply and demand are directly translated into activities at the core of an arts group's financial planning. Let's begin by looking at the area of demand.

The *law of demand* explains the relationship between the amount of a good or service a buyer both desires and is able to purchase and the price charged for the good or service. In other words, the lower the price charged for a good or service (the vertical axis, or y-axis, of the graph), the larger the quantity demanded will be (the horizontal axis, or x-axis). Conversely, the higher the price, the lower the quantity demanded. Figure 10–1 shows this relationship. The quantity demanded for concert tickets priced from $1 to $20 is plotted. The downward sloping line is called a *demand curve.*

This graph, a fixed demand curve, is based on an important assumption: only the price change affects the quantity demanded, and the change in quantity only occurs along the demand curve. Other factors that influence the ticket buyer, such as the prices of other goods and services, income levels, and individual tastes, do not change.

Demand Determinants

Figure 10–2 shows what happens to the demand curve when other factors are taken into account. Each factor may cause the entire demand curve to shift to the left or right.

Price of other goods *Substitute goods* or services may cause a shift in the demand curve if their prices change. In Figure 10–2, the curve to the left (QD-3) shows a drop in demand when another similar service was available to the consumer. The curve to the right of the original (QD-2) shows an increase in demand. To understand how these factors work, suppose that another group is presenting a classical music concert featuring the same Mozart symphony on the same night. Because the other group is charging $5 less per ticket, the demand curve for your concert might shift to the left (QD-3), meaning that the overall demand will go down. Likewise, if the other group is charging $5 more than your group, the demand may increase, as shown in the shift to the right (QD-2). In summary, when the price of a substitute good or service goes down, the demand curve shifts to the left. When the price of the substitute good or service goes up, the curve shifts to the right.

The curve will also shift if a complementary good is introduced into the interaction. A *complementary good*, which is defined as a good or service that is used jointly with the original good, can cause the demand curve to shift to the left or right, depending on whether

DEMAND CURVE
Ticket Sales per Price

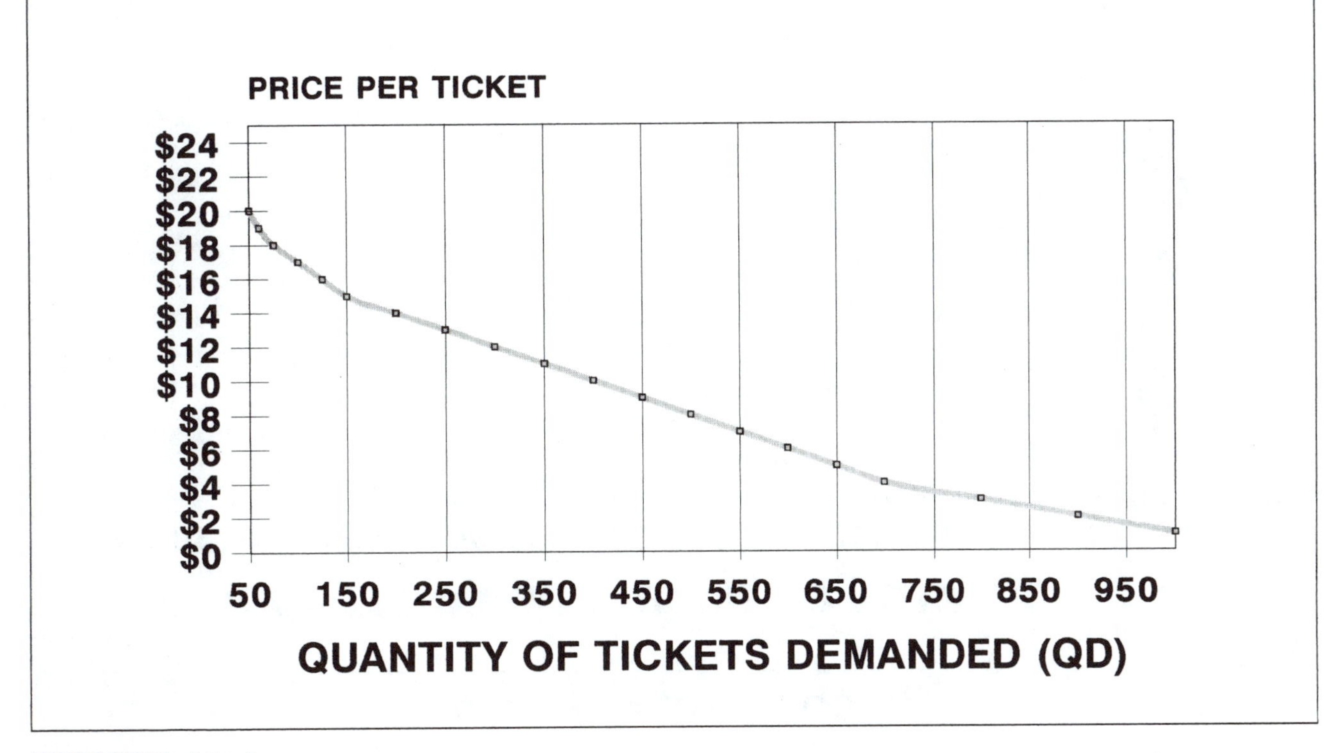

FIGURE 10-1

the price of a complementary good goes up or down. Such a relationship is illustrated by compact discs (CDs) and compact disc players. If the price of CDs declines, the demand curve for CD players should show an increase (represented by a shift to the right). If the price of CDs goes up, there should be a decrease in the number of CD players sold (a shift to the left).

An arts organization may feel the effect of this shifting demand curve in many ways. For example, if a large increase in parking fees occurs, some consumers may be discouraged from purchasing tickets.

Income Another factor that could affect the demand curve is individual income. If income increases, the demand for normal goods and services will also increase. Conversely, if income levels decrease, the demand for normal goods and services will decrease. *Normal goods and services* are defined as those things that people want as their income increases. *Inferior goods or services* are defined as those things that people will choose when their income decreases. For example, if there is a significant drop in income levels, people might choose less expensive forms of entertainment. Instead of buying a $300 season subscription to the symphony, people may shift their spending to a less expensive community orchestra series. The community orchestra may

SHIFTING DEMAND CURVES
Ticket Sales per Price

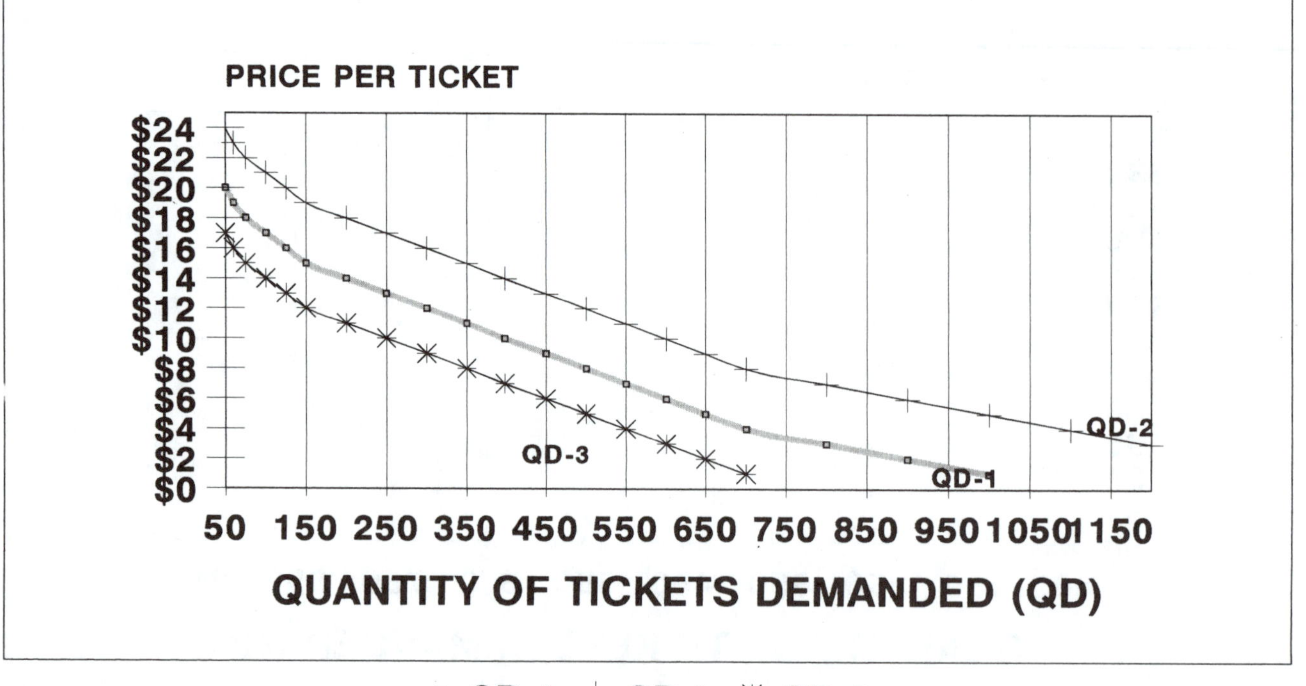

FIGURE 10-2

not perceive itself as an "inferior" good or service, but in economic theory, it fits the definition.

Expectations Demand is affected by individual expectations about the overall economic situation. If a consumer thinks that the price may go up for a good or service, they may make the purchase immediately. This expectation would shift the demand curve to the right. On the other hand, if the consumer thought that the price was going to drop, they may delay making the purchase. If enough people share this expectation, the demand curve will shift to the left. An arts manager might take advantage of this phenomenon by stressing in a subscription renewal campaign that prices will go up next season but that subscribers who renew by a certain date can save money. This tactic would help increase the renewal rate.

Tastes Personal consumer tastes also cause the demand curve to shift to the left or right. If a significant number of people shift their interests away from classical music and toward bluegrass, the curve will shift to the left. An increase in demand might result if a popular soloist is added to a concert performance.

Market Demand and the Arts

The *market demand* for a particular product or service is obtained by summing up all of the individual demand curves. When this is done, we must consider the overall size of the market as a factor that will shift the demand curve to the left or right. Overestimating the market demand for any product can lead to a series of financial problems for the organization.

Do these concepts have any place in the day-to-day planning activities of an arts manager? They do when the principles are applied realistically. For example, a dance group with the mission of bringing postmodern dance to the masses, would be wise to start with the expectation that there will be limited demand for the product. Although the community may regularly support dance performances by the regional ballet company—there is thus a market for the entertainment service of dance—there is no guarantee that this market will support an experimental dance troupe.

Advertising is one way to affect the demand for a particular product within an overall market. However, reaching individual consumers with the message about a particular dance organization and being able to make a sale are two different matters.

To understand the entire relationship of the arts organization to the economic environment, we must look at the supply of the product or service.

Law of Supply

We have seen that there is a relationship between price and quantity demanded for a product or service. Now let's look at the supply side of the theory. The application of these concepts to the operation of an arts organization requires some explanation.

The relationship between the amount of a product and its price is at the center of the *law of supply*, which states that "suppliers will supply larger quantities of a good at higher prices than at lower prices."[8] The graph in Figure 10–3, shows the supply curve for concert tickets. To maximize revenue, the supplier of tickets would like to sell 1000 tickets at $20 each. The supply curve shows how few tickets would be provided at $1. The supplier has an incentive to provide the $20 ticket because the costs of producing the event remain constant within the single performance. In fact, the supplier may want to provide multiple performances of the concert to bring in as much revenue as possible. For example, if you are the supplier of a symphony orchestra performance and your players are all under contract, it is to your benefit to offer this service as many times as possible.

The *supply curve* shown in Figure 10–3 shows a change in quantity supplied based on a change in one variable: price. Change in the quantity supplied therefore only occurs along this supply curve. Other variables will have an effect on the shift of the entire supply curve to the left or right. These variables are called *supply determinants.*

Supply Determinants

Price of resources If the price of resources used in creating the good or service rises or falls, the supply curve will shift to the left or right, respectively. Figure 10–4 shows examples of what a shift in the supply curve would look like. In economics, the term *factors of production* is used to describe the "inputs of labor services, raw materials and other resources" used to create the final product or service.[9] For suppliers, a

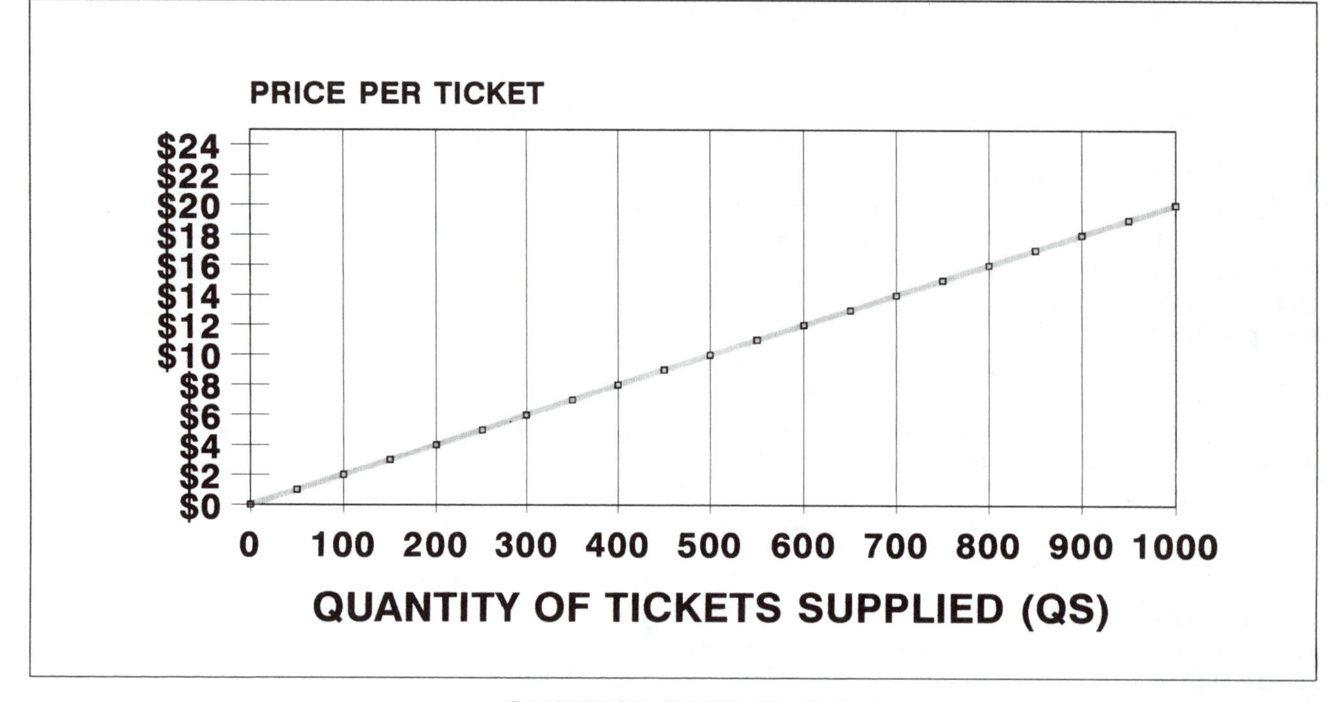

FIGURE 10-3

drop in the factors of production translates into greater profits. Therefore, suppliers have the incentive to supply more (S-3) because they can make more profit. Conversely, if the factors of production go up, suppliers have an incentive to supply less (S-2) because their profit margin goes down. Hence, suppliers reduce output to bring production costs into line with the quantity supplied.

In the case of arts organizations, we seldom see the behavior exhibited by the theoretical supplier of concert tickets. The law of supply, as applied to this example, does not translate into the ticket supplier reducing the quantity of performances supplied. The supplier is not motivated to provide as much of a good or service if the factors of production rise. If costs increase, the typical concert ticket seller would pass the cost along to the consumer in the form of a price increase. The other alternative is to go further into debt by not charging enough to cover the costs of production. Ironically, this is what many nonprofit arts organizations do.

Prices of other goods The fixed supply curve assumes that the prices of other goods remain constant. If the prices of other goods rise, the supply curve will shift to the right or left. In other words, a change in the price of one good produced by an industry may be expected to shift the supply curves of all other goods produced by that industry.[10] If, for example, the supply curve for the particular type of paper used for

SHIFTING SUPPLY CURVES
Tickets Supplied per Price

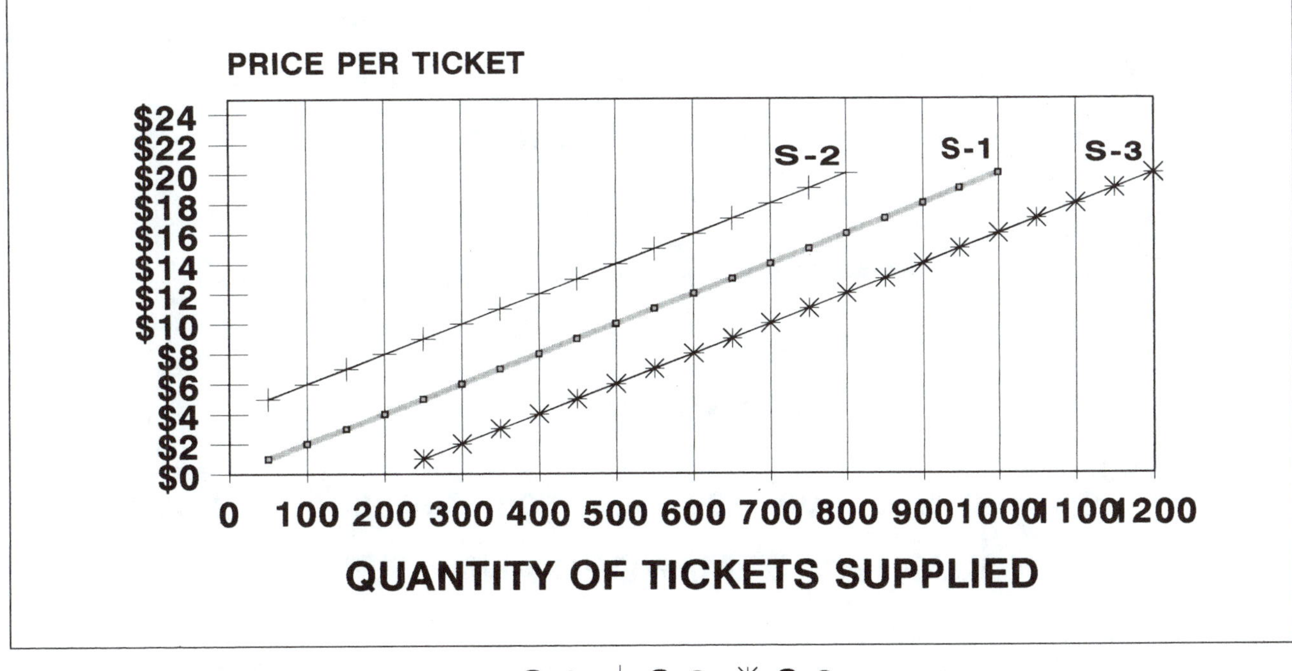

FIGURE 10-4

newspapers increases, or shifts to the right, the costs of paper for programs and brochures will also increase. Why does this occur? If paper producers increase the quantity of paper supplied for newspapers and cut back on the paper supplied for brochures and programs, the supply of the latter is decreased, and its price will rise. Such a change is demonstrated in Figure 10–4 when the supply curve shifts from S-1 to S-2. Printers will have to pay more for paper, and that price increase will be passed along to the organization. This in turn will affect the organization's overall operating budget. More funds will have to be allocated to printing, or a lower quality of paper will have to be selected.

Technology Advances in technology reduce the costs of production for suppliers and increase the productivity of the industry. This translates into lower costs for suppliers, and it is an incentive to supply more. The concert ticket seller may sell the performance rights to a television station with a new pay-per-view system. In addition to selling the one thousand tickets for the live performance, the concert supplier may now find a whole new distribution system for the product, and this may become an incentive to schedule more performances.

Number of suppliers When the number of suppliers in a product area increases, the supply curve shifts to the right, and conversely, when the number of suppliers decreases, the supply curve shifts to the left.

SUPPLY AND DEMAND CURVES

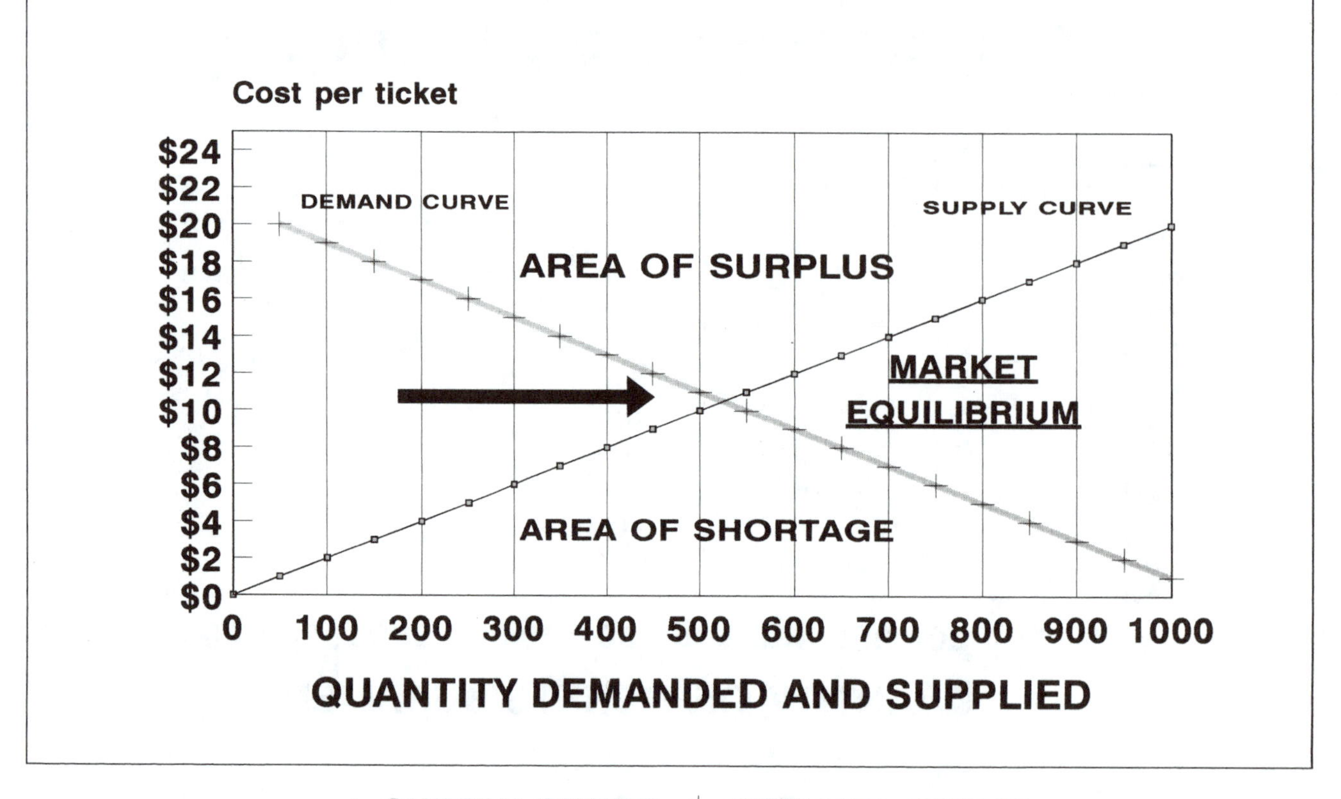

SUPPLY CURVE DEMAND CURVE

FIGURE 10-5

The entertainment industry, taken as a whole, generally has a great many suppliers at various prices. The supplier of classical music concerts might be able to increase prices if a great number of alternative forms of entertainment disappeared. On the other hand, if you have the only performance available of a specific work with a special performer, you may be able to raise the price despite the number of suppliers in the entertainment industry.

Suppliers' expectations When establishing the quantity to produce, suppliers consider many different factors. For example, the concert supplier expects that this will be the final performance before the famous soloist retires and thus increases the number of concerts.

Market Equilibrium: Supply and Demand Curves

Now let's see what happens when we bring supply and demand together.

Figure 10–5 shows the supply and demand curves brought together for the concert. The point at which the two curves intersect is called the *market equilibrium,* or the "price where the quantity of the good or service buyers demand and purchase is equal to the quantity suppliers supply and sell."[11] The area directly below the equilibrium point reflects a shortage of tickets, and the area directly above the equilibrium point represents a surplus of tickets. In other words, if all of the tickets were priced at $20, there would be excess supply. If, on the other hand, they were all priced at $1, there would be a shortage.

Figure 10–6

PRICE AND REVENUE MATRIX

PRICE	ESTIMATED DEMAND	REVENUE	NOTES
$20	100	$2,000	
$19	125	$2,375	
$18	150	$2,700	REVENUE DECREASES AS
$17	200	$3,400	TICKET PRICE INCREASES.
$16	250	$4,000	
$15	300	$4,500	
$14	350	$4,900	
$13	400	$5,200	
$12	450	$5,400	
$11	500	$5,500	* MAXIMUM
$10	550	$5,500	* REVENUE
$9	600	$5,400	
$8	650	$5,200	
$7	700	$4,900	
$6	750	$4,500	
$5	800	$4,000	
$4	850	$3,400	REVENUE DECREASES AS
$3	900	$2,700	PRICE DECREASES.
$2	950	$1,900	
$1	1000	$1,000	

Revenue maximization Figure 10–6 shows one application of how revenue maximization relates to demand. In this case, the theoretical demand generates the most revenue from sales at $10 or $11 per ticket. As we lower or raise the ticket price below or above the equilibrium point, the estimated revenue changes. Assuming that the demand moves along the curve estimated, we will make less money selling $15 tickets or $6 tickets than we would selling $10 or $11 tickets.

If we distribute the seat prices by location and charge more for the seats in the better viewing areas, we can overcome this problem and maximize the revenue. For example, if three seating areas are used and the prices are set at $10, $15, and $17, it might be possible to sell 1050 seats and collect $13,400.

By reducing the price, it would be possible to sell more tickets and make the same amount of revenue. It would be possible to make the same $13,400 by pricing tickets at $10, $6, and $4, if the quantity of tickets demanded is strong enough to sell a total of 2150. The price

In the News—Economic Impact

Spoleto Means Profits to Charleston Vendors
Joe Drape

The following article is an example of the *multiplier effect* in action.
CHARLESTON, S.C.—They come to this port city in search of avant-garde dance, startling opera stagings and the bonhomie shared by like-minded lovers of arias and teasing denouements. But Spoletogoers also dig hot dogs, die for lemonade, scarf up souvenir T-shirts and gulp brews with the same gusto as a baseball crowd.

In short, the French, Italians, Georgians and others who converged here beginning yesterday for the annual international arts festival are tourists.

And local hoteliers and hucksters look to the festival's 17 days of high-minded art as an economic opportunity akin to fishing in an aquarium.

"It's 17 days of extra cash and extra hard work," Meeting Street vendor Brenda Williams explained, squirting mustard on some hot dogs for an Italian couple outfitted in sailor caps with "Charleston" emblazoned on the front.

"All you got to do is be out on the street when the doors open and you've got it made."

Nearly 100,000 people are expected to attend the festival, celebrating its 15th year. They will spend almost $2 million on tickets to attend more than 120 performances. The spinoff economic impact of the festival, according to Spoleto and city officials, exceeds $50 million.

"It's a really big time for everyone," said Patty Downs, a desk clerk for the Planters Inn.

Innkeepers and restaurateurs are the biggest beneficiaries of the arts dollars. As with most inns and hotels in The Battery, all 41 rooms at the Planters Inn are booked throughout the festival, and have been for 5 months.

She-crab soup and other native seafood will be spooned out at more than double the rate at 82 Queen, a favorite downtown eatery. Nearly 150 lunches and 450 dinners per day will be served in the antique indoor dining rooms and balmy patios of the 19th-century estate.

Because this is a pedestrian-oriented city fraught with the opulence and charm of a Confederate yesteryear, retailers also experience dramatic profit increases.

SOURCE: Joe Drape, "Spoleto Means Profits to Charleston Vendors," Cox News Service.

reduction becomes a demand-increasing tactic that may have long-term benefits to an organization trying to build an audience. Economists refer to this as the *income effect*. A price reduction gives people more money to spend. When faced with a choice of paying a minimum ticket price of $4 or $10 for the same event, most people would rather make their dollar go further.

This is a simplification of the entire ticket pricing process, but it is also a reminder that fundamental concepts in economics are a part of the arts manager's job.

General Applications of Economic Theories

Throughout this chapter, examples are cited to show how economic theories can be applied to the arts environment. A few other general theories have a direct impact on the arts.

Economic Impact

One key area on which the arts have focused in the last few years is the economic impact of the organization on the community. Arts organizations expend funds; pay salaries, wages, taxes, and benefits; and use goods and services in the community. They also help stimulate the local economy through the *multiplier effect* of money. "Spoleto Means Profits to Charleston Vendors" provides evidence of the economic impact of the arts on a community.

The impact of the basic operation of the arts organization extends beyond the community. The salaries paid to staff members are used to pay for rent, make car payments, or buy groceries. The money is then used again by the property owner, bank, or store to purchase things or to make loans. In effect, the money that the arts organization puts into the local economy ripples throughout the region. In addition, when consumers buy tickets and make the journey to the performance, they may pay for a babysitter, gas for the car, a meal at a restaurant, parking, and a purchase at the arts center's gift shop. The $40 paid for the ticket may generate four or five times that amount in other goods and services.

Organizational Impact

Several basic economic principles relate to how an organization operates in the total economic environment. For example, calculating the impact of such things as total fixed costs, total variable costs, average fixed and average variable costs, and marginal costs helps the organization with financial and operational planning. The ideas related to the law of diminishing returns, long-term operational costs, and economies of scale also have some application to the arts. Let's take a brief look at some of these theories and laws.

Fixed, variable, and marginal costs All organizations must identify the fixed costs of operation to form the base operating budget. Expenses for such things as renting or leasing space, buying equipment, and repaying loans, must be carefully calculated as part of the *total fixed costs* (TFC) of operating. These costs will not change whether you do two or two hundred performances, and they are often called *overhead.* Salaries for a minimum or core staff and the various benefits paid to them might also be part of the fixed overhead of an arts organization.

When performers, designers, and technicians are hired for a given production (assuming a season of several shows), these expenses are added to the *total variable costs* (TVC). Materials purchased for the scenery and costumes, phone calls, blueprints, paint, and labor to produce the show are all part of the TVC. The *total cost* (TC) represents the total of the fixed and variable costs (TC = TFC + TVC). You could divide the TFC and TVC by the number of shows in the season to derive the average fixed costs (AFC) and average variable costs (AVC) per show. Added together, the AFC and AVC equal the average total costs (ATC).

Another key indicator of costs is the *marginal cost* (MC), which is defined as the cost of producing an additional quantity of the product. In the performing arts, the most obvious output increase would be the number of performances scheduled. For example, when you plan for an evening concert performance, you can estimate the total variable and fixed costs associated with that concert. The marginal cost of doing another performance that afternoon would probably be less than the marginal cost of scheduling another performance for the next day. One reason why the MC might be lower is that the rental of the performance space is based on a daily rate of eight hours. If you use the space for only four hours in the evening, you still pay for time you do not use the space. The matinee performance would not add to the variable costs of the hall.

Diminishing returns On the other hand, your payroll and production costs would rise with the extra performance. These variable costs would offset the gains from the decreased hall rental and would allow the organization to gain firsthand experience with the *law of diminishing returns*. The law states that "as more and more of a variable factor of production, or input (e.g., labor) is used together with a fixed factor of production, beyond some point the additional, or marginal, output attributable to the variable factor begins to fall."[12] To put it simply, at some point the costs rise enough to reduce the marginal gain from creating the additional output.

Profit-making ventures must constantly watch all costs related to making, selling, distributing, and advertising the product as well as the marginal costs. Otherwise, there will be no money left to call a profit. Nonprofit organizations may not try to generate revenue above costs for the owners or investors, but they must still carefully control fixed, variable, and marginal costs. In fact, a nonprofit organization may generate a surplus of revenue, as we will see in the next chapter. To generate a surplus or to break even, the arts manager must draw on a great deal of skill in controlling costs, setting prices, and estimating demand. Chapter 11 examines various fixed and variable costs in the process of designing budgets and managing cash flow.

Economies of scale Another issue related to overall cost is how economies of scale operate in an organization. The technical definition of *economies of scale* is "a decrease in the long-run average total costs (ATC) of production that occur when larger facilities are available for manufacturing a product."[13] These economies are achieved because the business is able to specialize production techniques, its labor force becomes more expert, volume discounts are available for materials used to produce the product, or the by-products of manufacturing reach a large enough quantity, to become salable.

There are limited applications of economies of scale in aspects of the production process. For example, instead of setting up and equipping its own scenery and costume production shop, an organization could develop a central production center to be used by all of the major arts organizations in the region. This production center could achieve costs savings from scale through construction techniques and bulk purchases. For example, by grouping various construction projects and buying in quantity, the organization may be able to achieve a large enough scale of operation to reduce overall costs.

Another example of economies of scale is a performing arts center with multiple spaces. The assumption that it is less costly to run three theaters under one roof than three theaters under three roofs, helped motivate the construction of these types of facilities. However, diseconomies of scale can also affect such centralized operations if management is not careful to control growth as the organization matures. A diseconomy of scale might be achieved by having to hire extra people and buy extra tools to take on the increased scale of production. This would increase the average total costs to produce the sets in the central shop. The savings realized by the scale of the operation would be quickly negated.

The Economic Problems Facing the Arts

Up to this point, we have examined many of the basic economic principles as they relate to the arts. A limited number of studies have been done and a few books have been written about the arts and economics. One that is required reading for all arts managers is William J. Baumol and William G. Bowen's landmark study, *Performing Arts—The Economic Dilemma*,[14] which was published in 1966. The book was the

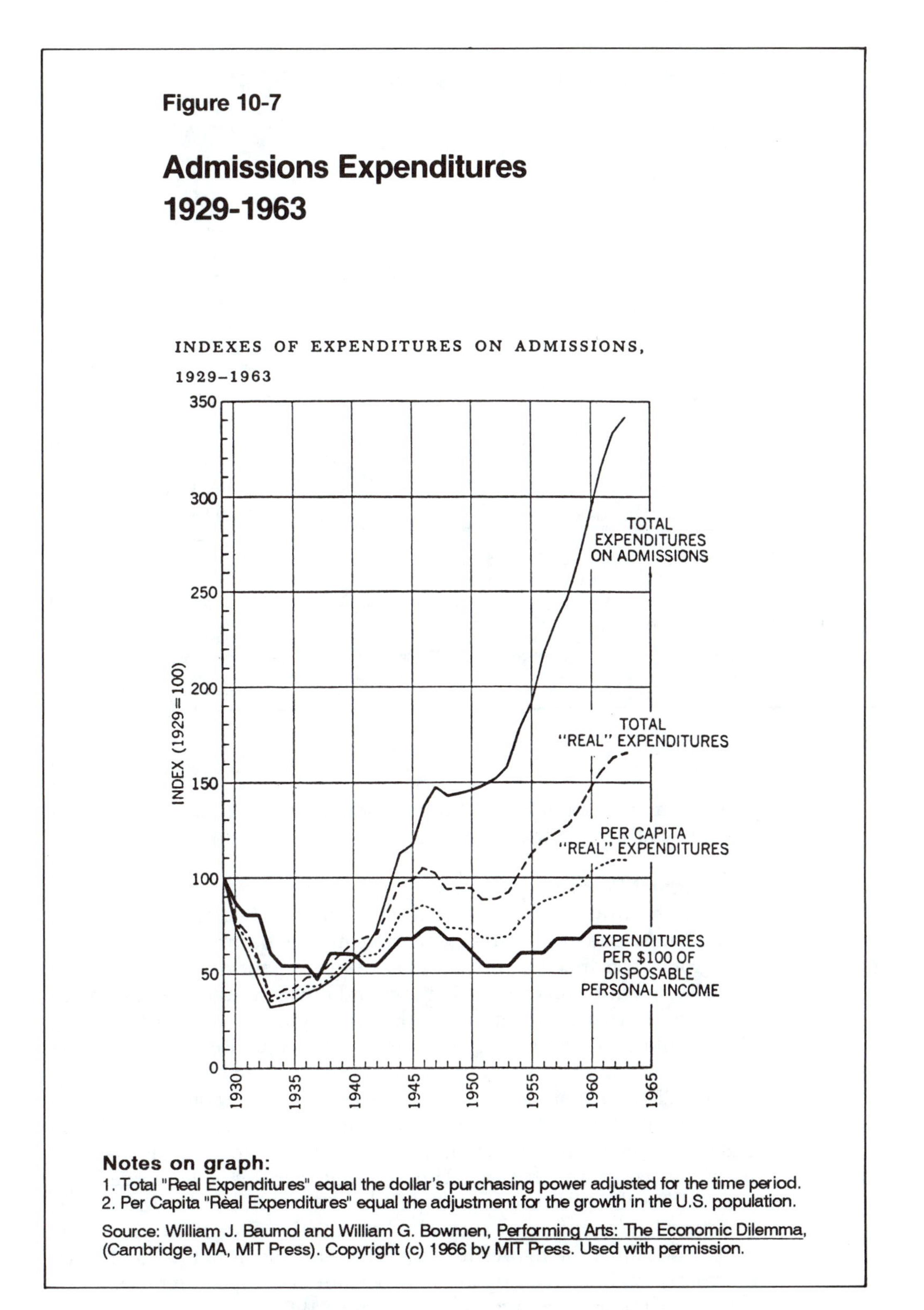

Figure 10-7

Admissions Expenditures 1929-1963

Notes on graph:
1. Total "Real Expenditures" equal the dollar's purchasing power adjusted for the time period.
2. Per Capita "Real Expenditures" equal the adjustment for the growth in the U.S. population.

Source: William J. Baumol and William G. Bowmen, Performing Arts: The Economic Dilemma, (Cambridge, MA, MIT Press). Copyright (c) 1966 by MIT Press. Used with permission.

first detailed analysis of the economic conditions of the arts. Other writers have also explored the relationship of the arts and economics. These sources are listed at the end of the chapter.

Although the data gathered by Baumol and Bowen are from the early 60s when a ticket to the Metropolitan Opera cost less than $9, the methodical approach they used to gather and assemble the informa-

Figure 10-8 Expenditure on Admissions 1930-1985

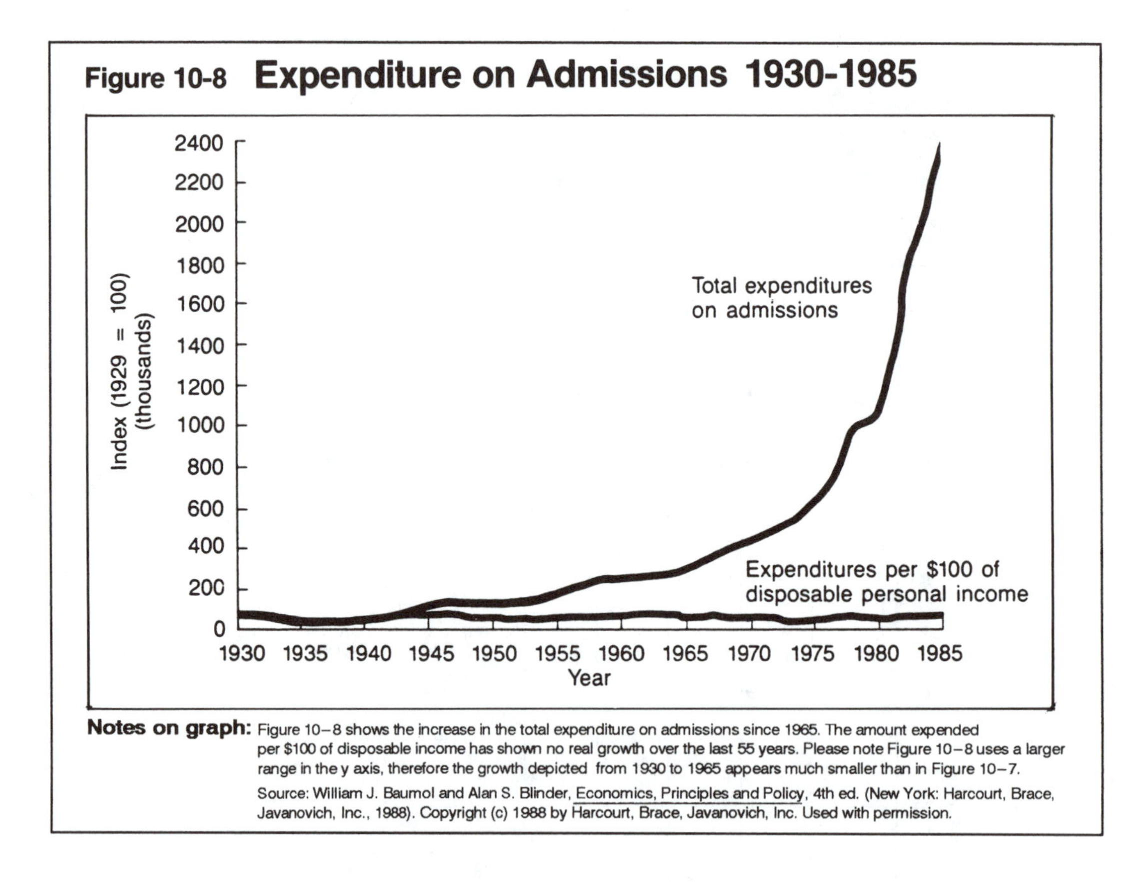

Notes on graph: Figure 10–8 shows the increase in the total expenditure on admissions since 1965. The amount expended per $100 of disposable income has shown no real growth over the last 55 years. Please note Figure 10–8 uses a larger range in the y axis, therefore the growth depicted from 1930 to 1965 appears much smaller than in Figure 10–7.

Source: William J. Baumol and Alan S. Blinder, Economics, Principles and Policy, 4th ed. (New York: Harcourt, Brace, Javanovich, Inc., 1988). Copyright (c) 1988 by Harcourt, Brace, Javanovich, Inc. Used with permission.

tion became a model for all future research on the topic. Let's take a brief look at each of their key findings.

The Cultural Boom

During the mid-60s there was much talk of a "cultural boom" in America. Attendance at arts events was up, the number of performing arts groups was increasing yearly, and regional performing arts centers were being built everywhere. Baumol and Bowen found that, although there was indeed a lot of activity, the actual growth of the arts in relation to all other factors was very modest. By adjusting and correcting the data for inflation, population increases, and income growth, they found a disturbing situation. Instead of growth, they found a decline in the amount of money expended per person on the arts. Figure 10–7 is a graphic representation of this depressing news, and Figure 10–8 updates the data through 1985. The line that represents what everyone was calling the "culture boom" (labeled "Total Expenditures on Admissions") does appear to increase dramatically after 1940. However, when you adjust the figures for disposable income, inflation, and wage increases, the line hovering near the bottom of the graph represents an actual decline in spending on the arts since 1930.

The Arts Audience

Before Baumol and Bowen's study, there was very little statistical data available about who was attending various theater, dance, opera, and concert performances in the United States. The authors used detailed audience surveys to gather data on income levels, education, age, gender, and preferences. They found that the "common man" was fairly uncommon among those who attended live professional performances. They found a predominance of white-collar occupations and high levels of education and income. Forty-five percent of the audience members were between 35 and 60 years of age. Their research led them to estimate that in 1966 about 4 percent of the U.S. population over 18 years of age attended some professional arts event. Taken at today's population levels, that would translate to around 10 million people.

In the 25 years since Baumol and Bowen's study, not a great deal has changed when it comes to audience composition. The audience for "high culture" events such as opera, theater, dance, and museum exhibits continues to be a relatively small segment of the population. However, some sources claim there was a higher attendance at arts events than at sporting events in 1990, indicating that, on the surface at least, the arts are doing much better than at the time of Baumol and Bowen's study. However, when the millions who watch sports on television are taken into account, it would probably be reasonable to assume that more people, representing a greater cross section of the population, would consider themselves to be sports, not arts, fans.

The apparent growth experienced by arts organizations in the 80s was very much a product of the baby boomer population increase driving up arts attendance and participation. The diversity of interests of the people educated in the 60s and 70s is now reflected in the many small arts organizations that specialize in more nontraditional work. However, the actual audience demographic profile has not changed significantly.

The Productivity Issue

Baumol and Bowen's detailed economic study uncovered data that support the conclusion that the financial difficulties that arts groups were experiencing would only worsen over time. The basis for this finding is directly related to how the entire economic system has seen slow, but steady, growth in productivity. In theory, new technologies and processes of production increase the quantity of work output for each employee, that is, they increase productivity. This, in turn, should reduce the cost per unit for production and increase profits. If profits can be increased, then suppliers will have an incentive to produce more, and the entire supply curve and all of the goods and services produced should shift to the right. At the same time, if businesses are able to achieve higher levels of productivity and profit, then the overall income level of the work force should rise. The net result, in theory, is economic growth as the demand curve moves to the right with the supply curve.

To stay ahead of *inflation*, which is the constant increase in the price levels of everything, a business must find ways to more productively use the labor force and any other inputs needed to produce the product or output. Without productivity gains, the producer would have to raise the price higher and higher to cover the increasing input costs. If other companies that make a similar product are able to increase productivity, the prices they charge could be kept lower. This would cause consumers to buy the less costly product and would eventually put the less productive company out of business.

Baumol and Bowen applied this theory to arts organizations and found the basis for the gap between their income and their expenses. The authors argued that the technology of presenting an arts event was subject to limited increases in productivity.

The time it takes to rehearse and perform a play, opera, dance, or symphony was not subject to increases in productivity. In addition, the supply of the product was limited. A live performance can only be repeated a certain number of times in a day. From that conclusion and from the income-expense gap data they collected over a 15-year period, Baumol and Bowen predicted that arts productivity would actually decline over time as other segments of the economy became more productive. The long-term effect of this is the ever-increasing income gap. They predicted that without regularly increasing donations, the income gap would continue to increase even with increasing ticket prices.

Productivity

The application of computers to almost all phases of American business has helped increase productivity. In fact, computers in the offices or production areas of arts organizations have helped make people more productive. However, no amount of computer technology will have a significant impact on shortening the time spent on rehearsing a scene or repeating a musical passage until it is perfect. The final product will take about the same amount of time to present. The result is that whatever gains in productivity are realized by computerizing the office will be offset by increasing costs to present the basic product.

Cost increases caused by inflation coupled with limited productivity and supply gains in operations further increased the income gap for the arts. In most cases, salaries, benefits, utilities, supplies, and so forth, are increasing each year. Unless ticket prices are increased, more performances are scheduled, or more funds are raised, arts organizations will be doomed to an ever-increasing income gap.

This gap, sometimes called the *Baumol and Bowen's Disease*, is a fact of life for arts organizations. Any arts group that attempts to ignore the implications of the income gap in its strategic planning, marketing, or fund raising efforts will eventually find itself going broke.

Other Findings

There were many other key findings in Baumol and Bowen's study. Here are just a few:

1. The costs per performance were rising faster than the consumer price index (CPI).
2. Increases in ticket prices often exceeded the CPI, but because these increases were below the increases in the costs per performance, there was no real gain in additional revenue.
3. Cost pressures and the demand to provide artists with a living wage placed many arts organizations in a state of almost constant budget crisis.

The problems Baumol and Bowen saw 25 years ago are still with us today. In Chapter 4, "Quiet Crisis in the Arts" noted that the financial problems facing arts groups today are worsening. The number of organizations facing economic difficulty seems to multiply daily.

Conclusion

What does Baumol and Bowen's study and this chapter on economics lead us to conclude? The optimist might conclude that the economic system has worked well enough to get us to the point where more arts organizations are operating in a wider geographical distribution than ever before. On the other hand, it is also possible to conclude that arts organizations have expanded too much in relation to the real market demand for their products. As a result of this overexpansion, there are too many arts groups chasing too few dollars.

Somewhere between these two viewpoints resides the realist, who believes that market forces, government subsidies, and private sources will provide enough support to keep a limited number of arts organizations alive. The realist would argue that culture is expressed in part through its artistic creations and activity. Society will continue to invest its resources in the arts because enough people want and need what the arts provide. The satisfaction gained from creating and consuming the arts ensure their place in the overall economic system. However, if the system is to work for the benefit of everyone, the artists and the arts organizations must take an active role in shaping public policy about the place of the arts in society.

The basic economic principles and theories discussed in this chapter have an application to the arts. The relationship of the organization to the macroeconomic environment and the conditions in the microeconomic environment in which the organization operates should affect planning. Understanding the supply and demand issues facing the organization are critical. Being able to predict and control fixed and variable costs accurately, and being able to predict marginal costs and gains can help keep the organization from falling into financial trouble. Finally, it is important for nonprofit organizations to accept that the income gap is a fact of life. Strict cost-control strategies and plans can be adopted to minimize the negative consequences of this underlying problem.

Summary

Arts organizations function in macro- and microeconomic environments that affect the long-term financial health of the organization.

The economy is the organization of the production, distribution, and consumption of all things people use to achieve a certain standard of living. Economics studies how societies employ scarce resources to produce goods and services and to distribute them among people. Goods are tangible things, and services are intangible things. The price for a good or service is based on how plentiful or scarce the good or service is and how much it is demanded by people in the economic system. Opportunity costs are the costs of giving up one thing in exchange for another.

Macroeconomics refers to the overall relationship between business, government, and households. Microeconomics deals with the sales and purchase activity that affects a good or service.

The law of demand states that there is a relationship between the costs of a good and what people are willing to pay. In most cases, the higher the price, the lower the quantity demanded, and the lower the price, the higher the quantity demanded. The entire demand curve may be increased (indicated by a shift to the right) or decreased (shifted to the left) if there are changes in the prices of other goods, or if consumer income levels change. In addition, consumer expectations and tastes may also cause a change in the demand curve.

The law of supply states that suppliers will provide larger quantities of goods and services at higher prices. Supply curves are affected by the price of production resources, the price of other goods, technological advances, the number of suppliers, and suppliers' expectations. Market equilibrium is said to exist when the quantity demanded matches the quantity supplied.

Organizations face fixed, variable, and marginal costs in all phases of operation. Calculating these costs helps to determine the organization's baseline operating budget and the costs of doing additional activities. Some aspects of the arts organization may benefit from economies of scale, which are savings derived by decreasing the average total costs of operation.

Baumol and Bowen found that the cultural boom in the United States was exaggerated, and when measured against real growth, there had been a decline in per capita expenditures on the arts. Their study also found that a small segment of the population, mostly white-collar, highly educated, and economically advantaged people, were regular consumers of the arts. The key finding of the study was that increased costs of production coupled with limited productivity gains created a growing income gap in the long run. They predicted that unless arts organizations regularly increased donated revenue, the gap would far outdistance their ability to raise revenue through ticket prices.

Key Terms and Concepts

Economic system
Macroeconomics
Microeconomics
Goods and services
Economic problem
Opportunity costs
Inflation
Gross domestic product (GDP)
Market
Law of demand
Demand determinants
Substitute goods
Complementary goods
Law of supply
Supply determinants
Market equilibrium
Fixed, variable, and marginal costs
Economies of scale
Productivity
Baumol and Bowen's disease

Questions

1. Define an economic system.
2. Define macroeconomics and microeconomics.
3. What are some of the opportunity costs that an arts organization might have to face?
4. Give three examples of the laws of demand and supply at work in the arts.
5. To what degree do public taste and expectations affect the demand and supply curves for arts events? Give examples.
6. Use graph paper to create a demand curve based on what you think would be the quantity demanded for tickets for a symphony or-

chestra concert in your area if the prices ranged from $5 to $35 in $5 increments. Assume there are 2000 seats in the performance hall.

7. Based on this demand curve, what would be the equilibrium point in demand and supply?

8. How much revenue do you think this event would generate if you had four different prices for the tickets (for example, $10, $20, $30, and $35)?

9. Give two examples of economies of scale that could be applied to an arts organization's daily operation.

10. Summarize the key findings in the Baumol and Bowen study of the economics of the arts.

Case Study

The following article raises some key economic issues about supply, demand, and the costs of production on Broadway.

Broadway Adopts Plan to Cut Costs and Ticket Prices

Mervyn Rothstein,

NEW YORK—Broadway producers, theater owners, craft guilds and unions announced final agreement today on a plan to cut costs and ticket prices to encourage the production of more plays and to develop new audiences.

Under the agreement, each of the three major Broadway theater owners—the Shubert and Nederlander Organizations and Jujamcyn Theaters—designates one of its theaters for such productions. Dramas or comedies are eligible, but musicals are not. Royalties, many salaries and other costs for plays at those theaters would be cut by an average of at least 25 percent. In return, ticket prices at those theaters would range from $10 to a maximum of $24, a significant reduction from the present range for nonmusicals on Broadway of about $25 to $45.

Early details of the plan, which was formulated during a year and a half of negotiations and which was formally known as the Broadway Alliance: A New Plan for Play Productions, were first reported last December. The plan is to be in effect for 18 months, from Sept. 1, 1990 to Feb. 29, 1992, and will be subject to periodic review.

"This is an auspicious occasion," Harvey Sabinson, the executive director of the League of American Theaters and Producers, said today in announcing the "collaborative effort" at a news conference at the Holiday Inn Crowne Plaza. Mr. Sabinson then read a statement prepared by Peter Stone, the president of the Dramatists Guild.

'Broadway Could Only Watch'

"The Broadway community can no longer stand helplessly by as many of America's finest plays are being produced everywhere but on Broadway," Mr. Stone's statement read. "Most of these plays, in the face of growing costs and dwindling audiences, lacking a big star, a major prize or a lavish production, have been relegated to regional and Off Broadway theaters. Such acclaimed works as 'Driving Miss Daisy,' 'Steel Magnolias,' 'Frankie and Johnny in the Clair de Lune,' 'Other People's Money' and 'The Cocktail Hour' have achieved national recognition and large motion picture contracts while Broadway could only watch and sadly remember a time, no more than 10 years ago, when *every one* of these fine plays would have been lighting up Broadway houses."

Much mention was made at the news conference of the need for attracting new audiences, and a number of speakers referred to the high price of tickets as a prime factor keeping customers away from Broadway. The top Off Broadway ticket is about $30. "This plan addresses that problem and makes theater available," said Paul Libin, the producing director of Circle in the Square. "It embraces those people we've lost and helps attract young people."

The theaters designated for the plan are the Walter Kerr, which is owned by Jujamcyn and is on West 48th Street; the Lyceum, on West 45th, or the Belasco, on West 44th, which are Shubert properties; and the Nederlander, on West 41st. They are to be booked on a rotating basis, and the theater owner has the right to reject a play, as well as to book a regular attraction and drop out of the rotation while that attraction is running.

Guidelines for Plays

A producer wishing to present a play under the plan must submit it to the Broadway Alliance for approval—proposals are currently being accepted—and is subject to strict guidelines. First the play must be produced for $400,000. A serious play presented under normal Broadway rules these days costs from $800,000 to $1 million to produce, though some have been done for more and some for less. Weekly operating costs must be kept roughly between $60,000 and $65,000, also less than half those for a typical play. And if the play turns into a hit, the producer is not allowed to switch it to a full Broadway contract, raise ticket prices and transfer it to a different theater in New York City.

"The fundamental trust is the $24 ticket," said Cy Feuer, the president of the league. "Any movement out of the plan would increase that price and violate that trust."

One major element of the plan is profit-sharing: all those working for reduced salaries are to share in 10 percent of any profit after the play has earned back its original investment. On weeks in which the show loses money, producers are to receive payments for limited office expenses only, but they will not have to pay rent for the theaters until the investment has been recouped. There are also reductions in advertising commissions and other fees.

The theaters designated have not been used in recent years, though the newly refurbished Walter Kerr (formerly the Ritz) is the home of "The Piano Lesson," by August Wilson, and cannot take part in the plan as long as that Pulitzer Prize-winning play continues to run. That fact limits the startup of the plan to two theaters, as only one of the Shubert theaters may be designated.

In addition, all the theaters involved are ones that were previously designated endangered, for reasons that include their setting on the periphery of the Broadway district and second balconies that are difficult to fill at high prices. In the last decade, the craft guilds and unions had agreed to various concessions at those theaters, including lower wage scales and more flexible work rules. The producers reduced ticket prices by about 20 percent and agreed to sell tickets for 499 of the theater's seats.

That plan did not succeed in attracting many new plays. But Rocco Landesman, the president of Jujamcyn, said that the new plan had a much better chance of succeeding.

"There's a huge difference between the plans," Mr. Landesman said in an interview. "The old plan saved a production between $4,000 to $4,500 a week, and involved only a couple of unions. This plan is across the board and involves the entire theater community. The costs are much lower."

Sample budgets provided by the league show that even at reduced ticket prices, producers would be able to recoup investments fairly quickly if they succeeded in attracting audiences. "You can make a profit under this program," said Alan Eisenberg, the executive secretary of Actors' Equity and head of the committee that prepared the plan.

The kind of budget figures involved means that most plays likely to take advantage of the plan would be small, with one set and small casts.

The theater league is contributing $250,000 as startup money, to be used for administrative expenses and for lighting systems in the designated theaters to help reduce production costs. A separate fund is being established to help producers complete the financing of their plays if needed.

"There is a loyal audience for plays, and it goes where the plays are," Mr. Stone said at the news conference. "We want to bring that audience back to Broadway."

SOURCE: Mervyn Rothstein, "Broadway Adopts Plan to Cut Costs and Ticket Prices," *New York Times*, June 27, 1990

Questions

1. Restate, in economic terms, the objectives defined by the parties to the agreement noted in the first two paragraphs of the article.
2. Based on the law of supply, do you think the stipulation that a hit show may not increase prices or increase the number of seats available will make sense to producers? Explain how the law of supply relates to this plan.
3. If there is a "loyal audience for plays," as asserted in the article, what other economic reasons besides price might explain the decline in audiences?
4. Do you think this plan will work? Explain.

References

1. Roger N. Waud, *Economics*, 3d ed., New York: Harper and Row, 1986) 4.
2. Ibid.
3. Ibid., 12.
4. Ibid., 13.
5. Ibid., 24.
6. Ibid., 25.
7. Ibid., 33.
8. Ibid., 70.
9. Ibid., 72.
10. William J. Baumol and Alan S. Blinder, *Economics - Principles and Policy*, 4th ed. (New York: Harcourt Brace Jovanovich, 1988), 63.
11. Waud, *Economics*, 77.
12. Ibid., 503.
13. Ibid., 514.

14. William J. Baumol and William G. Bowen, *Performing Arts: The Economic Dilemma* (Cambridge, MA: MIT Press, 1966). This section of the chapter summarizes Baumol and Bowen's key points.

Additional Resources

The following list contains articles and books related to the topic of economics and the arts.

Blaug, M., ed. *The Economics of the Arts.* Boulder, CO: Westview Press, 1976.

Hanasman, H. "Nonprofit Enterprise in the Performing Arts." *Bell Journal of Economics* 12 (Autumn 1981).

Passell, Peter. "Broadway and the Bottom Line." *New York Times,* December 10, 1989.

Vogel, Harold L. *Entertainment Industry Economics: A Guide for Financial Analysis.* Cambridge, England: Cambridge University Press, 1986.

Financial Management

11

In the last chapter, we examined the basic theories and principles from the world of economics, and we studied some of the ways in which the economic environment affects an arts organization. The topics of supply, demand, utility, and variable, fixed, and total costs were studied in an attempt to build a foundation for understanding the financial management of an arts organization. In *Managing a Nonprofit Organization,* Thomas Wolf notes that "financial management is, for many, one of the most forbidding aspects of the administration of nonprofit organizations."[1] This is due in part to phobias that people have about anything having to do with quantitative thinking. When faced with an income statement and a balance sheet some board members develop a glazed look in their eyes and suddenly lose the ability to reason. Even worse, they approve budgets and accept financial statements without understanding the numbers. Unfortunately, comprehension usually comes to the board of directors when it is too late to correct a financial problem that has been staring them in the face for months.

This brings us to the goal for this chapter: to unravel some of the mystery about financial management and to provide a clear picture of the relationship of planning, organizing, leading, and controlling to the financial health of the organization. We will look at an overview of financial management in profit and nonprofit organizations and then move on to the financial management information system. We will look next at budgeting, cash flow, and record-keeping systems. Finally, we will integrate the whole system into the balance sheet and statement of account activity.

Overview of Financial Management

In many arts organizations, the responsibility for financial records, budgets, payroll, and money management falls to a business manager who works under the supervision of a general manager or managing director. In a smaller organization, an accountant or bookkeeper may do the processing and record keeping. This person reports to the artistic director and the chair of the board of directors. As we saw in Chapter 9, many arts organizations do not start up with a management information system. Instead, a system evolves as the organization grows. The same is often true for the financial management of the organization. If it is to be effective, a *financial management information system* (FMIS) must be a comprehensive investment, reporting, control, and processing system that helps managers realize the financial objectives of the organization. The financial management of the organization is one of the manager's most important responsibilities. The long-term health of the enterprise depends on the arts manager's vigilance in monitoring the revenue, expenses, and investments of the organization.

Let's look at the responsibilities and evolution of financial management.

For-Profit Organizations

In the profit sector, a vice-president of finance usually supervises the controller and treasurer of the organization. The controller usually supervises the accounting system and the tax operations of the organization. The treasurer manages the firm's investments and cash and oversees the management of the capital resources (the plant and equipment) and planning.[2]

The financial manager's biggest responsibility in a for-profit business is to plan for the acquisition and use of funds to maximize the overall value of the business. Therefore, the financial manager makes decisions focused primarily around three areas: forecasting, investing, and control.

The financial manager's forecasting and planning role involves interacting with other executives who are charting the strategic course for the organization. Growth or change usually means that funding will be required. The financial manager must be involved in advising and directing how the organization will pay for its plan.

The financial manager also actively works to raise money to finance the growth of the organization or to pay for equipment or the acquisition of other businesses. The financial manager must be able to tap into the stock and bond markets, and the banking community to pay for the future of the organization.

Finally, the financial manager must be able to help the company operate as efficiently as possible with the resources it currently possesses. It is safe to say that all business decisions have financial implications. For example, if marketing decides to expand into a new product line, finding the money needed to pay for the new product line will be the responsibility of the financial manager.

Profit in the arts Commercial theater, popular music presentations, films, television, and the entertainment industry are very concerned with maximizing revenue and creating a profit. For example, Stephen Langley's text, *Theatre Management and Production in America*,[3] contains very clear examples on how to calculate the costs, profit, and break-even points for Broadway and Off Broadway productions. Techniques for budgeting, cost control, and cash flow analysis, which are discussed later in this chapter, can also be applied to profit or nonprofit organizations.

Nonprofit Financial Management

Because the primary objective of most nonprofit arts organizations is not focused on increasing the wealth of the owners or stockholders, the financial manager's job is somewhat different than in the profit sector. A nonprofit organization is still a business, though, and it must collect as much revenue as it expends, or it will go out of business. The financial manager's job in a nonprofit organization is critical to planning and using the limited resources available. For example, if the artistic director wants to add another show to the season, do a world premiere, or take a production on tour, how will the organization pay for this activity? Understanding the cash flow, current debt load, fund balances, and so on, is the first step in analyzing what can and cannot be done.

A good financial manager maximizes the use of the available resources. For example, funds raised through ticket sales or donations should be invested to generate further revenue. If an organization collects $200,000 from subscription sales for next season's shows, that money should be invested in interest-earning accounts before it is needed for the new season's operating budget. A financial manager also actively seeks out ways to minimize costs for insurance (health and life) and to reduce other operating costs.

Financial Management Information System

In Chapter 9, we learned that the management information system is responsible for gathering data, formulating information, and distributing it throughout the organization. The financial management information system is a key part of the overall MIS. Figure 11–1 shows a schematic drawing of a FMIS.

The FMIS must ultimately provide the board of directors and the upper management with accurate, timely, and relevant information based on the data that the system gathered. Without this information, the planning process will not work. The questions are usually very simple: How are we doing? Did we spend more than we budgeted? Why? Did we raise less than we budgeted? Why? Did we increase or decrease our debt? Will we have sufficient resources to continue operating?

If the system is working properly, the answers to these questions pose no problem for management. For the system to work properly, the operating system noted in Figure 11–1 must accurately gather the data needed to process the records. These data become the information used in reports and in the analysis of the current financial health of the organization. At the same time, the organization has specific legal responsibilities to report on its financial activities to various state and federal agencies. The FMIS must be capable of gathering and reporting this information if the organization is to retain its nonprofit legal status.

Before delving into the details of the FMIS and the accounting system, it is important to clarify the legal status of the organization. The FMIS serves two important purposes. It provides information about the fiscal health of the organization to people inside the organization, and it reports to external agencies, such as the IRS, and to granting agencies, such as the NEA and state and local arts councils.

Legal Status and Financial Statements

When a business or arts organization starts up, it may be owned and operated by one individual. However, once the operation grows to the point that a staff and office space are required, it may be time to incorporate the enterprise. Individual artists may also incorporate themselves to gain some specific tax advantages. In many areas of the country, the Volunteer Lawyers for the Arts help individuals and organizations with the incorporation process.

Incorporation

The major reason why an individual or organization decides to legally incorporate is to provide protection for the people who operate the business. Without the protection of incorporation, the owner is legally responsible for all debts incurred and may be personally sued. A legal

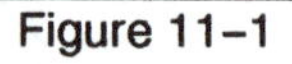
Figure 11-1

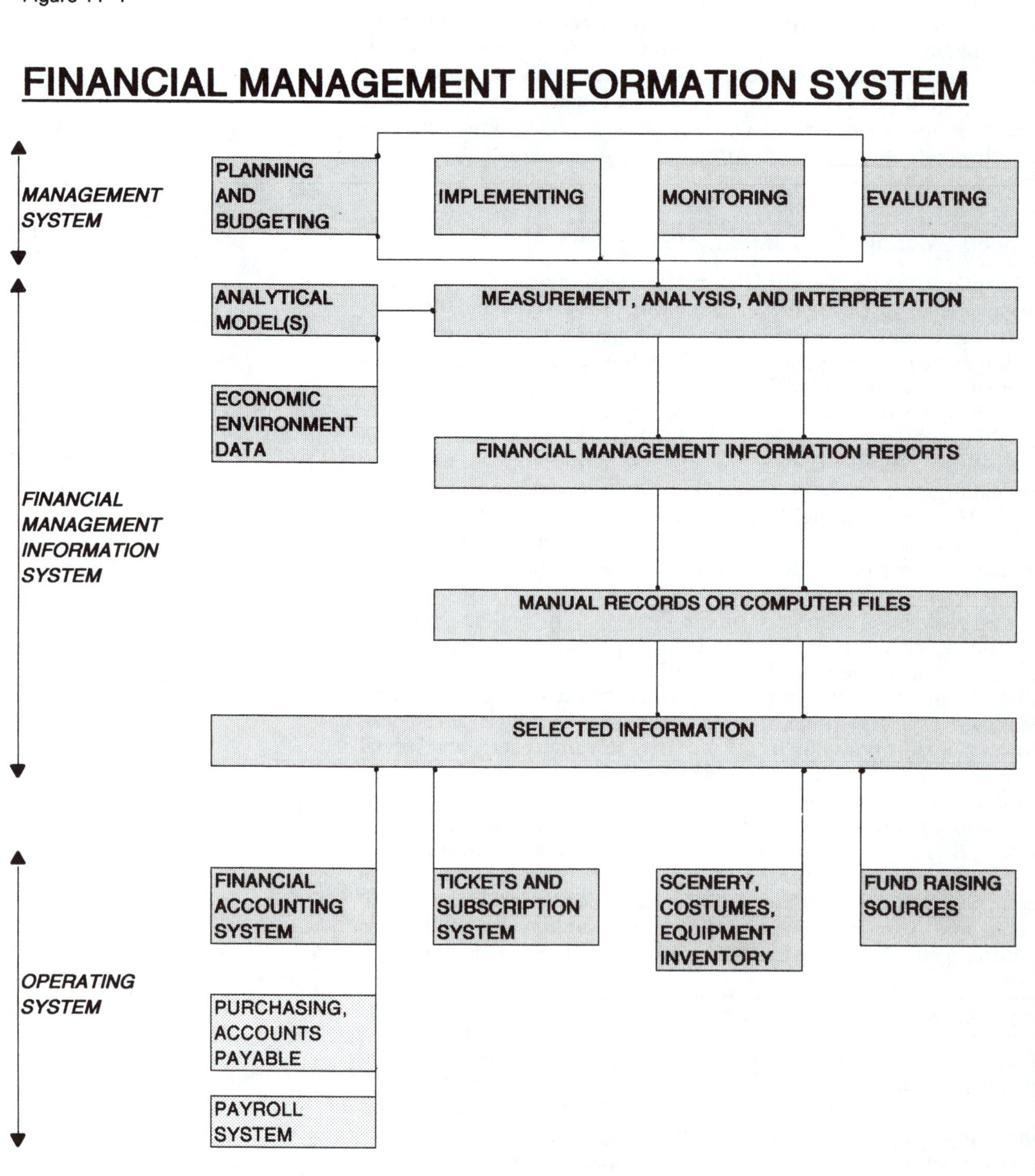

Source: Frederick J. Turk and Robert P. Gallo, Financial Management Strategies for Arts Organizations, (New York: American Council for the Arts, 1984). Copyright (c) 1984 by The American Council for the Arts. Used with permission.

settlement against an individual might mean that all personal assets could be sold to pay the organization's debts.

In the case of most arts organizations, filing for incorporation as a nonprofit business is fairly straightforward. Provided that the proper papers are filed, the state bestows upon the organization the legal right to operate. Filing for exemption from state and local taxes requires additional paperwork. Filing for incorporation is usually covered under the operational procedures established by the Secretary of State and generally requires the following:

- the official name of the organization
- the purpose or purposes of the organization
- the scope of activities; if you are filing for tax exemption, this limits what you can and cannot do
- membership provisions (if any)
- name of the person registering the incorporation and the place of business
- the names and addresses of the incorporators and the initial board of directors (if any)
- how any assets will be distributed when the corporation is dissolved

Additional legal regulations may affect nonprofit corporations, including business or occupation licenses and state or local charitable-solicitation licenses. Incorporation and nonprofit status, if accompanied by tax exemption, empowers the organization to raise funds. Vending licenses may also be required if you plan to sell items through a gift shop.

Tax Exemption

As noted, exemption from local, state, and federal taxes does not automatically come with nonprofit incorporation. The Internal Revenue Service (IRS) Code, section 501(c)(3), exempts charitable organizations and public and private foundations from paying taxes on net earnings. However, an organization must still pay some taxes. For example, a sales tax must be collected if the organization operates a gift shop. The IRS has many tax-exempt categories, covering social welfare organizations, such as the League of Women Voters (501,c,4), and even cemeteries (501,c,19).

When applying for tax-exempt status, financial data for the current fiscal year and the three preceding years will be requested. If the organization is just getting started, the current year's budget and a proposal for the next two years will be accepted. A form that fixes the organization's fiscal year—for example, July 1 to June 30—is also required.

To qualify for tax-exempt status, the organization must be operated for a charitable purpose. The exemption status is bestowed upon organizations that fulfill some of the following purposes: religious, charitable, scientific (research in the public interest), literary, educational, or testing for public safety. In addition, there are restrictions pertaining to making a profit from enterprises not directly related to the exempted purposes of the organization. These activities will be subject to the *unrelated business income tax* (UBIT). For example, if an arts organization starts acting as a travel agent and sells bookings for cultural cruises, the IRS might rule that this is unrelated to the organization's stated mission, and any surplus revenue from this activity would be subject to income taxes. Certain lobbying and propaganda activities are also prohibited.

A 501(c)(3) organization is not restricted from making a profit. As long as the profit making relates to the stated purpose of the organization, net earnings (profit after deducting expenses) may be accrued and retained. However, these earnings may not be distributed to members of the organization. Net earnings are usually placed in endowment funds or a restricted account.

As should be expected, the rules and regulations contain a lot of fine print. Hiring a lawyer, using legal services donated by a board member, or contacting the Volunteer Lawyers for the Arts is a prerequisite to filing the incorporation papers. Once an organization has attained the legal status to operate, it is obligated to provide reports and documentation to local, state, and federal agencies. The organization will also be required to file forms related to Social Security taxes and withholding taxes. The organization will also have to file tax forms with the IRS that list revenues, expenses, and changes in *fund balances* (the nonprofit organization's equivalency of worth). The details of the organization's liabilities, assets, programmatic activities, revenues, donations, and expenses for the previous four years must be filed every year.

A financial management information system and a person designated to oversee this important area become vital once the organization reaches the level of legal incorporation. The preparation of required reports—such as a balance sheet, a statement of account activity, and a financial statement of the worth of the organization—will be required. In addition, the organization's finances must be in order to the degree that an outside auditor could come in and analyze the financial operation. A complete audit, which could be very costly, is often required.

(Barbara Singer's *Nonprofit Organizations* was the source for much of this information. I recommend this book to anyone who is interested in a more detailed description of the legal issues involved in starting a nonprofit organization.)

Budgeting and Financial Planning

To attain a clear understanding of the entire financial management system, let's return to Figure 11–1. The top left box represents planning and budgeting as a key component in the FMIS. The budget is the framework around which the entire operation is organized. The decisions about how the organization will establish its budget and manage its income, expenses, and investments in a fixed time period are critical if artistic goals and objectives are to be met.

The Budget

As noted in Chapter 9, a budget is a quantitative representation of the organization's plans for a given period of time. Budgets usually cover what is called a *fiscal year* (FY), which can be any designated 12-month period of expense and revenue activity. Most financial planners suggest that organizations set their fiscal year around the programmatic profile of the organization. The IRS tax year, for example, is January 1 to December 31. However, a performing arts group with an eight-month season of October to May would probably find that a fiscal year of July 1 to June 30 better fits their programming and expense patterns.

The *summary budget* usually works well for smaller operations with a limited number of account lines. Figure 11–2 shows part of a proposed budget for a small theater group. This budget lists all of the income and expenses for the organization. In a typical fiscal year, the income and expense lines form the most active part of the budget. The budget does not reveal whether the organization is financially healthy. It simply tells what is expected in revenue and what is planned in expenses.

The *detailed budget*, shown in Figure 11–3, goes into more depth. Account numbers are assigned to the individual line items in the budget. The account numbering system helps identify the type of transaction that took place, and the system can be further broken down into account subcodes to provide as much detail as possible. For example, under the marketing budget (3000), there is a line item for a seasonal brochure (3010) and a spring brochure (3011). Figure 11–3 also shows in each line expenses incurred to date and the percentage of the budget expended. This information is very important to the person responsible for issuing budget reports in the FMIS. For example, the marketing area has used 68.8 percent of its budget to date. The information in the revenue lines shows that the theater company has only completed the first three shows of the season (see revenue for lines 0301-0303). Thus, the marketing department may not have enough left in its budget to finish the season. A quick review of the expenses for programs (3040) indicates that almost 80 percent of the budget has been expended only halfway through the season. In this case, the budget serves its purpose as a financial management information tool. Decisions must now be made to solve this budget problem in the marketing department.

Another way to organize a budget is along project or department lines. A *project budget* distributes revenue and expense centers across the organization. Figure 11–4 shows one way the theater company could express its budget by using this approach. In this example, the account titles are distributed across six major operational areas: regular season, touring, education, building fund, fund raising, and administration, and rents. This new distribution gives much more information about how the organization distributes its income and expenses. For example, it is possible to see that the regular season, touring, and fund raising are operating with a deficit (see account totals). A manager analyzing this budget as an FMIS tool would immediately notice that the organization is spending more than it raises on its fund raising activity. This should lead to a reorganization of the department and a reallocation of the budget to lower the costs of raising money.

The project budget is an excellent way to help with the fund raising needs of the organization. For example, the touring area could make a case for an additional $11,200 to achieve its objectives. If the $11,200 could be raised from outside sources, funds could be shifted to help pay for other priority projects that the organization has established in its planning process.

Another useful element of the project budget is its ability to show the proposed distribution of resources across the entire organization. The salary line for regular staff presents a clear picture of how much of the budget has been allocated to support the regular season or to fund raising. The FMIS becomes a tool for helping the manager evaluate whether the budget is best serving the organization. For example, in looking at this budget, the manager might wonder why so much of the organization's resources is required to put on the shows. The distrib-

Figure 11–2

SUMMARY BUDGET -- THEATER COMPANY

INCOME	90–91 ACTUAL	91–92 BUDGET	92–93 BUDGET	92–93 $CHANGE	92–93 %CHANGE
Ticket Sales	121,324	124,500	132,000	7,500	6.02%
Donations	68,459	73,000	77,100	4,100	5.62%
Foundations	18,500	12,600	13,250	650	5.16%
Arts Council	9,598	10,000	8,000	(2,000)	-20.00%
Concessions/Merchandise	4,521	4,500	5,000	500	11.11%
Program Advertising	9,365	11,000	12,500	1,500	13.64%
Total Revenue	231,767	235,600	247,850	12,250	4.94%

EXPENSES	90–91 ACTUAL	91–92 BUDGET	92–93 BUDGET	92–93 $CHANGE	92–93 %CHANGE
SALARIES					
Regular Staff	45,234	52,750	58,500	5,750	10.90%
Acting Co.	42,000	40,000	42,500	2,500	6.25%
Production	24,789	24,750	23,000	(1,750)	-7.07%
Subtotal Payroll	112,023	117,500	124,000	6,500	5.53%
Rents and Insurance	29,328	32,000	33,600	1,600	5.00%
Supplies	12,378	8,000	8,400	400	5.00%
Travel	4,234	2,500	2,750	250	10.00%
Entertainment	2,345	1,500	1,500	0	0.00%
Marketing	28,000	20,000	22,000	2,000	10.00%
Telephone	5,089	3,200	3,200	0	0.00%
Production	83,458	45,000	46,500	1,500	3.33%
Equipment Maintenance	2,748	5,000	5,000	0	0.00%
Truck	1,254	900	900	0	0.00%
Subtotal Expenses	168,834	118,100	123,850	5,750	4.87%
TOTAL PAY AND EXPENSES	280,857	235,600	247,850	12,250	5.20%
ACCOUNT TOTALS	(49,090)	0	0	$0	

uted budget shows $32,000 for regular staff salaries and another $18,000 for the production staff. The combined total of $50,000 is more than 23 percent of the entire budget. Further investigation may lead to the discovery that a staffing shift to the building fund raising campaign could help activate that department.

The project budget may also be broken down in various graphic representations to help make the FMIS more effective. This is especially important when a manager must work with a board of directors or a finance committee that is not close to the daily operational needs of the organization. The pie graphs (shown in Figures 11–5 though 11–7) illustrate how the revenue, expenses, and salaries are distributed by amount and percentage.

Figure 11–3

DETAILED BUDGET -- THEATER COMPANY

SAMPLE INCOME ACCOUNTS

ACCOUNT NUMBER	ACCOUNT TITLE INCOME	BUDGET	INCOME TO DATE	BALANCE	PERCENT OF INCOME
0300	SUBSCRIPTION SALES	52,000	58,000	(6,000)	111.5%
0301	SINGLE TICKETS SHOW 1	5,000	4,235	765	84.7%
0302	SINGLE TICKETS SHOW 2	8,000	7,980	20	99.8%
0303	SINGLE TICKETS SHOW 3	23,000	22,567	433	98.1%
0304	SINGLE TICKETS SHOW 4	6,000	1,234	4,766	20.6%
0305	SINGLE TICKETS SHOW 5	12,000	890	11,110	7.4%
0306	SINGLE TICKETS SHOW 6	9,000	567	8,433	6.3%
	SHOW REVENUE SUBTOTAL	**$115,000**	**$95,473**	**$19,527**	**83.0%**
0500	TOURING	17,000	0	17,000	0.0%
0600	INDIVIDUAL DONATIONS	43,500	17,689	25,811	40.7%
0700	FOUNDATIONS	13,250	0	13,250	0.0%
0800	ARTS COUNCIL	8,000	4,000	4,000	50.0%
0900	CONCESSIONS AND MERCHANDISE	5,000	1,678	3,322	33.6%
0950	PROGRAM ADVERTISING	12,500	12,500	0	100.0%
	OTHER REVENUE SUBTOTAL	$99,250	$35,867	$63,383	36.1%
	TOTAL REVENUE	**$214,250**	**$131,340**	**$82,910**	**61.3%**

SAMPLE OPERATING ACCOUNTS

ACCOUNT NUMBER	EXPENSES	BUDGET	EXPENSE TO DATE	BALANCE	PERCENT OF EXPENSES
3000	MARKETING DEPARTMENT				
3010	SEASON BROCHURE	6,200	7,359	(1,159)	118.7%
3011	SPRING SEASON BROCHURE	3,500	0	3,500	0.0%
3020	NEWSLETTER	2,950	1,545	1,405	52.4%
3030	POSTERS	1,000	467	533	46.7%
3040	PROGRAMS	3,500	2,780	720	79.4%
3050	FLYERS	450	250	200	55.6%
3060	PHOTOGRAPHY	2,000	1,267	733	63.4%
3100	ADMIN. SUPPLIES	1,000	769	231	76.9%
3200	MARKETING RESEARCH	1,400	700	700	50.0%
	SUBTOTAL – MARKETING	**$22,000**	**$15,137**	**$6,863**	**68.8%**
	TOTAL EXPENSES	**$214,250**	**$156,432**	**$57,818**	**73.0%**
	ACCOUNT TOTAL	**$0**	**$25,092**	**($25,092)**	

NOTE: ACCOUNT TOTAL EQUALS EXPENSES MINUS INCOME

Figure 11–4

PROJECT BUDGET -- THEATER COMPANY

ACCOUNT TITLE	REGULAR SEASON	TOURING	EDUCATION	BUILDING FUND	FUND RAISING	ADMIN. AND RENTS	TOTAL
INCOME							
TICKET SALES	115,000	17,000	0	0	0	0	$132,000
DONATIONS	10,000	5,000	2,500	10,000	16,000	33,600	$77,100
FOUNDATIONS	5,000	0	8,250	0	0	0	$13,250
ARTS COUNCIL	2,500	0	5,500	0	0	0	$8,000
CONCESSIONS	5,000	0	0	0	0	0	$5,000
PROGRAM ADS	8,500	4,000	0	0	0	0	$12,500
OTHER INCOME	0	0	0	0	0	0	$0
TOTAL INCOME	**$146,000**	**$26,000**	**$16,250**	**$10,000**	**$16,000**	**$33,600**	**$247,850**

EXPENSES SALARIES	REGULAR SEASON	TOURING	EDUCATION	BUILDING FUND	FUND RAISING	ADMIN. AND RENTS	TOTAL
REGULAR STAFF	32,000	5,000	4,500	1,000	12,500	3,500	$58,500
ACTING COMPANY	32,500	10,000	0	0	0	0	$42,500
PRODUCTION	18,000	3,000	0	0	0	2,000	$23,000
RENTS AND INSURANCE	5,600	5,600	5,600	5,600	5,600	5,600	$33,600
SUPPLIES	1,250	500	450	200	3,000	3,000	$8,400
TRAVEL	1,000	250	0	0	1,500	0	$2,750
ENTERTAINMENT	500	0	0	500	500	0	$1,500
MARKETING	14,000	5,000	0	2,000	1,000	0	$22,000
TELEPHONES	700	150	150	200	1,000	1,000	$3,200
PRODUCTION	38,500	6,500	1,000	500	0	0	$46,500
MAINTENANCE	2,000	500	0	0	0	2,500	$5,000
TRUCK	200	700	0	0	0	0	$900
SUBTOTAL EXPENSES	**146,250**	**37,200**	**11,700**	**10,000**	**25,100**	**17,600**	**$247,850**
ACCOUNT TOTALS	**($250)**	**($11,200)**	**$4,550**	**$0**	**($9,100)**	**$16,000**	**$0**

From the Budget to Cash Flow

The budget shows the manager the planned revenues and expenses for the fiscal year. The budgeting process that took us to the point of preparing a detailed budget still does not tell us if we have enough resources to operate the organization during the year. To make the budget work, a cash flow chart must be developed. It will provide a detailed look at the budget before the organization begins to spend money.

Cash flow statement It is possible to predict periods during the fiscal year when the organization may not have enough cash to pay its bills. Figure 11–8 shows the theater company's budget distributed over a fiscal year that begins in July and ends in June. The performances will take place October through December and February through April. The touring activity is scheduled for January and May.

In this example, the expense portion of the statement shows that the company will spend more than it makes in sales and donations for eight months out of the year. However, by carefully planning the use of cash reserves, the company should be able to make it through the year. For example, if the company starts the year with a cash reserve of

REVENUE DISTRIBUTION - THEATER COMPANY

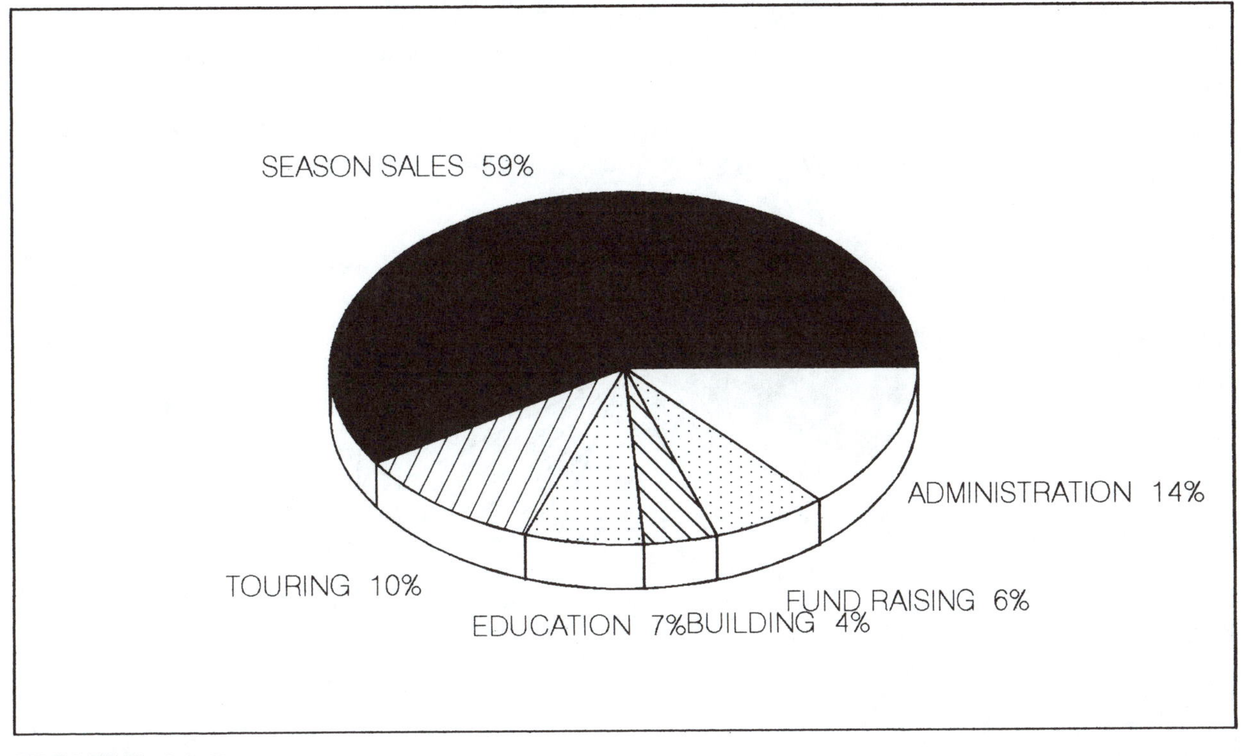

FIGURE 11-5

$29,600 (see Cash Projections section), it will have enough to get through the season. The bank account would zero out in October and be overdrawn in November by $4,775, but sales from the annual Christmas production and spring season subscriptions will quickly put the company back into a positive cash flow.

It is easy to see from this simple example how an arts organization can get into financial difficulty. What would happen, for example, if the actual ticket revenue for September and October came in $10,000 lower than projected? The company would be short $14,775 by November. A short-term loan from the bank would help the company meet its expenses, but it may cost the company another $300 to borrow the $14,775 for the month. The company will owe the bank $15,075, which it can pay back. By the end of the year, however, the cash reserves will be reduced.

Once an arts organization begins to find itself in a cycle of borrowing to make up cash flow and paying very high short-term interest rates, the erosion of the fiscal foundation begins. It may take two or three years, but overestimating revenue in combination with overspending will eventually lead to a deficit that could bankrupt the organization.

Base-line budgets In the previous chapter, the concept of fixed and variable costs was introduced. The total operating costs of the organization were derived from the combination of these two figures. The theater company's operating budget is an expression of these figures in the designated account lines. However, an arts organization is a type of

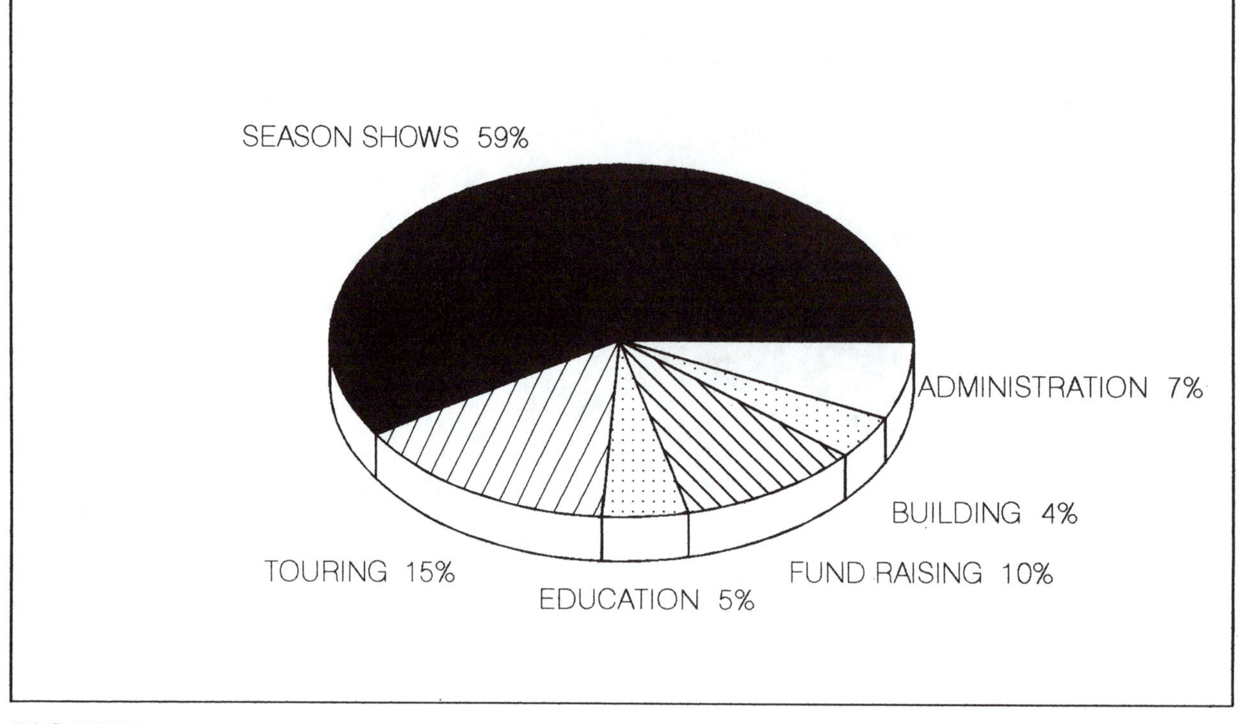

FIGURE 11-6

service business. Therefore, the distribution of fixed and variable costs will differ from that of a manufacturing enterprise. Identifying fixed costs becomes the starting point in the financial manager's quest to establish how much it will cost to run the arts organization.

Fixed and variable costs In economics, a *fixed cost* is "the cost of the inputs needed to produce any output and it is a cost that does not change when the outputs are changed."[5] For example, the theater company will have a fixed annual cost for leasing office space. It does not matter how many productions the company mounts, it will still have to pay the rent. The same applies to the shop space it rents to build the scenery and costumes. A case can also be made for budgeting a core staff, payroll taxes, insurance, minimal office supplies, telephone lines, and specific equipment charges as part of the fixed costs of the organization. Without these core elements, it would be impossible to operate the organization. It is obviously in the organization's best interests to control these core costs carefully.

After the organization has designated the fixed costs, the *variable costs* are any other costs required to run the company. The variable costs relate directly to the myriad of decisions that the theater company made about the season titles, the scale of production, the marketing campaign, and so forth, as part of the planning process. If the company decides to produce six shows, all of the variable costs related to supporting that effort should be reflected in the budget.

Figure 11–9 demonstrates the relationship between fixed and

DETAILED EXPENSE DISTRIBUTION - STAFF SALARIES

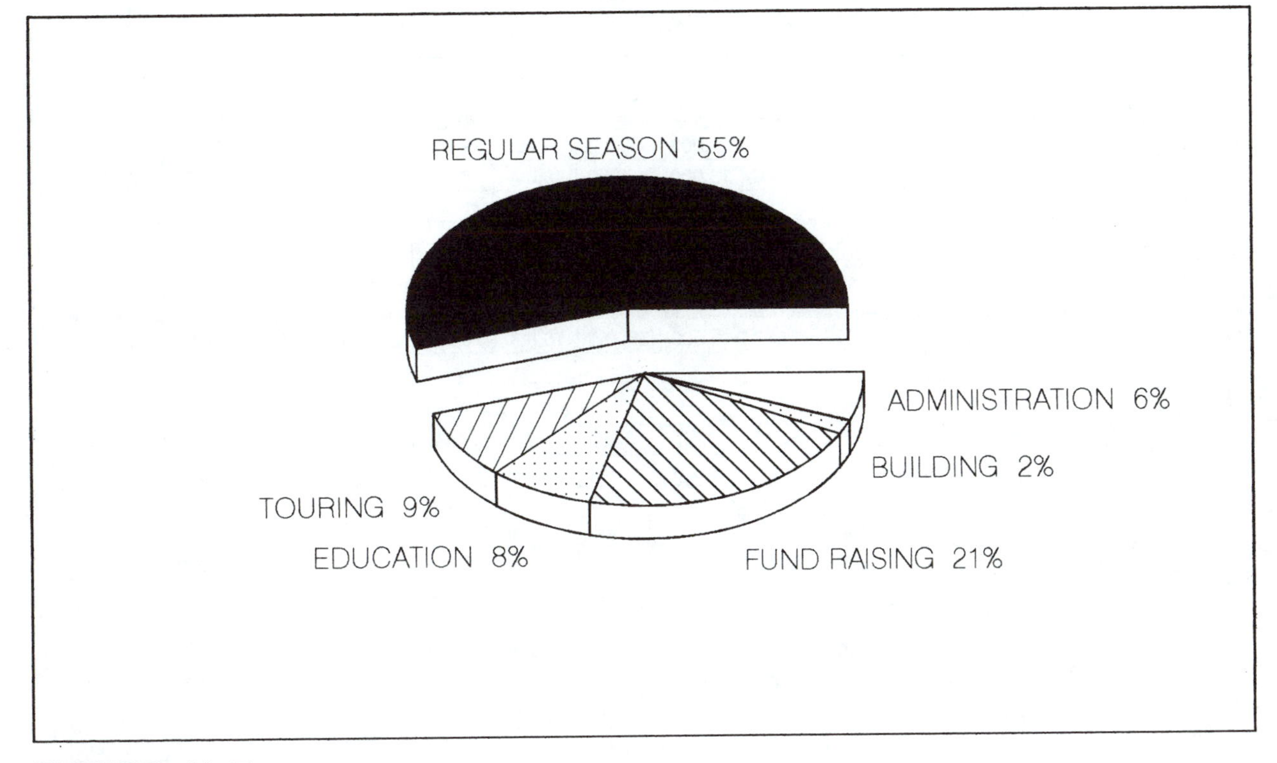

FIGURE 11-7

variable costs. In this case, the fixed cost base is set at $60,800. As each production is added to the season, the variable costs rise. By the time we reach the sixth show, we have accumulated combined fixed and variable costs of $247,500.

This chart is intended for illustration only. The details of working out fixed and variable costs would be much more complicated. In fact, the financial manager would probably be expected to calculate the variable and fixed costs if the company added an extra performance of a show.

Until an organization has a clear picture of all of its fixed and variable costs and understands how to adjust those figures to budget realistically, the financial management system will experience recurring problems. In the long run, the chances are good that the company will find itself growing into a larger deficit each year. Once an organization reaches its debt threshold, it becomes very difficult to pay back the borrowed money and continue to meet its stated objectives. The news items in this chapter are examples of the very real impact of inadequate financial management.

Developing a Financial Management Information System

The financial management information system illustrated in Figure 11–1 requires that data from the operating system be transmitted to the records system. These data are then assembled in various reports used by the different departments in the operation. To report accu-

Figure 11–8

CASH FLOW PROJECTIONS -- THEATER COMPANY

INCOME

TITLE	JULY	AUG	SEPT	OCT	NOV	DEC	JAN	FEB	MAR	APR	MAY	JUNE	TOTAL
TICKETS SALES	11,000	15,000	17,000	10,000	8,000	23,000	17,000	6,000	12,000	9,000	500	3,500	132,000
DONATATIONS	2,800	3,200	2,500	3,200	8,500	13,000	2,200	8,800	8,000	12,000	7,900	5,000	77,100
FOUNDATIONS	0	0	0	0	0	0	13,250	0	0	0	0	0	13,250
ARTS COUNCIL	0	0	0	0	2,500	0	0	0	5,500	0	0	0	8,000
CONCESSIONS	0	0	0	500	500	2,500	0	500	500	500	0	0	5,000
PROGRAM ADS	0	0	0	0	0	7,000	4,000	1,500					12,500
TOTAL INCOME	13,800	18,200	19,500	13,700	19,500	45,500	36,450	16,800	26,000	21,500	8,400	8,500	247,850

EXPENSES

TITLE	JULY	AUG	SEPT	OCT	NOV	DEC	JAN	FEB	MAR	APR	MAY	JUNE	TOTAL
STAFF	4,875	4,875	4,875	4,875	4,875	4,875	4,875	4,875	4,875	4,875	4,875	4,875	58,500
ACTORS	0	0	3,000	4,000	8,000	8,500	5,000	4,000	6,000	4,000	0	0	42,500
STAFF	0	2,400	2,400	2,400	2,400	2,900	2,400	2,400	2,400	2,400	900	0	23,000
INSURANCE	2,800	2,800	2,800	2,800	2,800	2,800	2,800	2,800	2,800	2,800	2,800	2,800	33,600
SUPPLIES	2,000	2,000	500	500	225	200	200	200	250	325	1,000	1,000	8,400
TRAVEL	0	0	0	2,000	0	0	0	0	0	750	0	0	2,750
ENTERTAINMENT	0	0	0	500	200	200	0	200	200	200	0	0	1,500
MARKETING	4,000	2,000	2,000	2,000	1,000	1,000	2,000	1,500	1,500	1,000	2,000	2,000	22,000
TELEPHONES	200	200	200	200	300	300	300	300	300	300	300	300	3,200
PRODUCTION	0	15,000	8,500	6,000	4,000	0	5,000	5,000	3,000	0	0	0	46,500
MAINTENANCE	500	500	400	400	400	400	400	400	400	400	400	400	5,000
TRUCK	75	75	75	75	75	75	75	75	75	75	75	75	900
SUB TOTAL	14,450	29,850	24,750	25,750	24,275	21,250	23,050	21,750	21,800	17,125	12,350	11,450	247,850
BALANCE	(650)	(11,650)	(5,250)	(12,050)	(4,775)	24,250	13,400	(4,950)	4,200	4,375	(3,950)	(2,950)	0

CASH FLOW PROJECTIONS

RESERVE	29,600	28,950	17,300	12,050	0	(4,775)	19,475	32,875	27,925	32,125	36,500	32,550	END
BALANCE	28,950	17,300	12,050	0	(4,775)	19,475	32,875	27,925	32,125	36,500	32,550	29,600	BALANCE
	JULY	AUG	SEPT	OCT	NOV	DEC	JAN	FEB	MAR	APR	MAY	JUNE	

FIXED AND VARIABLE COSTS -- THEATER COMPANY

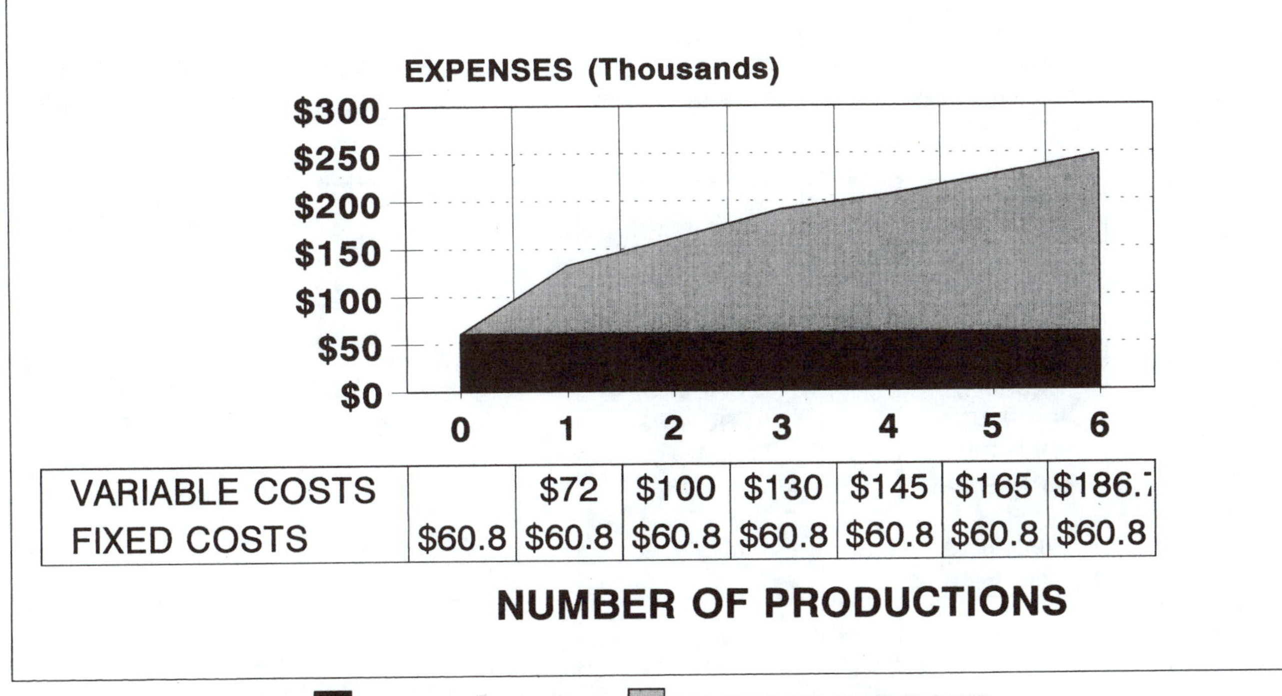

	0	1	2	3	4	5	6
VARIABLE COSTS		$72	$100	$130	$145	$165	$186.7
FIXED COSTS	$60.8	$60.8	$60.8	$60.8	$60.8	$60.8	$60.8

FIGURE 11-9

rately on the fiscal activity of the organization, an accounting and record-keeping system must be in place. Figure 11–10 shows one version of an accounting system. In this case, the flow is from the top to the bottom of the page.

The personal computer and inexpensive accounting software has had a tremendous impact on arts organizations. Inexpensive systems are available to help organizations quickly enter data and print reports. However, due to the nature of nonprofit businesses, different reporting formats are required when discussing how much the "business" is worth and how it is using its assets.

Accounting

Accounting is defined as identifying, collecting, analyzing, recording, and summarizing business transactions and their effects on a business. A *transaction*, which is a key element in this definition, is an exchange of property or services. *Bookkeeping* involves the clerical work of recording the transaction. In a sense, the accountant begins where the bookkeeper leaves off by summarizing and interpreting the records or books. In many cases, the bookkeeping and accounting are done by the same person.

The Financial Accounting Standards Board (FASB) oversees the practices of the profession and regularly updates what are referred to as the generally accepted accounting principles (GAAP).

Figure 11–10

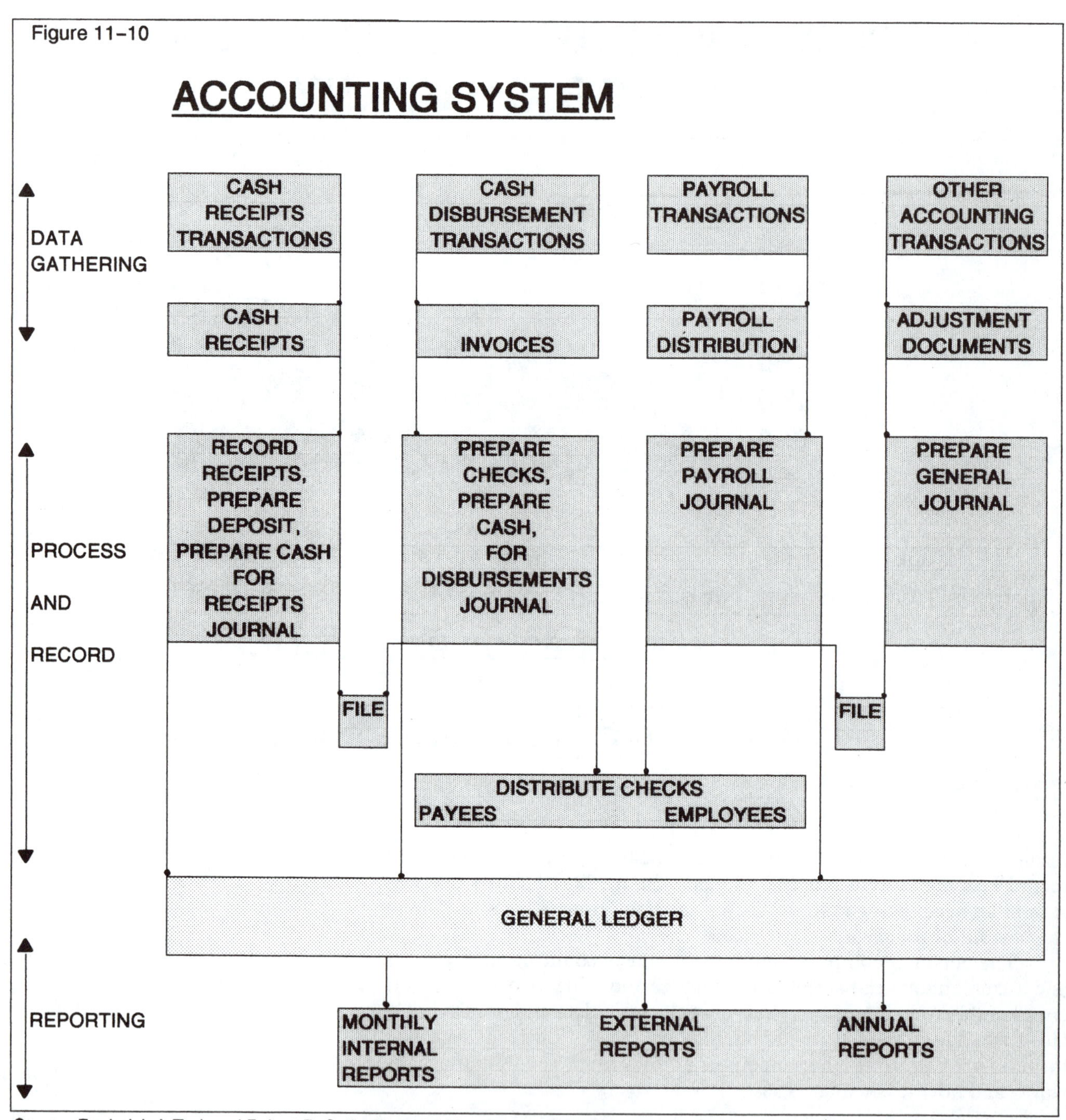

Source: Frederick J. Turk and Robert P. Gallo, Financial Management Strategies for Arts Organizations, (New York: American Council for the Arts, 1984). Copyright (c) 1984 by the American Council for the Arts. Used with permission.

Accounting has evolved its own specialized language to describe transaction activity. Let's look at a few key terms.

Cash-based accounting A personal checking account is one example of a cash-based accounting system. You make a deposit, you write checks, and at the end of the month, you have a positive or negative balance. The major problem with the cash-based system is that it gives you no information about how much you are worth, how much you owe, or how much is owed by others.

Accrual-based accounting The accrual system recognizes expenses when they are incurred and income when it is committed. The primary advantage of this system is that it shows future commitments and how much money you owe. When you charge something to a credit card, you are using an accrual system. You will have to make a future payment to an ongoing account in your name.

An organization typically opens accounts with several different vendors. Office supplies, lumber, and paint are purchased with the understanding that the organization will pay for the materials when they are billed. For example, when $250 worth of office supplies are purchased from a local supplier, the $250 becomes a *payable*, and when the bill is paid, it becomes an *expense* item in the budget. A foundation grant for $5000 awarded to the organization is called a *receivable*. When the check arrives and is deposited, it becomes *revenue*. The accounting system deducts the payables, adds the receivables, and arrives at the balance with which the organization will operate.

Fund-based accounting Because nonprofit organizations receive money in such forms of income as grants and gifts, the FASB has a special set of standards for nonprofit organizations to use to report the overall account activity. Fund accounting is a system of classifying resources into activities or projects so that each area can accurately report account activity during a fiscal year.

Frederick J. Turk and Robert P. Gallo note that "the fundamental principle in fund accounting is stewardship," and the "funds are accounted separately according to restrictions established by donors" or the board.[6] In Figure 11–4, for example, the activities related to the regular season, touring, education, and the building fund could all be established as fund balance accounts. Each account becomes a separate entity and may be designated as either a restricted or an unrestricted fund. A *restricted fund* is designed to control how the funds are used from a specific account. For example, if an organization has an endowment fund, the money received or expenses incurred will need special approval. In the project budget shown in Figure 11–4, the building fund would be a restricted account because gifts and expenses would be intended for a specific purpose. The regular season fund, on the other hand, would be an *unrestricted account*, meaning that funds could be expended for a wide variety of organizational activities, as approved by the board.

The Accounting system To extract the information required for an effective FMIS, the accounting input must be as detailed as possible. Figure 11–10 shows how a typical system takes the input of cash receipts, invoices, payroll, and other documentation and processes the data to produce monthly external and annual reports.

In the News—Financial Reality and Dance

Here is an example of a dance company that does not have enough money to finish the season or to make the payroll.

Dancers Shun Pay, Hope to Save Ballet
Associated Press

PHILADELPHIA—The Pennsylvania Ballet, two weeks away from closing, hopes to get $2.5 million in donations to keep alive.

The troupe's dancers and staff said they would work without pay and appeal to audiences for donations rather than accept trustees' decision to suspend operations.

Board Chairman Edwin E. Tuttle said trustees made their decision Friday. The dancers said they voted unanimously Sunday to waive payment for the next production, which opens tomorrow. They were joined by the musicians, stagehands and other staff, about 200 people in all.

SOURCE: Associated Press, "Dancers Shun Pay, Hope to Save Ballet." March 13, 1991.

And More . . .

Despite the positive spin the artistic director tries to put on the announced season in the following article, cutting the company size by more than 25 percent and reducing the budget by $1.4 million represent significant changes brought on by financial management problems.

Ballet to Pare Season, 10 dancers
Wilma Salisbury

CLEVELAND—Cleveland Ballet will reduce the size of the company and present fewer performances at the State Theater next season in order to operate within the constraints of a $5.1 million budget.

The 38-member company will be cut to 28 dancers. There will be no new productions and no new choreography by Artistic Director Dennis Nahat.

The season will consist of two repertory programs, a full-length abstract ballet and "The Nutcracker." All will be on the subscription program. . . .

"I don't think of it as downscaling," Nahat said of the plans he announced yesterday for the 1991–92 season. "We are not pulling back on quality. We are not moving in another direction. The audience will not be getting less. They will be getting the real roots of the company. . . .

Cleveland Ballet began its 1990–91 season with a budget of $6.5 million and an accumulated deficit of $2.4 million. Under pressure from funders, the company announced in October that it would cut its budget across the board by $1 million. The Ohio Chamber Orchestra musicians volunteered to play at half-salary for one week. The dancers agreed to give up two weeks of their 38 week contract.

SOURCE: Wilma Salisbury, "Ballet to Pare Season, 10 Dancers," *Cleveland Plain Dealer*. May 10, 1991.

When an organization establishes its accounting system, specific account numbers and designated subcodes are used to identify *liabilities* (money owed or funds committed) and *assets* (property or resources owned by the organization). When expenses are incurred, they are recorded as liabilities that reduce the organization's overall worth. When money is received in the form of revenues or gifts, it is classified as an asset that adds to the organization's overall worth.

The accounting system shown in Figure 11–10 identifies a data-gathering area where cash is received, checks are written (disbursement transactions), payroll is processed, and all other account activity takes place. A general ledger processes the four other journals (receipts, disbursements, payroll, and general journal) and allows reports to be assembled that provide the feedback required by the management and the board. Computers and software automate the bookkeeping to the point that the journal and ledgers are modules within the software. However, all accounting systems still generate a tremendous amount of paperwork. Check stubs, invoices, receipts, purchase orders, requisitions, credit card slips, and so forth, can quickly overwhelm an organization. Organizations are required to keep all of this documentation—it is sometimes called a *paper trail*—to complete successfully a full audit or a simple review by an outside accounting firm.

Accounting formula The record-keeping system used in accounting is based on the simple formula that assets must equal the liabilities plus the fund balance (A = L + FB), and the fund balance must equal the assets minus the liabilities (FB = A - L). In Figure 11–11, an individual starts with $5000 in cash for a nonprofit consulting business. The accounting formula expresses this relationship as $5000(A) = $0(L) + $5000(FB). The assets total $5000 and because no debts have been incurred, there is $0 in liabilities and $5000 in the fund balance. (In the profit sector, the fund balance would be called *equity*, and the formula would read A = L + E.)

The consultant buys a computer system for $2000 on credit. The business takes on a liability it must pay. Figure 11–11 shows the relationship between assets, liabilities, and the fund balance after purchasing the computer. The cash assets are reduced by $2000 and the liabilities are increased by $2000. To recalculate the fund balance of the business, we apply the formula. In this case, we find a fund balance of $3000 by subtracting liabilities from assets (FB = A – L).

Financial Statements

So far, we have been looking at accounting documentation from the FMIS that was designed for internal use by the staff and the board. These various budget reports do not tell us about the overall fiscal health of the organization. Does the theater company have any money in the bank? How much does it owe in short- or long-term debt? To locate this information, we need to review two reporting formats in the accounting system: the balance sheet and the statement of activity.

Balance sheet The balance sheet shown in Figure 11–12 is an example of the A = L + FB formula. The top half of the balance sheet details the current and previous years' assets as of June 30. The bottom half of the sheet shows the organization's liabilities and fund balances. On this day, June 30, 1992, the organization had $29,600 in cash, $4292 in accounts receivable, and so forth. In comparison to the previ-

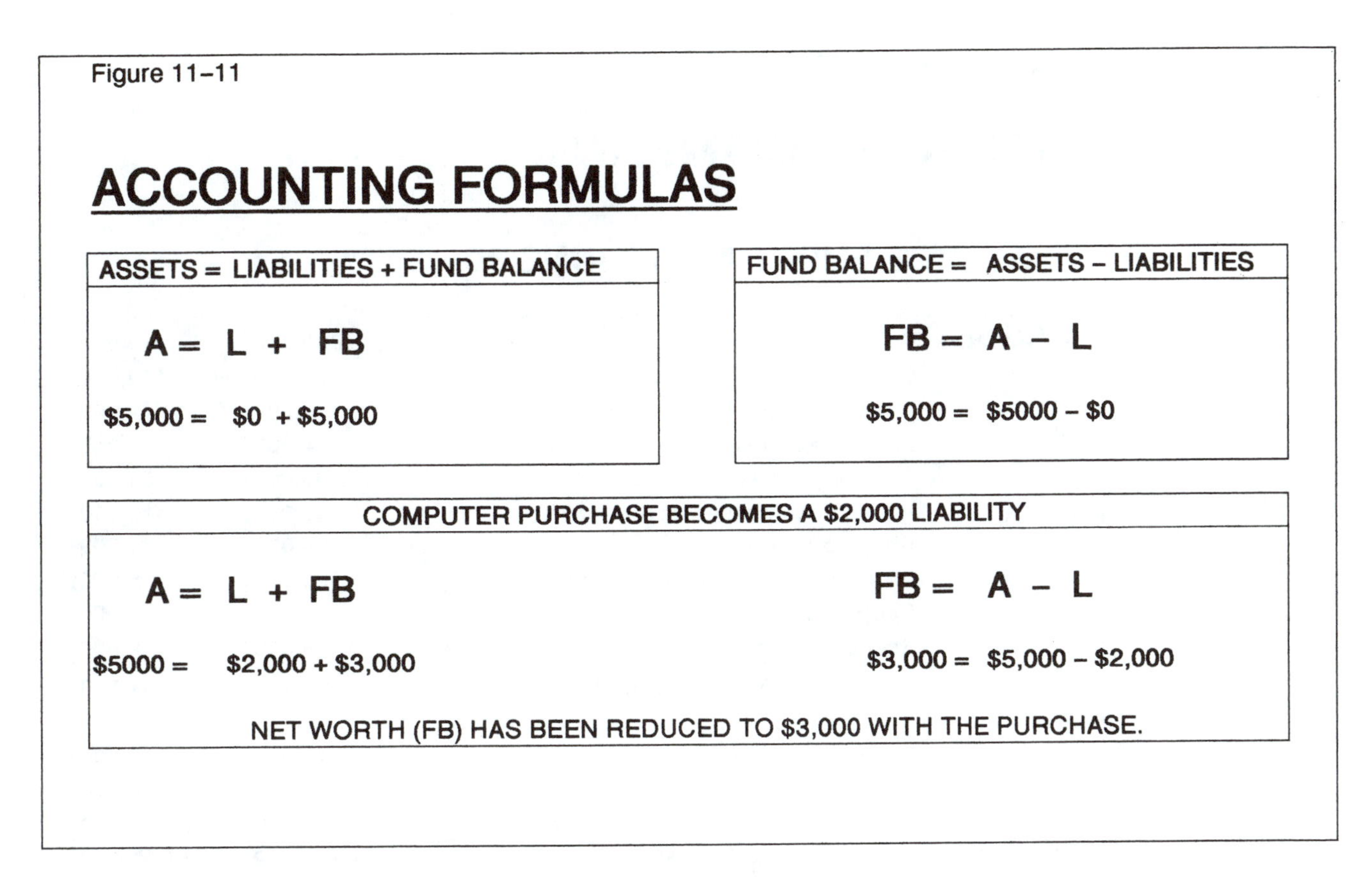

ous year, the organization has more in total assets in 1992 ($385,292) than it had in 1991 ($378,410). However, the organization's liabilities also increased by $23,500. Accounts payable has increased by $29,500 this year. The balance sheet also tells us that the organization decreased its operating deficit from $49,090 to $26,618 this year. Unfortunately, the deficit reduces the organization's overall fund balance by this amount. The theater company must use its operating endowment as a cash reserve to cover the deficit.

The good news is that the company managed its funds well because the endowment fund earned $27,000, or almost 8 percent this year. The bad news is that the $27,000 was used to cover the deficit. The company did manage to reinvest $15,000 in the endowment, but the deficit and the reinvestment still translate to a net loss of $12,000 to the fund. Therefore, the overall financial worth of the organization has decreased.

Statement of activity The second of the financial reports is the statement of activity and changes in fund balances. The example shown in Figure 11–13 makes it clear where the deficit came from and what changes took place in the fund balances.

The revenue section tells us how much money came into the organization from various sources. In comparison to last year, the overall revenue rose by $5499, or a modest 2 percent, and expenses fell by $16,973, or 6 percent. Unfortunately, the theater company still spent $26,618 more than it made.

The operating fund balance decreased as a result of the previous year's debt. The beginning balance for the operating endowment in

Figure 11–12

THEATER COMPANY

BALANCE SHEET AS OF JUNE 30

ASSETS	6/30/92	6/30/91
Cash	$29,600	$12,000
Accounts receivable	4,292	2,445
Prepaid expense	4,800	3,245
Office equipment	38,000	34,000
–Less accumulated depreciation	(15,000)	(11,000)
Shop equipment	34,000	31,000
–Less accumulated depreciation	(20,000)	(15,000)
Shop inventory	13,400	10,000
Scenery and costumes	26,500	19,500
Subtotal Assets	**$115,592**	**$86,190**
Endowment earnings	27,000	22,000
Investment Funds	242,700	270,220
Subtotal	**$269,700**	**$292,220**
TOTAL ASSETS	**$385,292**	**$378,410**

A = L + FB

FB = A − L

LIABILITIES	6/30/92	6/30/91
Accounts payable	$105,500	$76,000
Notes payable	15,000	25,000
Endowment reinvestment	15,000	11,000
TOTAL LIABILITIES	**$135,500**	**$112,000**

FUND BALANCES	6/30/92	6/30/91
Operating excess (deficit)	($26,618)	($49,090)
Operating endowment	245,910	295,000
Building fund	10,000	5,000
Restricted gift fund	12,000	8,000
Education fund	8,500	7,500
TOTAL FUND BALANCE	**$249,792**	**$266,410**
TOTAL LIABILITIES AND FUND BALANCE	**$385,292**	**$378,410**

Figure 11–13

STATEMENT OF ACTIVITY AND CHANGES IN FUND BALANCES

REVENUES

	6/30/92	6/30/91
TICKET SALES	$125,436	$121,324
DONATIONS	75,478	68,459
GRANTS	20,500	28,098
CONCESSIONS	4,867	4,521
ADVERTISING	10,985	9,365
TOTAL	$237,266	$231,767

EXPENSES

	6/30/92	6/30/91
SALARIES	$121,786	$112,023
RENTS/INSURANCE	34,500	29,328
SUPPLIES	8,965	12,378
TRAVEL	1,965	4,234
ENTERTAINMENT	4,865	2,345
MARKETING	29,327	28,000
TELEPHONE	6,876	5,089
PRODUCTION	49,654	83,458
EQUIPMENT MAINTENANCE	4,790	2,748
TRUCK	1,156	1,254
TOTAL	$263,884	$280,857
EXCESS (DEFICIT)	($26,618)	($49,090)

FUND BALANCE CHANGES

FUND BALANCE CHANGES	6/30/92	6/30/91
Operating fund balance	$245,910	$295,000
Excess (deficit)	(26,618)	(49,090)
Other Endowments and Restricted gift funds	30,500	20,500
Transfers/earnings	27,000	22,000
END OF YEAR FUND BALANCE	$276,792	$288,410

Figure 11 – 14

RATIO ANALYSIS

1. Ratio of Expendable Assets to Total Liabilities

EXPENDABLE ASSETS (A) / TOTAL LIABILITIES (B)	
	1:1 OR GREATER MEANS MORE ASSETS THAN LIABILITIES
	0.3:1 OR LESS MEANS CASH FLOW PROBLEMS

CURRENT FISCAL YEAR			LAST FISCAL YEAR		
29,600	CASH		12,000	CASH	
4,292	ACCTS RECEIVABLE		2,445	ACCTS RECEIVABLE	
4,800	PREPAID ACCOUNTS		3,245	PREPAID ACCOUNTS	
$38,692			$17,690		
$38,692 / $135,500	(A) / (B)	EQUALS A RATIO OF 0.28:1	$17,690 / $112,000	A / B	EQUALS A RATIO OF 0.15:1

IN BOTH FISCAL YEARS THE THEATER COMPANY DID POORLY IN THIS AREA.
THE RATIO PROBLEM BECAME WORSE IN THE CURRENT FISCAL YEAR.

2. Ratio of Expendable Fund Balances to Total Expenses

EXPENDABLE FUND BALANCES (C) / TOTAL EXPENSES (D)

NOTE:
IN THE CURRENT FISCAL YEAR THE RATIO DECREASED, INDICATING LESS RESERVES WERE AVAILABLE.

CURRENT FISCAL YEAR			LAST FISCAL YEAR		
245,910	OPERATING ENDOW		295,000	OPERATING ENDOW	
10,000	BUILDING FUND		5,000	BUILDING FUND	
8,500	EDUCATION FUND		8,000	EDUCATION FUND	
$264,410			$308,000		
$264,410 / $263,884	(C) / (D)	EQUALS A RATIO OF 1:1	$308,000 / $280,857	(C) / (D)	EQUALS A RATIO OF 1.09:1

1991 was $295,000. Because of the deficit of $49,090, the fund balance was reduced to $245,910 at the beginning of the 1992 fiscal year.

Ratio analysis Ratio analysis is a quick way of examining the organization's balance sheet and statement of activity. These ratios, and dozens of others, are detailed in Chapter 8 of Turk and Gallo's book.

Figure 11–14 compares two key indicators of the theater company's fiscal health. The ratio of expendable assets to total liabilities reveals that in the current year, the theater company has a problem. Turk and Gallo recommend that organizations have at least a 1:1 ratio of assets to liabilities. The balance sheet may show that assets equal liabilities and fund balances, but this ratio reveals that there is not enough cash on hand to pay the bills.

The ratio of expendable fund balances to total expenses for the current year is 1:1. The situation was not much better last year with a 1.09:1 ratio. Turk and Gallo indicate that ratios greater than a 0.3:1 are a reason to be concerned.

The operating fund balance represents the only large source of money available to keep the organization alive during difficult economic periods. Ideally, the theater company's operating endowment should be around $900,000 to provide the necessary cushion. That amount would give the company a 0.29:1 ratio, using the current year's expense totals.

Problem areas If, as a member of the board of directors of the theater company, you read the balance sheet and statement of activity, what would be your reaction? In this case, the financial management information system reveals that the organization has serious cash flow problems and a decreasing endowment. The reports also show that you have very little cash on hand and that most of your assets are not liquid. *Liquid assets* simply means cash or an asset that can be easily converted to cash. On paper, the organization has $385,292 in assets. However, $242,700 of that is money committed to long term investments, and $70,000 is committed to equipment, stock, and inventory. It is not easy to turn investments, the inventory of a shop, or a rack of period costumes into cash.

The immediate crisis facing the theater company will be meeting the accounts payable liability of $105,500 with $29,600 in cash and $4200 more expected in receivables. The balance sheet tells the board and the staff that there will be a shortage of $71,700 on the first day of the new fiscal year. On top of that, the company still owes $15,000 in accounts payable to the local bank (see "Notes Payable," Figure 11–12).

The other problem area is the shrinking operating endowment. Without this endowment the company has no future. If it spends more than it earns each year, it will eventually zero out the account. As we shall see in Chapter 13, having a large endowment fund is not a hindrance to raising money. In fact, funders like to see organizations with reserves.

Solutions The theater company has its work cut out for it. The FMIS has made it clear that the company has difficulty keeping to a budget. By comparing the proposed 1991-1992 budget shown in Figure 11–2 with the statement of activity shown in Figure 11–13, we see that although revenue was $1666 higher than budgeted, expenses were

$28,284 higher than budgeted. The previous year's deficit was even worse: $49,090. A very high priority for theater management and the board will be to get the cost centers under control. The protection and development of the operating endowment fund will also be required. The discipline of controlling costs and a more aggressive reinvestment policy should rebuild the endowment.

Investment

One of the most important responsibilities of a financial manager in a nonprofit arts organization is to work with the board of directors to develop investment strategies that will help ensure the survival of the enterprise. Earlier in this chapter, we discussed the responsibility of the stewardship of the funds for an organization. This *fiduciary responsibility*, or trusteeship of the organization's resources, is the fundamental role of the board of directors. To exercise this fiduciary responsibility, the board must have accurate information about the assets and liabilities of the organization. At the same time, a finance committee should work with the financial manager to seek out ways to maximize the assets available. The management of the various fund balances of the organization must be monitored closely by the board.

In *Financial Management Strategies for Arts Organizations*, Turk and Gallo detail the best way to approach the *asset management* of the organization.[7] Their suggestions include developing a system for managing the cash available to the organization. Cash control can be realized through limiting access to cash accounts, timing the payment of bills, and closely monitoring the organization's accounts receivable. They recommend that the organization invest in a mix of short-term interest-earning instruments, such as certificates of deposits (CDs), treasury bills (T bills), and money market funds. The objective of short-term investment is to maximize the interest-rate return while maintaining quick access to the funds should an emergency arise.

The board of directors also needs a long-term investment policy. One strategy is to take a conservative approach and invest in a variety of well-known stocks and bonds. Higher risk investments will usually provide greater returns, but board guidance is necessary to ensure that the financial manager invests the organization's assets safely.

Finally, Turk and Gallo urge managers to monitor the fixed assets of the organization, including inventories and expensive capital equipment and property. Maintaining sufficient insurance for equipment, inventories, and buildings help minimize the financial risk to the organization if a disaster strikes.

Conclusion

It is important to remember that the FMIS is a tool designed to help the organization fulfill its mission. Maximizing revenue, building the endowment, controlling cash flow, and conducting ratio analyses are important, but these activities should not be the organization's controlling force. The primary objective of the FMIS is to support the organization in achieving its artistic goals. If the FMIS is working effectively, it should help, not hinder, goal achievement. It should also be remembered that the system is only as good as the people who use it. Inexperienced or poor management, not the FMIS, usually leads an organization into a financial crisis.

For example, if the artistic director insists that the only way to attain a high level of production quality is to build the costumes and sets in the company's own shops, it is the financial manager's responsibility to explain how much this will cost. If it can be done with the resources available in the short and long run and the board approves the idea, then it should be done. If the organization cannot afford to build its own costumes and sets, the financial manager has an obligation to say so. Unfortunately, this scenario is not always followed. Instead, either the manager cannot stand up to the artistic director, or the cost analysis is done without a real understanding of the organization's fixed and variable costs. When poor management is combined with a lack of understanding about the fiscal health of the organization, even the most sophisticated FMIS will not help.

Managing Finances and the Economic Dilemma

Without cash coming in, there is little chance that the show will go on or that the exhibit will open. The noble artists and staff can come to the rescue by working for free or for half pay once in a while, but in the long run, trying to operate an arts organization by using a crisis financial management technique will spell doom. Limited productivity and ever-increasing labor costs means that all arts organizations must adapt a flexible plan to keep up with the income gap. For example, if we take the hypothetical theater company's budget for 1992–1993 of $247,850 and project the numbers through 1999–2000 at an inflation rate of 5 percent, a budget of $348,750 would be required to do the same six-show season. In reality, the cost increases that the company will face will probably exceed 5 percent per year. Salaries, artist fees, construction materials, rents, and so on, will all increase at rates that are very difficult to project accurately.

Without careful planning, the organization will be susceptible to this familiar cycle: a growing deficit leads to a budget crisis, followed by a high visibility fund raising campaign and a last-minute rescue by a group of donors.

These realities make it all the more important for arts organizations to have at least a five-year plan that establishes expense and revenue projections. Longer-range plans make sense for organizations with very high fixed costs, such as museums. The changing economic conditions will require that arts managers and board members revise the financial plans at least every year. In times of recession or when inflation rates begin to exceed 6 percent per year, organizations may need to revise the long-term plans every quarter.

Looking Ahead

In the next two chapters, we will study various approaches to maximizing the financial resources needed to realize the organization's goals and objectives. We will look at how the key areas of marketing and fund raising relate to the FMIS of the organization.

Summary

Financial management is usually supervised by a business manager under the supervision of a general manager or managing director. The financial management information system is a comprehensive investment, reporting, control, and processing system designed to realize the organization's objectives.

In the profit sector, the financial manager is responsible for forecasting, investing, and control. He or she is primarily concerned with increasing owner and shareholder wealth.

Nonprofit financial management implies a fiduciary responsibility for the fund balances of the organization. The term *fund balance* is the nonprofit sector's term for the equity, or the financial worth, of the organization.

A nonprofit organization must incorporate to be legally recognized in a state. Tax exemption will be granted only if the organization can meet the legal conditions set down by the federal and state governments. Even tax-exempt organizations must still pay some payroll and social security taxes. Any revenue earned from business operations that are not directly related to the main mission of the organization is also taxable.

The key to a successful FMIS is financial planning through a realistic budget. Budgets for small organizations simply list revenues, expenses, and the balance or deficit that remains after operations. Larger organizations usually distribute the budget across revenue and expense centers. Each area is budgeted as if it were an independent fund balance.

After a budget is created, the organization estimates the cash flow required to meet the budget during the fiscal year. The organization must establish its baseline budget of fixed costs and then analyze the variable costs of operation.

The FMIS depends on accurate data processing of all revenue and expense transactions. Arts organizations are expected to follow accepted accounting practices so that they may be audited by outside accounting firms.

Most organizations use accrual-based accounting, rather than cash-based accounting, to keep track of their transactions. The accrual system records receivables and payables, which are later recognized as revenues and expenses. The system allows financial statements about the organization to be developed. The accounting system is composed of three main components: assets, liabilities, and fund balances. The system is based on a simple formula: The assets of an organization equal the liabilities plus the fund balance.

The accounting system allows data to be collected for internal and external reports. The balance sheet shows what assets, liabilities, and fund balances the organization has. The statement of activity shows in more detail how the revenue, expenses, and fund balances were used during the fiscal year. A ratio analysis helps pinpoint problems in such areas as the relationship of assets to liabilities and of fund balance to expenses.

Organizations should develop short- and long-term investment plans to manage their assets. Certificates of deposit, treasury bills, and, money market funds are typical short-term investment tools. Stocks and bonds can be used for long-term investments.

Key Terms and Concepts

Financial management information system (FMIS)
Incorporation
Tax exemption
Section 501(c)(3) of the IRS Code
Unrelated business income tax (UBIT)
Account number
Account line items
Cash flow statement
Base-line budget
Fixed and variable costs

Cash-based accounting
Accrual-based accounting
Fund-based accounting
Payables
Receivables
Restricted and unrestricted funds
Liabilities
Assets
Accounting formula
Financial statements
Balance sheet
Statement of activity
Ratio analysis
Liquid assets
Fiduciary responsibility
Asset management
Short- and long-term investments
Fixed assets

Questions

1. What are the key components of an FMIS?
2. What are the three major responsibilities of a financial manager in a for-profit business?
3. What are two advantages of incorporating?
4. What is the importance of Section 501(c)(3)?
5. Define the term *budget*.
6. What is the major advantage of the project, or department budget?
7. What is the main purpose of a cash flow statement?
8. Why is it important to establish a base-line budget?
9. Define the term *accounting*.
10. Are there situations in which both cash-based and accrual-based accounting systems might be used in an organization?
11. Why is it better for organizations to have unrestricted, rather than restricted, fund balances?
12. What is the accounting formula, and what does it reveal about an organization's fiscal health?
13. Solve for the ratio of expendable assets to total liabilities and expendable fund balances to total expense and explain whether the organization has a financial problem. The organization has $250,000 in expendable assets, $125,000 in total liabilities, $1.25 million in expendable fund balances, and $225,000 in total expenses at the end of a fiscal year.

Case Study

Dance Company Financial Report

The *Wing and a Prayer Dance Company* just finished its third season of operation. The company performs four weekends distributed across one weekend in October, December, February, and April. They perform a Thursday to Saturday schedule in a thousand-seat theater. The single ticket prices are $10 and $16, and the series subscription prices to the four weekends are $32 and $52. During this last season, 60 percent of the subscriptions were sold at $52. The single ticket revenue was distributed through sales of 55 percent for the $16 tickets and 45 percent for the $10 tickets.

This year, the company was able to generate $99,950 in revenue. They sold $30,000 in single tickets and $39,855 in subscriptions. They received a grant from the state arts council for $10,000 and donations of $13,545. They raised $6550 from a benefit dinner.

The expenses for the year totaled $96,450. The total was distributed in the following manner: $25,500 for salaries and benefits, $30,000 for guest artist fees, $2400 for the office, $6450 for travel, $7800 for marketing, $15,000 for production, $1800 for utilities, $6500 in mortgage payments, and $1000 for miscellaneous expenses.

The dance company started with a fund balance of $11,500 for the year.

According to the bookkeeper for the company, the year ended with $53,500 in assets. The assets were made up of the following: $3100 in cash, $300 in accounts receivable, $1500 in an endowment fund, $1200 in prepaid expenses (deposits), and $47,400 in land and property values. The company liabilities were listed as $500 in accounts payable, $15,000 in a loan due to the bank, and $23,000 left on the mortgage. The total liability was $38,500.

Questions

1. Prepare an annual budget report for the dance company, showing the information above in the standard form of a balance sheet and a statement of activity.

2. Based on the balance sheet you created, what is the dance company's fund balance? Did the fund balance increase or decrease this year?

3. Assuming that cash, accounts receivable, and prepaid expenses are totaled to make expendable assets, what is the ratio of assets to total liabilities? What does this figure tell you about the company's financial condition?

4. Based on a total of 12,000 seats (a thousand-seat theater times four weekends times three performances per weekend), what was the average number of tickets sold per performance?

5. What is the percentage of total income earned through sales versus the total amount from donations, grants, and other sources?

References

1. Thomas Wolf, *Managing a Nonprofit Organization* (Englewood Cliffs, NJ: Prentice-Hall, 1990), 139.
2. J. Fred Weston and Eugene F. Brigham, *Essentials of Managerial Finance*, 8th ed. (New York: Dryden Press, 1987), 3–24.
3. Stephen Langley, *Theatre Management and Production in America* (New York: Drama Book Publishers, 1990), 317–26.
4. Barbara Singer, *Nonprofit Organizations: Operations Handbook for Directors and Administrators* (Wilmette, IL: Callaghan, 1987), 19–36, 63–98.
5. William J. Baumol and Alan S. Blinder, *Economics—Principles and Policy*, 4th ed. (New York: Harcourt Brace Jovanovich, 1988), 506.
6. Frederick J. Turk and Robert P. Gallo, *Financial Management Strategies for Arts Organizations* (New York: American Council for the Arts, 1984), 102.
7. Ibid., 141–54.

Marketing and the Arts

An arts manager must plan, organize, implement, and evaluate various marketing and fund raising strategies in an effort to maximize revenue to meet the organization's established objectives. There can be an enormous satisfaction in seeing a full house, a packed museum, or the ground breaking for the new building made possible by the efforts of a well-designed marketing or fund raising campaign. However, we will also see that no amount of managerial brilliance or sophisticated marketing efforts will amount to much if the basic product does not meet the needs of the consumers for whom it is intended. It is important to remember that, like a management information system or a computer, marketing and fund raising are nothing more than tools. Marketing and fund raising cannot make a bad script good or a weak performance strong. At best, marketing and fund raising can help support a long-lasting relationship between the individual consumer and the organization. If properly managed, this relationship can evolve, and the consumer can grow from a single ticket buyer to a subscriber or member and finally to an annual supporter. Unfortunately, this operational objective is much easier said than done.

In the United States today, it is almost impossible to avoid the efforts of someone trying to sell you something every day. We are bombarded with thousands of messages every week in the form of television commercials, newspaper and magazine advertisements, flyers and letters in the mail, or phone calls from total strangers. Thousands of new consumer products are released in the market every year. Billions of dollars and millions of hours of labor are expended on product research, design, and distribution.

Promotional activities related to the profit sector of the entertainment industry relentlessly let us know that a new film is opening, a new book is coming out, the ice show is in town, or a new ride is starting at the theme park. The escalating mixture of media blitz, promotional hype, and advertising competitiveness used to get the consumer's attention does not leave much room for local arts organizations to make much of an impact. For example, it is not unusual for a movie studio to spend more money advertising one new film than a major regional arts organization has in its operating budget for an entire year.

As we have seen, the economic environment in which arts organizations must function requires constant effort to find the resources to survive from year to year. The need to retain and expand the number of subscribers, ticket buyers, members, or donors also places an enormous amount of pressure on the arts manager. Arts managers with the expertise and a successful record of managing marketing and fund raising campaigns are very much in demand. However, because organizations depend so heavily on revenues generated from sales, when there is a decline in income, once-successful managers may find themselves suddenly unemployed.

An Event in Search of an Audience

No matter how lofty the aesthetic aims of an organization, without the regular support of an audience, patrons, or members, there will not be enough money coming in to keep the enterprise alive. In other words, there must be enough demand for the product, or the enterprise will be out of business.

Before the advent of "marketing," arts organizations had a fairly standard set of activities that they undertook in an effort to create enough demand for a show or an exhibit. A press release announcing the upcoming event was sent to the local papers (a photo or two may have accompanied the release), posters were put up wherever they were allowed, flyers were sometimes distributed, a few small advertisements were placed in the paper, and a brochure was mailed out to names on the mailing list. If the organization was lucky, a preview article might appear in the arts section of the local paper. Organizations with larger budgets placed bigger ads in the paper, and they sometimes ran a few radio or television commercials.

Many managers and board members wondered why after the term *marketing* came into vogue, arts organizations continued to do the same things but spent twice as much money to get the same audience. What happened was that organizations were really not engaged in marketing. They were still trying to sell events to an ill-defined public. As a result, they wasted a great deal of money trying to convince people to buy their product without really knowing to whom they were selling. Spending more on advertising, in this case, was wasted effort.

As we will see in this chapter, real marketing requires that an organization adapt and change its fundamental perceptions about its relationship to consumers. Marketing requires the adoption of a customer-orientated perspective that is often perceived as being incompatible with the fundamental mission of high-culture arts organizations. Selling, on the other hand, which is what most organizations still do, means that the organization tries to get the consumer to buy the product because it believes the product is inherently good and would be beneficial to the consumer.

A Means to an End

Whatever term is applied to the energy and resources used to find, develop, and keep an audience or membership base, all of this activity is still only a means to an end. Philip Kotler and Alan Andreasen noted in their definitive text, *Strategic Marketing for Nonprofit Organizations*, that we must "view marketing's role as one supporting the organization in achieving its goals."[1]

Let's look more closely at the definition of marketing and many of the key concepts inherent in this vital area of study.

Marketing Principles and Terms

A key part of any arts organization's strategic plan is how it plans to market itself. The marketing plan normally forms a major section in the foundation of an organization's strategic approach to its long-term growth. We will see how the term *marketing* is often used incorrectly to describe various promotional activities that organizations undertake.

The American Marketing Association's definition of marketing, is "the process of planning and executing the conception, pricing, promotion and distribution of ideas, goods, and services to create exchanges that satisfy individual and organizational objectives."[2]

Needs and Wants

The marketer strives to achieve a match between human wants and needs and the products and services that can satisfy them. Theoretically, the better the match, the greater the satisfaction. Marketers define a *need* as "something lacking that is *necessary* for a person's physical, psychological, or social well-being."[3] Charles D. Schewe's text notes that food, shelter, and clothing are universal needs. Psychological needs—such as knowledge, achievement, and stability—and social needs—such as esteem, status, or power—are shaped by the overall value system of the culture.

A *want* is defined as "something that is lacking that is desirable or useful."[4] Wants are intrinsic to an individual's personality, experience, and culture. You may have a need for knowledge, but you want to pursue an idea from a specific book. You need to eat, but you want a particular brand of pizza.

When you have needs and wants to satisfy, two other marketing principles come into play: functional and psychological satisfaction. When we purchase an item like a refrigerator, we achieve a *functional satisfaction* because of the tangible features of the product. When we purchase a car, we may satisfy a functional need, but a particular make and model may provide an intangible *psychological satisfaction* for recognition or esteem.

Obviously, functional and psychological satisfaction are not neutral terms. Americans have attitudes about products that have been shaped by the advertisements on television and radio and in print. According to Schewe, the "goal of (the) marketers (is) to gain a competitive edge by providing greater satisfaction."[5] Unfortunately for many consumers, the idea that a fine arts event could provide a degree of satisfaction is foreign. In many cases, arts organizations would also probably not see their mission as providing satisfaction to customers. After all, in the minds of those inside an arts organization, a symphony concert is not a mass consumer product like soft drinks or toothpaste. However, the reality is that arts organizations function in a highly competitive entertainment market. Ultimately, if the symphony concert does not provide some degree of satisfaction to audience members, they will not continue to purchase the product.

Exchange Process and Utilities

Wants and needs are satisfied through the process of *exchange,* which occurs when "two or more individuals, groups, or organizations give to each other something of value in order to receive something of value. Each party to the exchange must want to exchange; must believe that what is received is more valuable than what is given up; and must be able to communicate with the other parties."[6]

For example, suppose that you want to hear a piano recital, and the pianist wants to perform. You believe that the time you are spending to listen and the money you give up for the ticket are worth the exchange. The pianist believes that the fee and the satisfaction derived from playing will be personally rewarding. The performance and the recognition of applause form the communication to complete the exchange process. Performers sometimes forget just how important this final communication really is for the audience. The level of satisfaction felt is greatly diminished when the performer walks off the stage without acknowledging the audience.

The exchange process depends on four utilities that marketers have identified as form, time, place, and possession. The utilities interact as part of the exchange process in ways that promote or hinder the final exchange or transaction.

The *form utility* simply means the "satisfaction a buyer receives from the physical characteristics of the product."[7] Attributes such as style, color, shape, and function affect the exchange. Arts organizations that have gift shops must be very sensitive to this utility because the customers usually have fairly sophisticated tastes, and filling the shop with cheap products will do more harm than good for the organization. Unique, high-quality items may provide the organization with a chance to build a strong bond with the discriminating buyer.

Except for the printed program, a performance does not offer any form utility. The live performance is, as we all know, an intangible event. However, the psychological satisfaction gained from the event can form a powerful bond between the audience and the organization. The memories that trigger emotional and intellectual responses in relation to a particular performance, or exhibit, can help build a lifelong relationship between the arts organization and the consumer.

The *time and place utilities*, which involve "being able to make the products or services available when and where the consumer wants them,"[8] have a direct impact on arts organizations. Arts organizations usually have little flexibility when it comes to time and place. The customer has the choice of either coming to the performance at a specific time and a specific place or not seeing it at all. Experimenting with different performance schedules or locations or different exhibit hours, may offer arts organizations occasional opportunities to increase consumer access to their products. However, the live performing arts, by their very nature, will always be limited in their manipulation of the time and place utilities. The advent of television and home videotaping offers a way of partially overcoming the inherent limitations of the live performance. *Live from the Met*, for example, has provided a way for opera to reach audiences that would never be able to attend the production in New York City. Art museums have experimented with different programming and exhibit schedules. (See "Art Museum Thrives with Marketing.")

The *possession utility*, which refers to "the satisfaction derived from using or owning the product,"[9] has some application in the live performing arts. The tangible items offered by the organization can create a degree of consumer satisfaction in much the same way that the form utility did. For example, long-time subscribers often view the seats they regularly sit in as their possessions. For two or three hours on a given night, they do indeed possess those seats. Allowing subscribers to keep their seats each year can be a powerful tool for maximizing on the possession utility. It is also possible to reinforce the experience of having attended through the secondary means of selling souvenir programs or other related material.

As we have seen, the exchange process for consumers of arts products and services fits within the theoretical framework of basic marketing principles. As part of an arts organization's core strategic planning, it makes sense for the staff and the board to spend time asking very fundamental questions about exactly what they are offering to the public. For example, how does the organization's corporate structure and philosophy affect its relationship with its audience? Do its programs and activities satisfy the wants and needs of the audience? What mechanisms are in place to get feedback from the audiences about the organization's programming?

If the organization is to survive, it must be able to adapt to changing conditions in the marketplace. Schewe notes that "Strategic market planning is a managerial process of developing and implementing a match between market opportunities (i.e., unsatisfied wants and needs) and the resources of the firm."[10] This process is not exclusive to the profit sector. One need only look at the changes that nonprofit hospitals have had to make in the mix of services they offered in the 80s to see how essential organizational adaptability is.

Evolution of Modern Marketing

Marketing has moved through three eras in its evolution. It is important to note that although these phases represent a progression, many organizations still hold to attitudes and beliefs about their product or service that have not changed much in 75 years. As a result, there are no clean break lines in this evolutionary development.

The first era is tied to the production and manufacturing techniques that began with the industrial revolution in the eighteenth century.[11] Schewe and others note that the main emphasis up through the beginning of the twentieth century was on fulfilling the basic needs of consumers. Mass production techniques dictated an approach of deciding what the consumer wanted and then manufacturing the product in the most cost-effective way. The assumption was that consumers would buy whatever was manufactured.

During the second era, more attention was focused on sales of the mass-produced products. The rise of the salesperson as a dominant figure in a system of getting goods to consumers is a part of the American myth. The period after the Civil War was marked by growth and expansion. Masses of immigrants came to the United States, which also fueled rapid growth. Thousands of salespeople spread out over the country, trying to sell products to people whether they wanted them or not.

The marketing era, the third era, is an outgrowth of the diversification of consumer wants and needs that resulted from the demands of unprecedented growth in the economy after World War II. More companies began to pay attention to what consumers were saying about the available products. The idea of a consumer-driven economy meant that companies needed to consider their basic relationship with the consumer. The research and testing of products and the application of psychological theories about purchase behavior led to a greater emphasis on developing a long-term relationship with the consumer.

Modern Marketing

By the 1990s, the concepts of marketing had been applied in just about every segment of profit and nonprofit business in the United States, including the use of marketing as a tool for electing candidates to office. The use of computers to store massive amounts of information about consumer preferences and to provide almost instant feedback to companies about what is selling has revolutionized the marketing industry. The ability to track sales via *point-of-purchase* systems offers marketers immediate access to information about what people are buying. The ubiquitous bar code now gives the store and the suppliers up-to-the-minute sales information about what people are buying.

The proliferation of products designed to satisfy consumer needs and wants has led to an explosion of specialty goods and services. A journey to the supermarket provides evidence of products designed to meet special health and nutritional concerns. In fact, the reality of global marketing has led companies to use satellite communications to monitor worldwide sales and to make adjustments in production much more rapidly.

Marketing and Entertainment

A recent news article entitled "Off the Street and Into the Audience: Tourists Help Pick Fall TV Schedule"[12] provided an accurate description of the degree to which the commercial entertainment industry is committed to consumer feedback. People were asked to watch pilots of television shows and rate them with a hand-held counter that registered whether they liked what they were seeing. Writers and producers are not happy with the prospect of their shows being subjected to this simplistic evaluation system.

The film industry uses test screenings and similar audience feedback methods to find out what people like or dislike. Endings may be edited or even reshot if there are negative reactions to a film at a test screening. Theme parks do extensive surveying to get feedback about rides, exhibits, and services. The economic pressure to have a hit in the entertainment industry will no doubt lead to the application of more intensive prescreening and testing.

It is easy to see why there is so much suspicion about the place of marketing in the high-culture industry. As we will see, there are limits to how much consumer feedback is practical and how consumer oriented an organization can be. For most arts organizations, being totally consumer driven in the choice of programs and presentations remains a totally alien concept.

A Marketing Approach

A company attempting to make a profit usually has different values and goals than a local nonprofit health care center or symphony, but both rely on establishing a positive relationship with individual consumers and the general public. Both private and public sector companies make plans and state their missions based on satisfying the public's wants and needs. The mission statement is the source of the organization's goals and strategic plans. The planning process includes defining the function of marketing in the organization. Kotler and Andreasen identify three functional orientations that parallel Schewe's historical eras: the product, sales, and customer orientations.

Product Orientation

Kotler and Andreasen define the *product orientation* as one in which "the major task of the organization is to put out products that it thinks would be good for the public."[13] They cite as examples colleges and universities that continue to offer courses that are evaluated as being below standard or for which there is very low enrollment and museums that feature specific works of art from the collection even when there is little public interest in the exhibit. Product-orientated organizations are "in love with their product."[14]

Sales Orientation

The organization with a *sales orientation* thinks that "its main task is to stimulate the interest of potential consumers in the organization's existing products and services."[15] Most arts organizations engage in sales activities that they misidentify as marketing. Rather than make any changes in the product or how it is presented, they increase the resources allocated to advertising, direct mail, or telephone solicitations. These efforts usually result in short-term gains in audience. However, because it does not really adopt the consumer's perspective, the sales-oriented organization constantly has to replenish a large number of nonrenewing subscribers or members.

Customer Orientation

All of the marketing texts seem to agree that organizations that have evolved or start with a customer orientation have the best chance of competing in the world market today. An organization with a *customer orientation* "determines the perceptions, needs, and wants of target markets and then goes about designing, communicating, pricing and delivering appropriate and viable products or services." Kotler and Andreasen point out the following:[16]

> Adopting a customer orientation does not, as many nonprofit managers fear, mean that the organization must cater to every consumer whim and fancy. It doesn't mean that a symphony conductor or theater manager must give up his or her artistic integrity. . . . To restate: It means the marketing planning must *start* with customer perceptions, needs, and wants. It means that, even if an organization can't or ought not change certain aspects of the offering, the highest volume of exchange will always be generated if the way the organization's offering is described, "priced," "packaged," and delivered is fully responsive to what is referred to in the current jargon as "where the customer is coming from."[17]

As this quote should make clear, an organization that takes a customer's perspective would, for example, use text to describe an upcoming performance in terms that an audience can respond to rather than in the jargon of the profession. If a potential ticket buyer believes that arts events are only for the wealthy and everything the organization does with its promotional activity (ads, brochures, and so on) only reinforces this image, the arts promoter should not be surprised if the consumer feels reluctant to enter into the exchange process. On the other hand, arts organizations believe that they shouldn't have to describe a play like *Hamlet* as a "gut-wrenching tale of a family caught up in an whirlwind of lust and murder" in order to sell tickets. To discover the language that makes the most sense, the organization must engage in some basic consumer research. Research may show that a more dramatic description would make sense in their market.

The key to successfully adopting a customer orientation resides in the organization's having done research on its community. What are people's attitudes about music, opera, theater, dance, and art? Based on that research, the customer-oriented arts organization would have sev-

In the News—A Customer Orientation

This article on the Milwaukee Art Museum provides a practical example of the successful adoption of a customer orientation in marketing.

Art Museum Thrives with Marketing
David I. Bednarek

Through aggressive marketing, the Milwaukee Art Museum is attracting more visitors and members at a time when attendance at art museums nationwide is declining.

Since 1984, attendance at art museums in the US has gone down almost 5%, according to a Lou Harris poll, and is putting some museums on shaky footing because the higher cost of art and insurance demand greater attendance.

In the face of that national trend, attendance at the Milwaukee Art Museum has risen dramatically—from 129,000 in 1984 to 197,000 in 1989, an increase of 53%. Membership went up 56%, from 6,500 to about 10,000.

The museum's success with marketing stands out as a model for other arts groups and similar organizations as leisure time becomes more scarce.

Rebecca Turner, director of marketing for the museum, attributed the increases to a decision in the early 1980s to get into marketing instead of simply relying on public relations to keep the museum going.

In addition to the increases in attendance and membership, Turner said the number of people taking classes at the museum went up from 771 to 2,000, the number attending special events increased from 9,000 to 49,000, and the number taking museum tours rose from 25,759 to 53,000.

Since 1984, the museum also has attracted more docents, the volunteers who work as guides.

In deciding what to do to sell the museum, the marketers first found out why people did not go and then set up programs to counter those reasons.

To the response, "I have no time for arts," for example, the museum set up mini-lectures on gallery nights, First Friday events with live jazz, "Bagels and Bach" on Sunday mornings and lunchtime lectures.

To those who say, "I don't know enough," the museum set up audio tours of exhibitions, using tapes to tell about the arts on display, the "Bluffer's Guide to Art," and Master of the Month gallery talks.

And to the complaint, "I won't fit in," the museum organized or helped organize Senior Days, Grandparents' Day, Free Days, Lakefront Festival of the Arts and Music in the Museum.

SOURCE: David I. Bednarek, "Art Museum Thrives with Marketing," *Milwaukee Journal*, April 11, 1990.

eral different approaches to communicating with the different audiences in the community. In some cases, the promotional campaign might be targeted to educating people about a new work or a new author. In other cases, the organization may focus on the strong emotions that a story or a piece of music conveys. For some potential audiences, *Hamlet* may spark their interest if described in more lurid terms. The arts marketer must of course be careful about crossing a line that distorts or debases the product. On the other hand, the risk of offending the sensibilities of a small number of the old guard patrons may prove worthwhile if it brings in new customers. However, unless the organization has a method for tracking the impact of different advertising tactics, these efforts will be wasted.

The problem that most customer- or audience-oriented arts organizations face when it comes time to communicate effectively about the product is a lack of money. The cost of multitarget promotional campaigns is usually well beyond the reach of most groups. However, a marketer would argue that this is money well spent because the objective is to build up long-term audience support and consumer identification with the product. Unfortunately, many arts organizations take a middle ground and ultimately communicate a bland image by trying to straddle too many marketing perspectives in their brochures and publications.

Marketing Management

The Four P's

Using these principles of marketing now allows us to move into the process of marketing management. To market its products or services effectively, an organization must carefully design its marketing mix. Schewe defines *marketing mix* as "the combination of activities involving product, price, place, and promotion that a firm undertakes in order to provide satisfaction to consumers in a given market."[18] Each of these elements will have an effect on the exchange process.

The *four P's*, as they are often called, can be manipulated as part of the organization's overall strategy. For example, if you have a product with a brand name, such as the Metropolitan Opera, you may be able to manipulate the price based on the customer's perception of quality while stressing the place with its crystal chandeliers and red carpet in your promotional material.

The promotional aspect of the marketing mix is the most visible element, and it is usually divided into a further mix of types of advertising: newspaper, magazine, radio, television, direct mail, raffles, and other public relations activities (for example, getting a soprano on a local television talk show or radio program).

The overall marketing strategy for the organization may have several different marketing mixes. Depending on the target audience, you may stress price or product. For example, a group sales flyer sent to a retirement center may be accompanied by a letter that stresses price first and then product. The same group flyer may be accompanied by a different letter that stresses product first and then price when sent to a college or university drama department.

Market Segments

A marketing manager is expected to have a good grasp of the overall marketplace. As we discussed in Chapter 10, there are many markets in the system of supply and demand, and within the large markets, there are smaller markets for goods and services. Marketers use the term *market segment* to identify "a group of buyers who have similar wants and needs".[19] Once a market segment has been identified, the marketer begins the process of *target marketing* by "developing a mix of the four p's aimed at that market."[20]

In planning the marketing mix, information is the key ingredient in designing a successfully targeted campaign. For example, if you buy a mailing list from the state arts council with the names of ten thousand people interested in the arts in the state, you have identified a broad market segment. If this list of names is to be useful to you, it will need further analysis. How many of these people attend particular types of performing arts events? Narrowing the list further, how many of these people are geographically close to your performance or exhibition space? After you finish narrowing down the list to people within a three-hour driving distance, are there enough names left to make it worthwhile trying to target this group?

Mailing lists, which are purchased all the time in profit-sector marketing, may be far too costly for many arts organizations. For these groups, the existing audience is the best and most cost-effective resource for additional customers. The marketer's assumption is that if you consume the arts product, your friends or colleagues may share similar values.

In the News—Price as a Marketing Tool

This Theater Tells Patrons: Pay What You Can Afford
Associated Press

This article demonstrates the marketing principle of price manipulation as a method for increasing audiences.

PROVIDENCE, RI —At a time when the price of theater tickets is soaring beyond the reach of many, Trinity Repertory Theater will try a one-night experiment Tuesday of letting patrons pay what they will.

"The objective is to give everyone an opportunity to come to the theater," said E. Timothy Langan, managing director of the theater company.

The normal ticket price for the preview performance of Maxim Gorky's "Summerfolk"—a story Langan described as one about the Russian Empire's version of yuppies before the 1917 revolution—is $24. Based on similar experiments in Baltimore and San Diego, Langan said he expected people would be willing to pay $3 to $4.

"But that's OK," he said. "The whole purpose of this is for someone to be comfortable in coming to the theater."

Trinity has set aside $5,000 from a grant to cover projected losses.

SOURCE: Associated Press, "This Theater Tells Patrons: Pay What You Can Afford." October 10, 1989.

Market Research

To engage effectively in target marketing, much detailed information about the potential arts consumer must be known. Understanding the *demographics* (age, income, education, gender, race) and having an informed *psychographic profile* (consumer beliefs, values, attitudes) of the potential consumer is crucial to designing the marketing mix for the target market.

Marketing researchers in the profit sector have been developing various behavioral and psychological models in an attempt to make target marketing as cost effective as possible. The thrust of this work is to divide consumers into lifestyle segments based on such things as activities, interests, and opinions. An example of the psychographic approach (based on a behavioral profile) to understanding consumer behavior can be found in Arnold Mitchell's *The Nine American Lifestyles*.[21] His research resulted in a more elaborate version of Maslow's hierarchy of needs. Mitchell developed a hierarchy chart representing segments of the population as a way to identify consumer behavior. Mitchell called his chart a Values and Lifestyles Segment or VALS distribution.

The Association of Performing Arts Presenters hired Mitchell in 1984 to conduct a study of arts audiences. In his report, *The Professional Performing Arts: Attendance Patterns, Preferences and Motives*,[22] he found that four groups, which at that time made up about 66 million people, were the primary market for arts organizations. He called these groups the Achievers, the Experientials, the Societally Conscious, and the Integrateds. Of these four groups, the societally conscious (12 percent of the population) were the best market per capita. Mitchell also found that among these four lifestyles, the most common reason cited for attending an arts event was to see a specific show, performer, or group. He also found that even among these targeted groups, large percentages admitted that they never attended arts events. For example, an average of 28 percent never attended music concerts, 40 percent never attended theater productions, and 68 percent never attended dance events. His research found that lack of leisure time (30 percent), preferences for other leisure activities (34 percent), and not wanting to commit to season or series purchases (33 percent) were the primary reasons given for not attending.

Another approach to target marketing—one designed to help businesses connect with the consumer—is detailed in *The Clustering of America* by Michael J. Weiss. Weiss examines the work of a market research company. Claritas Corporation developed a system that uses a vast mix of census data to produce information that marketers buy to locate the people who might be disposed to buy their product. The Potential Rating Index for Zip Markets (PRIZM) system uses a zip code analysis of various neighborhood types. For example, Claritas's research has identified the top five clusters for classical music and named them Urban Gold Coast, Blue Blood Estates, Young Influentials, Bohemian Mix, and Money & Brains.[23]

The objective for an arts marketer using the Claritas system is to develop a database of the zip code distribution of its list of current subscribers and, at the same time, to gather information about the zip code distribution of the single ticket buyers and to compare the data with the neighborhood types. At this point, the marketer could determine which areas the organization has not reached. Buying a list of labels from the local utility company would allow the organization to send targeted mailings to households in the zip code neighborhoods that the organization has identified as potential customers.

Ultimately, a system such as the one Claritas has developed, should allow an arts marketer to target potential audiences by very narrow segments. After all, why should an arts organization waste its limited resources doing mass mailings when carefully targeted mailings to "the right people" will yield much more cost-effective results?

Demographic studies by Baumol and Bowen and recent studies by the NEA have contributed to this profile of the average arts consumer. The audience is equally split in gender distribution, has a median age of 35, and a median income of more than $20,000 (in 1980). The average consumer works as a professional, manager, or teacher or is part of a clerical or sales staff and is highly educated.[24]

Other arts research The PRIZM and VALS approaches to market research can be very expensive. Although Mitchell's study for the Arts Presenters goes into great detail about such things as reactions to different types of advertising, audiences will react unpredictably to various marketing plans. Other sources to consider for marketing information include the research division of the National Endowment for the Arts regularly publishes useful data, and the *Journal of Arts Management and Law.* Articles in a special issue of the *Journal* ("Consumer Behavior and the Arts," vol. 15, no. 1, Spring 1985) cover the general topic of marketing and provide detailed information on price theory. For example, one article applies psychologist Clare Graves's theories of personality to audiences. Another closely examines the variable of price on "discretionary income purchases." Another source for marketing ideas is the comprehensive anthology published in 1983 by FEDAPT entitled *Market the Arts!* In addition, a recent study by Lynne Fitzhugh provided a comprehensive update of audience surveys and audience purchase behavior.

Arts organizations should regularly survey people in their community for feedback on new programs and on problems with existing operations. A properly designed survey can give an organization the opportunity to adjust and change its marketing mix.

Marketing Ethics

Whatever approach is used in marketing research, the goal is to find out what the consumer thinks about the product or service. Marketers believe that with the right information they could better predict which combination of product, price, place, and promotion is needed to complete the exchange process on a regular basis with consumers. To bridge the information gap, marketers look to even more sophisticated applications of computers in their work. As a result, the line between market research and invading people's privacy has grown very thin today. The selling of vast amounts of information about consumers is a fact of life today. Michael Weiss points out that "the top five credit-rating companies have records on more than 150 million Americans. And Federal data bases contain some 288 million records on 114 million people, with 15 agencies mixing and matching data."[25]

Arts organizations, which depend on the sales of tickets and subscriptions for 60 percent or more of their operating budget, face a dilemma. How intrusive should they be when trying to reach potential arts consumers? Arts organizations want to identify and target these people who are most likely to be long-term consumers of their product. Techniques such as telephone marketing, if handled properly, can lead to direct contact with consumers. On the other hand, people resent phone calls and "sales pitches" that intrude into their private lives. Marketers for arts organizations must also face the ethical issue of selling information about their customers to commercial firms. The arts consumer is a prime target for the marketer of up-scale goods and services. Research has shown that arts consumers have more than the average amount of discretionary income and are therefore good targets for a wide variety of marketing assaults.

Strategic Marketing Plans

Now that the basic principles of marketing have been outlined, let's examine in more detail the critical planning process. As noted earlier in this chapter, if the marketing plan is to be effective, the entire organization must carefully consider how all phases of the operation relate to the dynamics of the marketplace. The simple fact facing all organizations is that new opportunities and new threats arise every day in the marketplace. An organization that can adjust to these changing conditions has the best chance of surviving in the long run.

Some board members may wonder why an organization such as a museum or some other well-established performing arts institution would need to worry about the changing dynamics of the marketplace. After all, won't people always go to the museum or to the symphony? Why should an organization spend time planning, reviewing its mission, devising strategies, and developing objectives when what it does is so obvious? Citing the examples of dance companies that have failed, museums that have had to reduce their hours and staff, and orchestras, theaters, and opera companies that have filed for bankruptcy should be enough to counter any argument that strategic marketing plans are a waste of time.

Planning Process

The organization's overall strategic plan (discussed in Chapter 5) incorporates the marketing plan. (See Figure 12.1). The organization's objectives drive the mission, goals, and objectives of its marketing plan. In addition, an analysis of opportunities and threats from the external environments, noted in Chapter 4 (economic, demographic, political and legal, social and cultural, technological, and educational) are weighed against the organization's strengths and weaknesses. Once the basic mission and objectives of the marketing plan have been defined, the core marketing strategy can be developed. The target markets and the proposed marketing mix can be articulated. The process now moves to the final stage by providing the system for carrying out the marketing plan, including what performance criteria will be used to monitor progress. In addition, the specific tactics can be created. Implementation plans and an evaluation system complete the process. The evaluation process feeds back to the core marketing strategy for long-term adjustments and directly back to specific tactics for short-term changes. For example, a short-term change might be to revise an advertisement in the paper when there is poor response to a particular offer. A long-term adjustment might be to evaluate all of the print media.

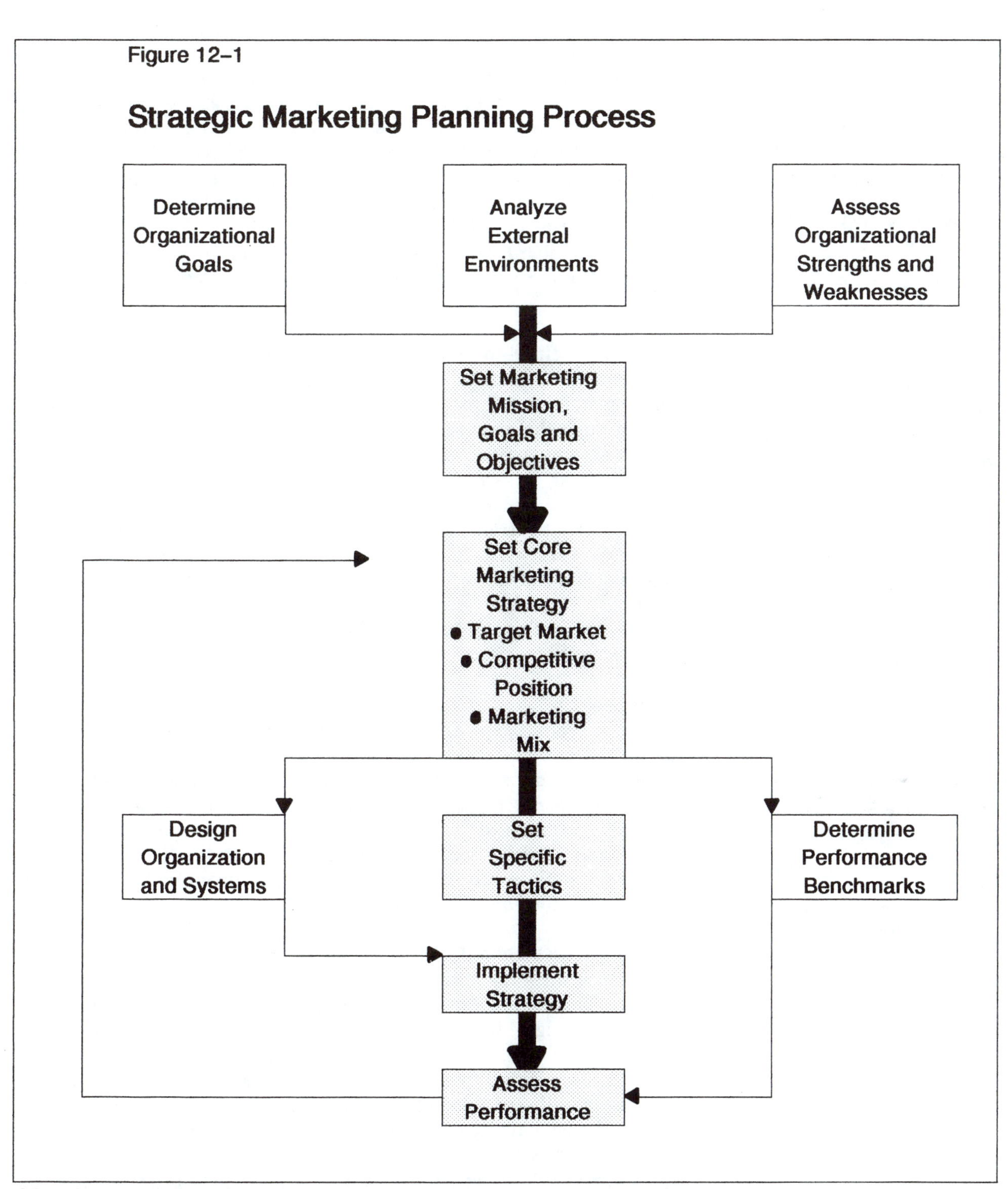

SOURCE: Philip Kotler and Alan Andreasen, Stratgic Marketing for Nonprofit Organizations, 4th ed. (Englewood Cliffs, NJ: Prentice Hall, 1991). Copyright (c) 1991 by Prentice Hall. Used with permission.

In the News—Marketing Strategies

This article shows how an arts organization can expand consumer awareness for a very modest investment.

Toasting Cabbies to Tout Visitors
Associated Press

CHICAGO—The Art Institute of Chicago hopes a free breakfast will go a long way toward steering visitors in its direction.

The museum treated about 100 taxi drivers to Danish pastry, juice, coffee and a tour last week in the hope that the steering-wheel philosophers will talk up the place to their fares.

"Chicago taxi drivers really are roving ambassadors," museum spokeswoman Eileen Harakal said. "Not only do they answer passengers' questions on Chicago's attractions, but they help steer people to them."

"It's a great idea," said cabdriver Bill Hogan as he contemplated Henri Matisse's 'Still Life With a Blue Tablecloth.' "When people ask what there is to see in Chicago I usually say the Sears Tower, the Museum of Science and Industry and the Shedd Aquarium. Now I'll mention the Art Institute."

SOURCE: Associated Press, "Toasting Cabbies to Tout Visitors." October 3, 1990.

Marketing Audit

One method that an organization might use to assess its ability to carry out a marketing plan is to do a marketing audit. Essentially, an audit consists of asking and answering a series of questions that explore the organization's markets, customers, objectives, organizational structure, marketing information system, and marketing mix. The Wolf Organization, a well-known arts consulting organization in Boston, devised the approach outlined in Figure 12–2. The audit gives the staff and the board a common ground on which to build a marketing plan that fits the organization's mission and function.

Consultants

It is usually helpful to get the perspective of an outside consultant when formulating any strategic or marketing plan. Someone with expertise in planning can save an organization a great deal of valuable time struggling through the planning process. As noted in Chapter 5, planning is hard work, and because of the pressing daily needs of keeping the organization afloat, this essential process is often relegated to a low priority by managers. A consultant, if used effectively, can go into an organization, shake up the status quo, and act as a catalyst to put planning at the top of the priority list. A word of caution, though: Consultants are not infallible, and they have been known to make mistakes. They can give bad advice and make recommendations that make conditions worse, not better. A background check of former clients is a requirement for organizations that want to protect themselves.

Strategies

The profit sector uses terminology borrowed from warfare when developing marketing strategies. Marketers use such terms as *frontal, encirclement, flanking,* and *bypass attacks* to describe marketing plans. Words such as *preemptive, counteroffensive,* and *contraction* are used to describe strategies.[26] Other options for organizations to explore include generic, market leader, market follower, and market niche strategies. Let's look briefly at the competitive environment facing many arts organizations and discuss strategy options.

The Competitive Marketplace and Core Strategies

When a community reaches the point in its growth where it has at least one professional arts organization from each of the major disciplines, the struggle for resources among these organizations can intensify. Arts people may carry on a cordial and friendly dialogue in public, but the simple fact of the matter is that there are only so many dollars that people will spend on subscriptions, tickets, memberships, and donations. As we have seen, arts organizations also face competition for the entertainment dollar from videotape rentals, films, television, and amusement parks. Therefore, in formulating a plan of attack, an arts organization might consider a *niche strategy*. Such a strategy focuses on the qualities that make a live performing arts event or a trip to the museum a unique activity. The niche strategy can be combined with a *differentiation strategy* in an effort to feature those things that are unique about the product. This strategy combination allows the organization to concentrate on what is special about their product while appealing to a targeted market.

If the organization's planning process leads to a decision to expand its audience base beyond the typical demographic blend, a strat-

Figure 12–2

Marketing Audit

1. MACROENVIRONMENT AUDIT

Demographic
What changes pose threats or opportunities?

Economic
What will be the impact of trends in income, savings, prices and credit?

Technological
What changes in technology pose a threats or opportunities?

Political
What local, state, and federal legislation or court decisions will affect the organization?

Cultural
How will changes in consumer lifestyles and values affect the organization?

2. TASK ENVIRONMENT AUDIT

Markets
- Market size, segments, growth, and distribution
- Describe major constituency -- how will it change?

Customers
- In each segment, what is volume of repeat business?
- How do current customers rate programs and services?
- How are customer interests represented in decisions about services?

Competitors
- Who are the major competitors? Strengths and weaknesses?
- How has market share changed regarding competitors?

Distribution
- What are the main distribution channels for the product? Are channels effective?

Publics
- What publics (financial, unions, media, government, citizen, local, internal) represent particular opportunities or problems for the organization?
- What steps have been taken to deal with the publics?

3. MARKETING OBJECTIVES & STRATEGY AUDIT

Organizational Objectives
- Is the mission clearly stated? Is it feasible given the opportunities and resources?
- What is the organization's overall orientation – Product, sales or marketing?
- What are the organization's major strengths and weaknesses?
- Are the objectives stated in market-oriented terms?

Marketing Strategy
- What is the core marketing strategy for the achieving objectives? Is it sound?
- What is the organization's marketing style (aggressive, passive, integrated)?
- Do marketing objectives directly relate to organizational objectives?
- Are enough resources budgeted to accomplish marketing objectives?

4. MARKETING ORGANIZATION AUDIT

Formal Structure
- Is there a high-level staff person with authority and responsibility over organizational activities that affect customer satisfaction?

Functional Efficiency
- Are there good communications and working relations between marketing and operations?
- Are there any groups involved in marketing that need more training, motivation, etc.?

5. MARKETING SYSTEMS AUDIT

Marketing Information System
- Is system producing accurate, sufficient, and timely information?
- Is marketing research being used?
- What information exists?
- Is system set up to measure post-purchase satisfaction?

Marketing Planning System
- Is planning system well-conceived and effective?
- Are sales forecasting and market potential measurements carried out?

Marketing Control System
- Are the annual plan objectives reviewed on a timely basis?
- Is provision made to analyze revenue production of different products or services?
- Are marketing costs periodically analyzed?

New Product/Service Development System
- Is organization organized to gather, generate, and screen new products/services?
- Is research and business analysis done before investing in a new idea?
- Is testing done before offering new product/service?

6. MARKETING FUNCTION AUDIT – The Marketing Mix

Products/Services

- Describe organization's products/services.
- Which should be phased out?
- Are there new products/services that should be added?
- Are there products/services that would benefit from improvements?
- For each product/service:
  - What goes wrong most often?
  - What are the major complaints
  - How vulnerable is it to competitive alternatives?

Pricing

- What are pricing policies and strategies?
- Are there pricing variations for different market segments?
- Do customers see the price as being in line with perceived value?

Distribution

- Is there adequate market coverage and service?
- Is service delivery decentralized (outreach)?
- Are information, reservation, or payment for services centralized?
- Does organization make good use of distribution, sales, or direct selling?
- When are products/services available?
- Are distribution decisions based on user preference or staff/volunteer convenience or custom?

Promotion

- What are the organization's advertising objectives? Are they sound?
- Is the right amount being spent on advertising? How is budget determined?
- Are advertising themes and copy effective? What does customer survey information say?
- Are advertising media chosen well?
- Are sales promotions used effectively?
- Is there a well-conceived public relations program?
- Is personal selling used effectively? Is sales force large enough?
- How is communication program designed for and directed to segments?

Source: Copyright (c) 1986 by The Wolf Organization, Inc., Cambridge, MA 02138. Used with permission.

egy to increase market share would be appropriate. In this *growth strategy*, the organization takes an aggressive advertising approach to reach new audiences. For example, if a theater company wanted to develop its market among African-Americans in the community, an advertising campaign using specific media publications and radio stations with a high ratio of minority consumers would make sense. Also, targeting group sales by using the local network African-American religious groups might prove successful. However, if the arts organization doesn't regularly offer a product that has some market appeal to members of various minorities, there is little chance that this strategy will succeed.

Whatever overall marketing strategy an arts organization selects, it is important to remember that it must fit with the mission of the organization. Care must be taken to avoid shifting the organization away from its mission to meet a market strategy. For example, museums are not in the gift shop business, but because these operations can become very healthy sources of cash, it is tempting to overstress their importance in marketing the organization.

Project Planning and Implementation

The details of preparing, budgeting, and implementing the marketing plan require careful attention. Decisions about where to put the limited marketing resources available to the organization can make or break a plan. The work done on researching the community and a detailed cost analysis of various media campaigns will pay off in the project planning stage. Organization and project management skills are required to prepare the overall schedule and budget distribution for the marketing campaign. *Market the Arts!* is an excellent resource book with examples of well-organized marketing campaigns.

Evaluation

After establishing the marketing mission, the core strategy, and specific tactics and implementing the plan, the organization must carefully evaluate and monitor how well its objectives are being realized. For example, if the costs of implementing the strategy exceed the budget, and the number of new subscribers or members is below the levels established for success, the organization must be able to adjust its tactics or to revise the entire strategy before it is too late. As noted in Chapter 5, the failure to abandon a plan that isn't working can lead to numerous problems.

Marketing Data System

Within the organization's overall management information system and financial management information system, you should find a *marketing data system* (MDS). Its purpose is parallel to that of the MIS. The MDS should be designed to gather and analyze data regularly and to issue reports on the success of current campaigns. Figure 12–3 shows a typical MDS for an arts organization. The four major sources of data are the sales system (box office), audiences, staff, and various external environments. If the system is working properly, the feedback provided to the marketing staff will arrive in time to make corrections and adjustments in the marketing plan. With some sources of information, such as surveys, the data gathered can be translated into information that is useful in planning future seasons or programs.

Computers can play a central role in gathering and processing data for the MDS. As we saw in Chapters 9 and 11, the MIS and the FMIS must be linked with the marketing data system if the organization is to monitor its operations effectively (see Figures 9–4, 9–5, and 11–1). It is essential that the organization have a network of computers that share data among the marketing staff and other members of the management team to enable the marketing plans to be successfully evaluated. The complexity of such a system may require hiring outside consultants to coordinate and advise the marketing manager. Without the ability to track all sales data quickly and accurately, the marketing manager's effectiveness will ultimately be seriously undermined.

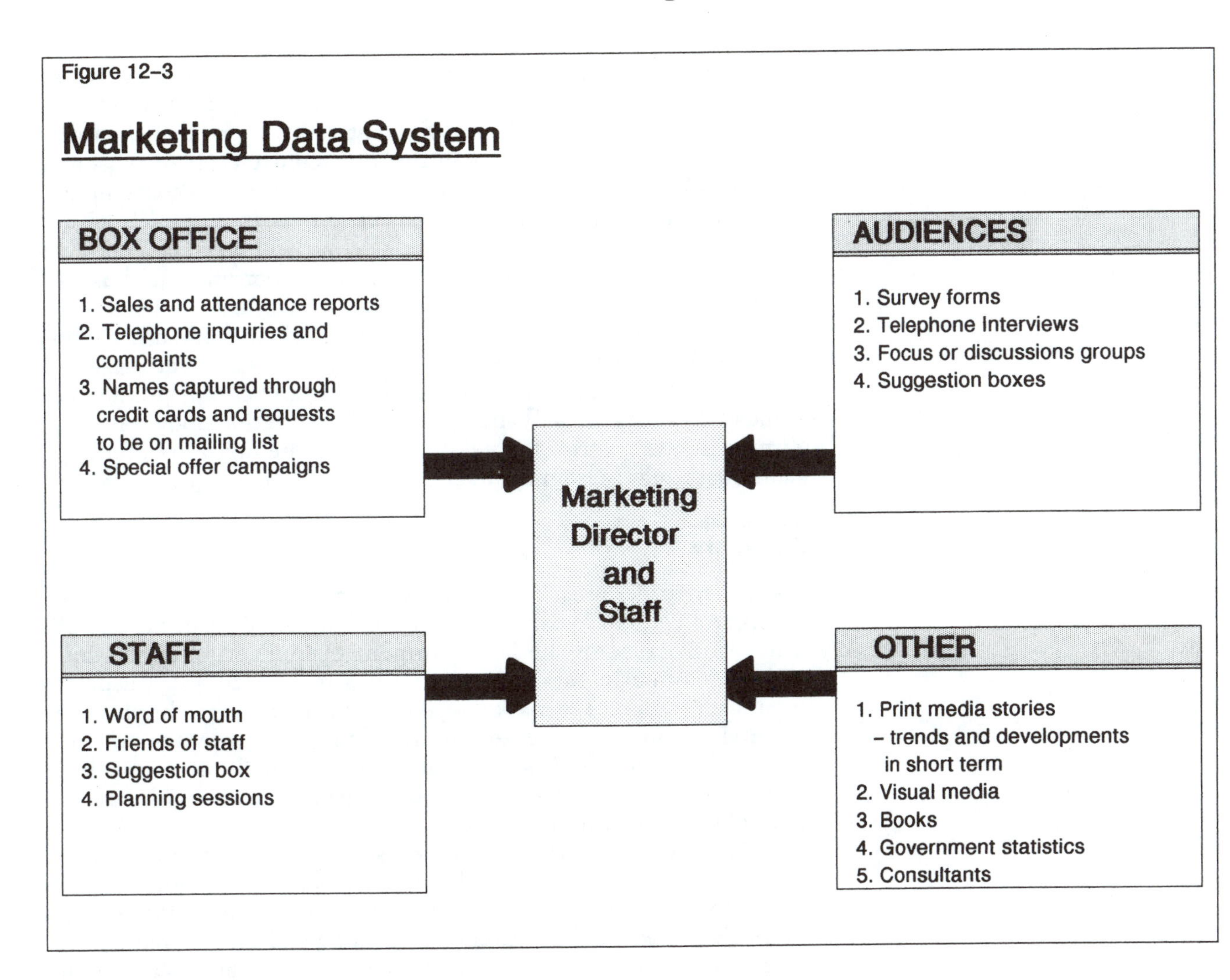

Conclusion

Marketing can be an effective tool for keeping an organization growing and evolving. However, marketing is also a long-term investment. A well-organized marketing campaign should be integrated with operational and long-term organization plans. It makes no sense to attempt a marketing plan without having first clearly defined the mission and goals of the organization. Like any tool, marketing can be misused. Thousands of dollars can be wasted on advertising campaigns or printed brochures that have little or no impact on sales.

Unfortunately, there is no guarantee that even the best plans will work. For organizations with limited resources, experimenting with different approaches to marketing is out of the question. However, finding the most cost-effective way to reach audiences when there are no funds for research also places impossible constraints on the marketing staff. It isn't surprising to find 100 percent yearly turnover in an arts organization's marketing staff. Impossible goals, limited resources, and poorly conceived plans take their toll on even the most ambitious people. Upper management involvement and commitment to planning and implementation can go a long way toward remedying the problems that arts marketers face.

The most daunting task facing any arts marketer is the development of future audiences. The simple reality of a very limited audience base coupled with the ever-increasing competition from other entertainment options makes for a difficult mix of circumstances. As most arts marketers know, unless you can establish a pattern of arts consumption at an early age, it is both difficult and costly to change people's leisure behavior later in life. The arts marketing effort will no doubt continue, and wherever appropriate, the arts marketer will borrow from the commercial marketing world those techniques that work.

For students interested in a challenging field of work, arts or nonprofit marketing has a great deal of potential. The use of language and images to express an idea or to convey an organization's mission demands a great deal of skill and creativity. Because this chapter is just a glimpse into the world of marketing, students are urged to explore the additional readings from the references cited in this chapter.

Summary

The arts manager must plan, organize, implement, and evaluate marketing strategies to maximize revenue and meet the organization's objectives. Because of the bombardment of marketing efforts by a multitude of businesses and causes, the arts manager must dedicate significant resources to marketing if the organization is to be visible in the highly competitive entertainment marketplace. Marketing is a means to an end, and it should therefore be thought of as one more tool available to the arts manager to be used in realizing the overall goals of the organization.

Contemporary marketing attempts to match the wants and needs of consumers with products and services. Needs are physiological and psychological things that are lacking and are necessary for people's well-being. Wants are things that are lacking and that people find desirable or useful. People can gain functional and psychological satisfaction from tangible and intangible features of products or services.

Marketing activity is designed to facilitate the exchange process. This process involves a transfer of something of value between two or more parties or organizations. The exchange process is successful to the degree that the utilities of form, time, place, and possession can be satisfied through the exchange. The arts exchange usually involves satisfying a psychological want through the intangible features of an experience, which is modified by the inherent constraints placed on the four utilities by the delivery system (the performance).

Marketing has evolved over the last three hundred years. The production era, which grew out of the Industrial Revolution, concentrated on satisfying basic needs. It was assumed that people would buy whatever was manufactured. The sales era, which concentrated on increasing demand, began sometime after the Civil War. More emphasis was put on customers' wants, but the manufacturers still dictated what would be available to purchase. The marketing era, which came to the fore after World War II, reversed the relationship between the consumer and the manufacturer. The consumer-driven market relationship starts with what the consumer wants, not the product.

Marketing today is classified along the same historical line of evolution. A company may have a product, sales, or customer orientation. The product-oriented company assumes that its product is inherently good and needs no changes. The sales-oriented company concentrates on trying to increase demand for existing products and services. The customer-oriented company determines the perceptions, needs, and wants of the market and goes about creating a product to fit those needs. Arts organizations can and do use these three orientations. The market-oriented arts organization thrives if it understands the market's perception of its product and describes, prices, packages, and delivers its product to reflect those perceptions. It does not mean that the organization must change the product to attract customers.

Marketing management is based on the organization's manipulation of the four P's—product, price, place, and promotion—or the market mix. The market mix can be adjusted to suit the target market. Market research has shown that people with various demographic and psychographic profiles react differently to various marketing mixes.

The entire marketing process is directly related to the organization's strategic plans. The main objectives of the strategic plan are incorporated into the marketing plan. An analysis of external environments and the strengths and weaknesses of the organization are also included. A detailed audit process may be used to assess the organization's capabilities to undertake an effective marketing campaign. From the marketing plan, specific strategies and detailed tactics can be designed to meet the defined objectives. The success of the marketing campaign depends on accurate and timely information gathered by the marketing data system.

Key Terms and Concepts

Marketing
Needs and wants
Functional satisfaction
Psychological satisfaction
Exchange
Form, time, place, and possession utilities
Production, sales, and marketing eras
Product, sales, and customer orientations
The four P's: product, price, place, and promotion
Marketing mix
Market segments
Target marketing
Demographic profile
Psychographic profile
Strategic marketing plans
Marketing audit
Niche strategy
Differentiation strategy
Market share strategy
Marketing data system (MDS)

Questions

1. Define the term *marketing*.

2. What are some of the wants and needs satisfied by the following: a brand-name soft drink, a meal at a French restaurant, and a visit to an art museum?

3. Does marketing make you buy things that you don't need? Explain.

4. Give an example of an exchange process in which you recently participated that was not satisfying. What went wrong in the exchange? What would you change to make the exchange satisfying?

5. What suggestions would you offer about form, time, place, and possession utilities to a museum and a children's theater company that are planning new outreach programs?

6. When you are considering the purchase of an arts product, which of the four P's is most important to you? Explain. Do you react differently to various marketing mixes? How?

7. What are some of the different market segments you would identify for theater, dance, opera, symphony, and museum organizations? How much attendance crossover do you think exists among the different segments? For example, do opera audiences go to the theater?

8. Do demographic and psychographic profiles of audiences match your perception of arts consumers? How do you think the profile of the audience will change over the next 20 years? How will changes in demographics affect arts organizations?

9. If you were managing a small modern dance company in a community with a well-established ballet company, what marketing strategy would you adopt to gain a market share?

For Further Discussion

According to "Art Museum Thrives with Marketing," what was the basic marketing strategy attempted by the Milwaukee Art Museum? Why was it so successful?

Case Study

The following article is intended to provide another perspective on the area of audience development. The issue for an arts marketer is reaching and developing new audiences. Although the topic area covers music and opera, it is fairly easy to extrapolate to theater, dance, or art.

Outreach Gimmicks Furrow a Skeptical Highbrow

Robert Finn

How did you first get interested in classical music?

Was it something you heard routinely at home?

Was there an inspiring teacher somewhere who unlocked that particular door for you?

Did you accidentally hear something on the radio one day that piqued your interest?

Did you have a friend whose enthusiasm was contagious?

Those used to be the standard avenues of approach. The rules seem to have changed (for the worse) recently, though, with less classical music heard (or made) in the home, the tragic decline of public school music education and the ghettoizing of the radio dial so that classical music is heard only on "classical" stations.

Everyone talks about "outreach"—going after new audiences with all sorts of gimmicks. This word was used in conversation by someone at the Cleveland Museum of Art last week to explain the rationale behind the museum's new Gallery Concert series. The feeling is that the seductive sounds of music coming from the armor court might interest casual gallery-goers who had come to the museum for non-musical reasons.

The museum has, of course, a regular concert series—but in order to attend those you have to enter Gartner Auditorium, a concert hall separate from the rest of the museum. Wednesday nights, you have to buy a ticket.

The magic word today is "crossover," and part of that strategy is to bring new audiences to classical music by having classical performers deal in popular or semipopular forms.

Does the tactic work? No one knows. Do people who come to symphonic pops concerts come back for full symphony programs? Do those who flock to concerts by the Philip Glass Ensemble end up as operagoers or symphony patrons?

My own feeling, admittedly subjective and anecdotal, is "crossover" in that sense is largely myth. Nor does it seem to work in the contrary direction: at Cleveland Orchestra's sold-out Christmas concert last month I did not see a single familiar face from the subscription audience.

Fine, you say, the orchestra is "reaching a different audience." But the question remains: How many of those Christmas concert patrons come back later in the season for the Mahler, Brahms and Prokofieff?

The business world long ago realized the power of classical music to sell its goods. That is why you hear snatches of opera or symphonic works in commercials for all sorts of products, and why we have two CDs on the market consisting entirely of operatic excerpts featured in recent movies.

One day last summer I had four or five phone calls from people who wanted to know what was the opera aria used in a champagne commercial on television (it was "O mio babbino caro" from Puccini's "Gianni Schicchi"). I told the callers as much as they cared to know about the opera—and wondered if any of them might be motivated to look into "Schicchi" itself, for it is a delightful show.

People cannot be coerced into a taste for classical music anymore than they can be coerced into a taste for stewed turnips or differential calculus. For more than half a century the Cleveland Orchestra has had one of the most extensive programs of children's concerts of any orchestra in the country—yet today it is just as worried as anyone else about where its future audience is to come from.

I have never forgotten hearing Louis Lane, in a speech many years ago, wonder out loud if all those well-intentioned children's concerts did not amount merely to "inoculating children against music."

A taste for classical music is something that has to sneak up on you and seduce you with its magic almost unawares. You hear something by accident somewhere; some friend puts a record on while you are visiting; someone recommends a record or hands you a concert ticket; you hear a snatch of something on television.

A taste for classical music will always be a minority taste, since the pop music industry is a large and profitable business that floods the marketplace with its product and makes a lot of people very rich. The classical music establishment, by contrast, lives a precarious existence, forever teetering on the edge of financial disaster. It cannot compete on anything like equal terms in the musical marketplace.

Critics and writers in classical music may be partly to blame for its minority status because we make this music that we love sound so formidable. We give the false impression that you have to have lots of education and know two or three foreign languages before you can "understand" what this music is all about. To read some people you would think that only those who took music appreciation 101 in college have any hope of enjoying a Tchaikovsky symphony. This, of course, is total hogwash.

I never took music appreciation 101. There was lots of music on the radio at home and a certain amount of piano playing, some of it my own. No one force-fed me. It was just there, and it ensnared me without my even knowing what was happening. When I began to show an interest, my parents, to their credit, took up the cause.

This does not seem to be happening now, which is why we have all these solemn projects for "outreach" and "crossover." These efforts are all praiseworthy, maybe even necessary for the survival of our musical life. But I have my doubts that they will succeed in the absence of the kind of silent, wonderful aural seduction that comes with simply having this music easily available.

SOURCE: Robert Finn, "Outreach Gimmicks Furrow a Skeptical Highbrow," *Cleveland Plain Dealer*, January 14, 1990.

Questions

1. How did you become interested in the arts?
2. Do you agree with the general premise that outreach programs are merely marketing gimmicks? Defend your argument.
3. Do you think the crossover program at the Cleveland Museum of Art has any merit? Explain.
4. Do you agree with the premise that "critics and writers in classical music may be partly to blame for its minority status" because they give the "impression that you have to have lots of education and know two or three foreign languages before you can`understand' what this music is all about"? Can you cite examples of how arts organizations have conveyed a similar message?
5. The author doubts that crossover and outreach programs will work because classical music is not easily available. What would you suggest be done to solve this problem?

References

1. Philip Kotler and Alan Andreasen, *Strategic Marketing for Nonprofit Organizations*, 3d ed. (Englewood Cliffs, NJ: Prentice-Hall, 1987) 61.
2. "AMA Board Approves New Marketing Definition," *Marketing News*, March 1, 1985, 1.
3. Ibid., 5.
4. Ibid.
5. Ibid., 7.
6. Ibid.
7. Ibid.
8. Ibid., 8.
9. Ibid.
10. Ibid., 19.
11. Ibid., 14–16.

12. Joy Horowitz, "Off the Street and Into the Audience: Tourists Help Pick Fall TV Schedule," *New York Times*, July 7, 1991.
13. Kotler and Andreasen, *Strategic Marketing*, 38.
14. Ibid.
15. Ibid., 39.
16. Ibid., 41.
17. Ibid.
18. Charles D. Schewe, *Marketing Principles and Strategies* (New York: Random House, 1987), 33.
19. Ibid., 36.
20. Ibid., 36.
21. Arnold Mitchell, *The Nine American Lifestyles* (New York: Warner Books, 1983) 13–24.
22. Arnold Mitchell, *The Professional Performing Arts: Attendance Patterns, Preferences and Motives* (Washington, DC: Association of Performing Arts Presenters, 1984), ES–1 to ES–4, and 21–24.
23. Michael J. Weiss, *The Clustering of America* (New York: Harper and Row, 1988), 130.
24. Lynne Fitzhugh, "An Analysis of Audience Studies for the Performing Arts in America," Part I, *Journal of Arts Management and Law 13* (Summer, 1983): 51–60.
25. Weiss, *Clustering of America*, 25.
26. Schewe, *Marketing Principles and Strategies*, 55.

Additional Resources

There is no shortage of books on marketing. A quick trip to a bookstore should turn up numerous titles on marketing. Listed below are some additional resources that will prove helpful for arts marketing.

Blimes, Michael and Ron Sproat. *More Dialing, More Dollars: 12 Steps to Successful Telemarketing*. New York: American Council for the Arts, 1984.

Morison, Bradley G., and Julie Gordon Dalgleish. *Waiting in the Wings*. New York: American Council for the Arts, 1987.

Melillo, Joseph V., *Market the Arts!* New York: Foundation for the Extension and Development of the American Professional Theatre, 1983.

Newman, Danny. *Subscribe Now!* New York: Theatre Communications Group, 1977.

Surveying Your Arts Audience. Research Division Manual, NEA., Washington, DC: 1985.

Market the Arts!, which was published in 1983, is available through sources such as the American Council for the Arts (ACA) in New York. The original publisher, The Foundation for the Extension and Development of American Professional Theatre (FEDAPT), is no longer in business.

13 Fund Raising

❑❑❑❑❑

Apart from the ballot box, philanthropy presents the one opportunity the individual has to express a meaningful choice over the direction in which our society will progress.

George G. Kirstein

The act of giving to good causes is well established in U.S. culture. The charitable system developed by various religious organizations to provide social services in the United States still depends on individual donations of funds, goods, and services. The intervention of direct government support in this system is a fairly recent phenomenon. In fact, government subsidies only became widely institutionalized after 1933 in the U.S. Today, the United States has a unique mixture of public and private support for health, education, social services, and culture. Government support of giving is also reflected in the tax benefits available when a person files a tax form and itemizes their expenses.

Organized fund raising by entities other than churches dates back to the nineteenth century. For instance, the International Red Cross operated the first disaster relief fund drives as early as 1859.[1] One source cites Lyman L. Pierce and Charles S. Ward as the fathers of modern fund raising, based on their work for the YMCA in the 1890s. The techniques they developed were used in 1905 to raise money for a new building in Washington, D.C., and these techniques made them pioneers of the major capital campaign. In fact, they may have been the first fund raising consultants, judging by the work they did assisting the U.S. government sell war bonds to help finance World War I.[2]

From these humble beginnings has risen a multibillion dollar industry. In 1990, for example, more than $11 billion was given to non-profit and charity organizations by corporations and foundations. Of that total, more than $1.4 billion was distributed to U.S. arts and culture organizations.[3]

Fund Raising and the Arts

Perhaps no area of managing an arts organization comes under closer scrutiny or is subject to more pressure than fund raising. For many organizations, 40 percent or more of the yearly operating budget may come from gifts or grants by individuals, foundations, arts councils, and corporations. If there is a decline in gifts from any of these sources, arts organizations with little or no cash reserves often find themselves in serious financial difficulty.

As we have discussed, the changing external environments (economic, political and legal, cultural and social, demographic, technological and educational) create opportunities and pose threats for arts organizations. Each of these environments may have an impact on the organization's fund raising efforts. A recession will probably signal a slowdown in giving because people feel they need to retain more of their discretionary income. In an election year, major donors may give more to candidates and less to cultural organizations. Changes in the tax laws could also affect giving. If people gain a benefit by making a donation to an arts organization and lowering their overall tax liability, they will become donors. However, the cause-and-effect relationship between these environments and donations is very unpredictable. Therefore, the arts manager must keep a watchful eye on the donation flow.

Direct government support of the arts in the United States still represents a very minimal commitment of resources even twenty-five years after the establishment of the National Endowment for the Arts. Direct government subsidies of the arts in many parts of the world are hundreds of times greater per capita than in the United States. The unique partnership of individual and private support for the arts defines the conditions in which all fund raisers must work.

Fund Raising Plans

Because fund raising is so closely linked to the overall fiscal health of the organization, management of fund raising activities must be thoroughly integrated into the strategic planning process. In fact, many arts organizations place marketing and fund raising under the control of a development director. This person hires specialists in each area of development to realize the objectives formulated in the short- and long-term organizational plans.

In organizations with inadequate staffing (which describes many arts organizations), one person may try to manage and implement annual giving, a capital campaign (for a new building, for example), foundation contacts, and local, state, and federal fund raising. It becomes very difficult to meet these diverse fund raising objectives effectively given this type of work setting. As we have seen in the chapters on planning, organization, and control, a manager needs adequate resources to carry out the organization's overall objectives. Because each of these fund raising areas requires a working knowledge of a vast amount of detail, it is unrealistic to expect one person to keep any sense of perspective with this impossible workload.

As we will see, much of the work involved in fund raising is research and writing. Much preparation is involved in carefully cultivating a match between the organization and the donor. On the other hand, a great deal of fund raising also involves "shmoozing"—having a friendly talk—with potential donors. Without the time and help to research and cultivate donors, the fund raiser's success will be very limited. The rewards are usually not immediate. Years may go by before an individual finally makes a donation to the organization. People who seek instant gratification will find development a very frustrating area in which to work.

On the whole, fund raising seems to be a growth industry. There is a constant high demand for people who can organize and effectively manage the fund raising activities of a nonprofit corporation. The downside of this high demand is the often-unrealistic expectations about how much money can actually be raised. The tendency to overestimate eventually leads an organization into a deficit operating mode. The net result is a high level of turnover in the development area in the nonprofit sector.

In this chapter, we will explore the requirements an organization

In the News—Fundraising in Action

Pennsylvania Ballet to Stay Open, After Fundraising Effort

PHILADELPHIA—The Pennsylvania Ballet said it completed a fundraising campaign that will enable the dance company to stay in business for at least one more season.

Major contributions from the Pew Charitable Trusts, the William Penn Foundation and the Annenberg Foundation, together with thousands of individual donations, brought in the $2.5 million the dance company needed to meet its budget deficit for the fiscal year ending July 31.

In March, the 27-year-old dance company was on the brink of going under, hurt by heavy debt and declining donations. But rather than sink, the company agreed to perform without pay and launched a "Save the Ballet" campaign.

After the initial campaign raised about $1 million, foundation and corporate donors agreed to kick in more money. The Pew Charitable Trusts donated $500,000 after the fundraising effort reached the $2 million mark.

With the immediate budget crisis behind it, the ballet company said it scaled back its budget for fiscal 1992. Moreover, it has cut back the number of performances to around 70 from more than 100, and its series to five from eight.

must meet before it tries to raise money. We will also discuss strategies to use in approaching different target donors and organizations that specialize in giving to the arts.

Preparing Fund Raising Plans

James Gregory Lord, a recognized expert in the field of fund raising, notes that "people give to people." He goes on to say, "People don't give to an institution. They give to the person who asks them. Often, a contribution is made because of how one person feels about another. The institution may be almost incidental. People also give *for* people—not for endowments or swimming pools."[4]

If fund raising managers keep this fundamental fact in the forefront of all planning and solicitation efforts, they will probably be successful in establishing in donors a lifelong pattern of giving. No matter what strategy an organization plans to adopt in its fund raising efforts, the bottom line depends on regular donations. Without the regular support of individuals, corporations, foundations, and government (even if it is only a tax break), most organizations would not be able to survive. Let's examine in more detail how to go about establishing a pattern of regular giving.

Strategic Planning

Most fund raising activity begins with a great deal of background work. Unless the organization happens to have a wealthy benefactor who hands out money with no questions asked, countless hours must be spent preparing to ask people for their support. The flow chart in Figure 13–1 depicts a typical system for organizing a fund raising campaign.

An organization's strategic plan normally contains a section outlining the proposed fund raising efforts. In Chapter 5, the concepts of the overall organizational strategy and the operational strategies for special areas were discussed. In Chapter 12, we saw how the marketing plan would be integrated with the strategic plan. Now we must consider how the fund raising needs would be integrated into the overall strategic plans.

Let's take the example of an organization that adopts a growth strategy. It is safe to assume that the fund raising management staff would need to address the issue of finding new sources of funds for the organization. This in turn requires that much time be spent on donor research. On the other hand, if the organization adopts a stability strategy, the fund raisers might concentrate their efforts on the current donor base. As with any planning process, multiple strategies can be incorporated into the overall master plan. However, the staff and budget resources required to support this approach can become burdensome.

The next step, determining the amount needed, is based on a careful analysis of the budget and the fiscal health of the organization. For example, if the strategic plan called for establishing an operating endowment fund to provide an annual income of $50,000, the fund raising goal could be as high as $1.25 million. This figure assumes that the $1.25 million would earn a conservative interest rate of 8 percent, thus yielding $100,000 per year. After deducting $50,000 for the operating fund, the remaining $50,000 would be reinvested in the endowment to overcome the annual effects of inflation, which is assumed to be 4 percent.

The actual campaign design involves formulating written mate-

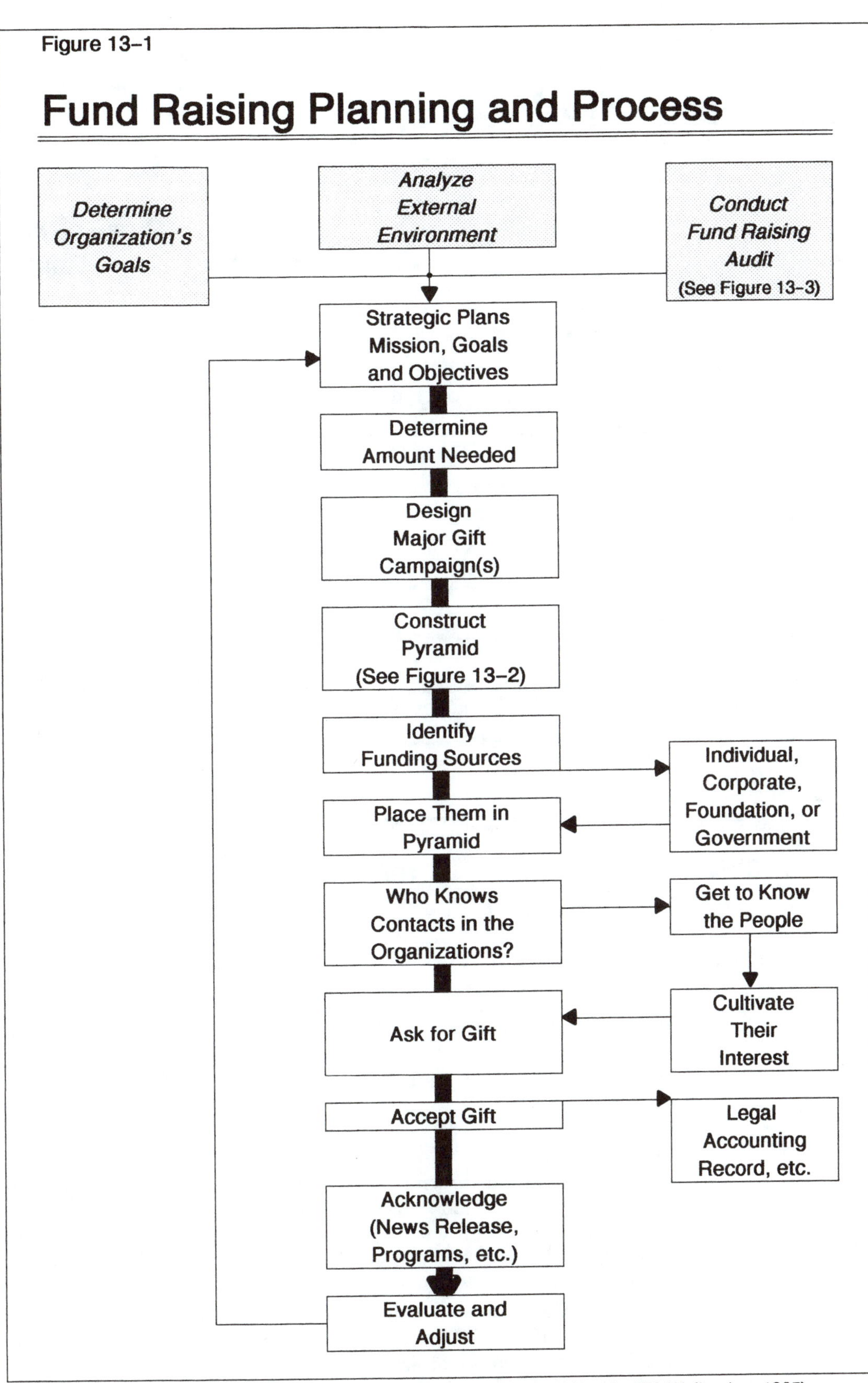

Source: Thomas W. Tenbrunsel, The Fundraising Resource Manual, (Austin, TX, Wellspring, 1985).

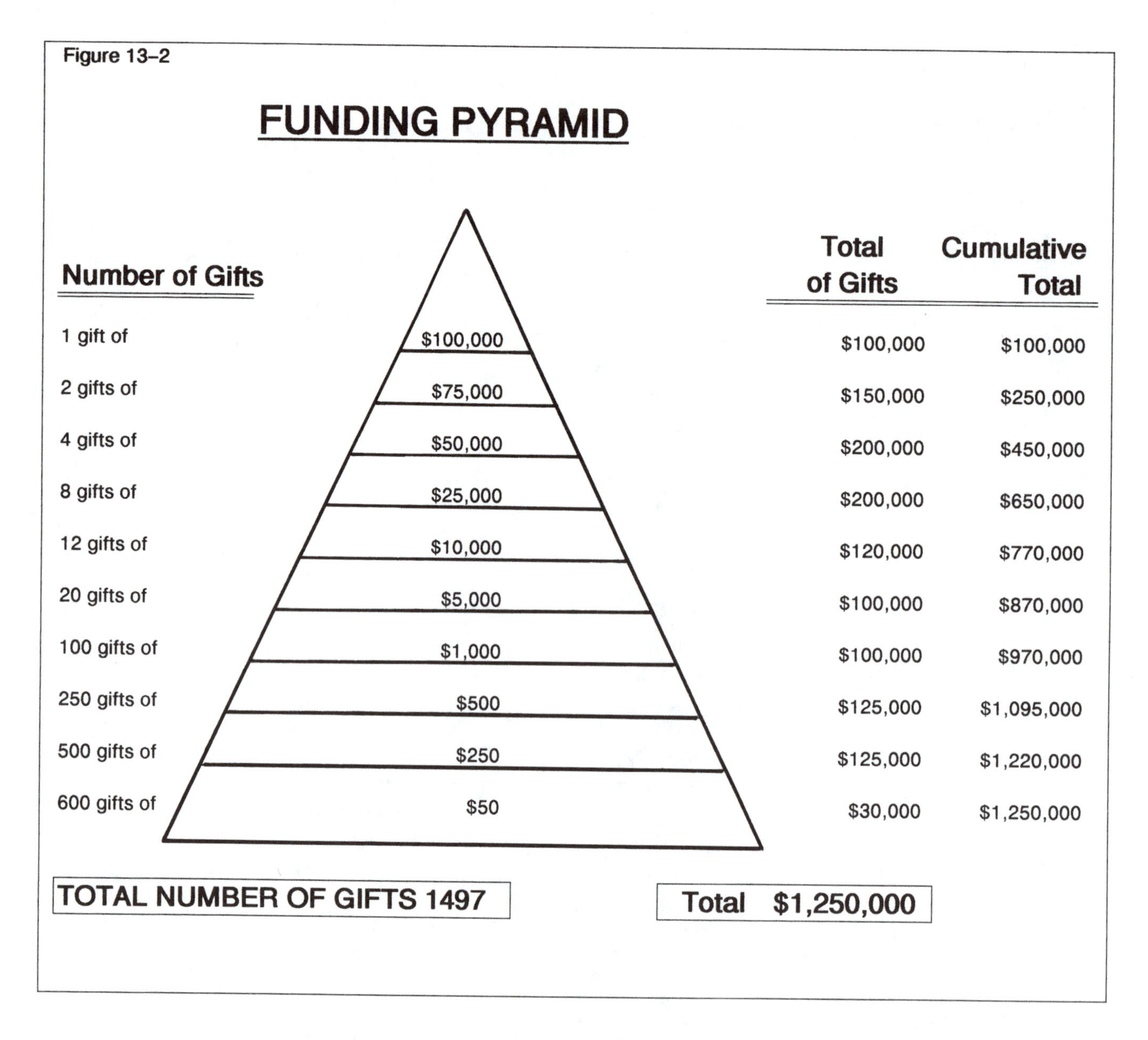

rial, creating the graphics and brochures to communicate the project or program, planning special fund raising events, (such as auctions, dinners, and costume balls,) and tactics such as telephoning donors to ask for support.

Funding Pyramid

The funding pyramid shown in Figure 13–2 is widely used to establish how many gifts at which amounts will be needed to meet a specific goal. A major campaign with a goal of $1.25 million usually concentrates its initial efforts on raising at least half of the money before publicly announcing the campaign effort. A lead gift of $100,000 and gifts of at least $10,000 are secured before the campaign is announced in order to build momentum. With at least half of the money raised, fund raisers can tell people, "Here's a project that others are willing to support."

To succeed in building the pyramid from the top down, the fund raisers must do their homework. Identifying possible funding sources, evaluating their giving potential, ranking them within the pyramid, and finding the right contact person could take a year or more. All of this work can amount to nothing if the wrong person asks for the gift. As James Lord says, "People give to people." It is critical that the fund raising staff educate board members and other volunteers about how and when to ask for support. As we will see later in this chapter, the entire fund raising effort is a marketing effort. The fund raiser tries to match the wants and needs of the donor with the goods and services of the organization.

Assuming success, the process is completed by legally accounting for the gift and acknowledging the support. As with the marketing campaign, evaluation and adjustment go on constantly in an effort to fine-tune and maximize the gift-giving program.

Fund Raising Audit

A fund raising audit is a useful approach to consider when preparing to raise money. Figure 13–3 shows an organization audit outline based on a format used by Thomas Wolf of the Wolf Organization, Inc., in his fund raising workshops. Each of the five sections addresses a major area that organizations must consider when preparing their fund raising programs.

Developing a clear mission, a positive image, and a strong case for support is the first step. Identifying the staff, board, and volunteer support system is next. Third, the organization must review its fiscal condition. Donors do not want their investment in an organization to go to waste as a result of financial mismanagement. The fourth area, activities and programs, asks important questions about what the organization is doing and how well it is serving the community. Finally, the audit coordinates the planning process by reviewing questions about mission, goals, objectives, process, and the evaluation system. As with the marketing plan, evaluating and adjusting the fund raising activities to fit changing conditions is very important.

Marketing and Fund Raising

An effective fund raising campaign requires the implementation of a well-organized marketing system. As in the marketing campaign, the goal is to achieve a close match between the donor and the funding need. In this case, the exchange process should make donors feel that the money, goods, or services they are donating will help solve a specific problem. For example, the money they donate to the operating endowment, if marketed properly, becomes a gift that supports any number of aspects of the organization. Finding the proper button for each donor—for example, the young artists' program or the museum arts classes for children—is a way to maximize the satisfaction of giving. The fact is, if the fund raising campaign is not donor driven in the way it is packaged and presented, it will probably not be effective.

As noted in Chapter 12, marketing and fund raising are usually less successful when presented in a product- or sales-oriented manner. A general appeal that says, "Give to us because we make great music," will not promote the exchange process. A letter mailed to a parent that says, "Your gift will help a child experience the wonder and joy of music," might prove to be more successful.

Fund Raising Management

Fund raising involves what by now should be the familiar aspects of project management: budgets, schedules, time tables, problem solving, and group leadership techniques. An individual with excellent group and project management skills is required for a successful campaign.

Sample Planning Document

The long-range plan for the San Diego Symphony Orchestra includes a section on fund raising. The material below demonstrates one way to articulate goals and objectives. Only one strategy for one goal is included in the excerpt. There are eight additional strategies for the first goal, and the entire development section of the five-year plan is 11 pages long.

Five-Year Plan, Fiscal Years: 1990–1995
San Diego Symphony Orchestra Association

Development

Goal: To maintain a stable and growing income stream from contribution programs (to include private sector, corporate, foundation and government support), which improve the organization's fiscal strength.

Objective 1: To generate and increase annual gift income to meet the needs of the annual operating budget.

Strategy: Maintain appropriate level of personal support from Board members for the annual fund and increase their awareness of their fund-raising responsibilities and the Development Department programs.
1990-91: President of the board solicits each Board member face-to-face each year, prior to the beginning of new fiscal year, with periodic follow-up when appropriate. Development Director meets individually with each Board member quarterly to review development needs and prospects; each Board member is to select six prospects for face-to-face solicitation.

Basis for Evaluation: Do annual comparison of percent of Board goal raised prior to start of new fiscal year.

SOURCE: San Diego Symphony Association, *Five-Year Plan, Fiscal Years 1990–1995*, November 1990, p. 60.

Figure 13–3

Fund Raising Audit

I. MISSION, IMAGE, CASE FOR SUPPORT

A. Mission
- What is the public service provided?
- Is it relevant? Current?
- Who is being served? How diverse is the constituency?

B. Image
- Is the organization known?
- What is its reputation?

C. Case for support
- What are the needs of constituency/clients?
- How well are they articulated?
- How varied a menu is offered to funders?

II. HUMAN RESOURCES

A. Board
- Giving and getting profile
- Evidence of responsible trusteeship

B. Staff
- Skills and qualifications
- Sufficient number of staff to carry out activities?
- Morale

C. Volunteers
- Numbers
- Experience
- Commitment level

III. FINANCIAL PROFILE

A. Fiscal accountability
- Do balance sheet and income statement show prudent financial management?
- Controls in place?
- Good budgeting systems? Program budgets available?

B. Income patterns

- Good balance between earned and unearned income?
- Sufficient availability of unrestricted funds?
- Not too dependent on single source or public sector?

IV. ACTIVITIES AND PROGRAMS

- Resource audit (what does organization do well?)
- What is the competition?
- Are programs really needed?
- How broad a constituency do they serve?
- What do the activities offer to donors?

V. PLANNING AND EVALUATION

A. Planning

- How well defined are the mission, goals, objectives, and action plans?
- How good is planning process? How involved is the Board, the constituency, and the staff?
- How far ahead has organization planned?
- Is there a multiyear financial plan?

B. Evaluation and accountability

- How are programs and activities evaluated?
- Are the evaluations based on quantitative targets?
- How is the overall effectiveness of the organization evaluated?
- How often are evaluations carried out and by whom?

Thousands of details must be coordinated into a unified whole if the organization is to reach its fund raising goals. Let's begin by examining the background work required for getting ready to ask for support. Then, we'll review the techniques and tools used to maximize the possibility of support from various funding entities.

Background Work

What Does the Organization Do?

After completing the fund raising audit, the process should shift to formulating what the organization can do to address a range of needs in the community. How does the symphony orchestra, museum, or

Sample Capital Fund Raising Campaign Document

The excerpt below illustrates how a case was built for a capital campaign to support the Hartford Stage Company in Hartford, Connecticut. The need to provide a stable base of support for a cultural resource and to maintain a high level of quality is emphasized. The full document explains the numerous ways people can support the campaign by taking advantage of "gift opportunities" that span such needs as the Visiting Director's Fellowships, a Student Subscription Fund (ticket subsidy), Senior Citizen Access Fund, and renovation projects in the lobby and offices.

Campaign for Hartford Stage:
A Critical Choice for Hartford Stage

Hartford Stage is one of the most exciting and influential theaters in America. Widely respected for the originality and quality of its productions, it has attracted the country's top artistic talents. Hartford Stage has been a vital force in the cultural, economic and social fabric of its city and serves as Hartford's cultural ambassador to the nation.

Over the last five years the theater has put its finances in order, increasing income and controlling expenses while maintaining artistic quality on stage. In the short term, austerity measures have enabled Hartford Stage to balance its operating budgets while managing to sustain artistic excellence.

But continuing austerity has also put the theater at risk for the future. Financial projections show that costs, some of which can no longer be deferred, will soon out pace revenues. The quality of its productions and the condition of its facility are in danger.

Hartford Stage must choose between sacrificing artistic quality or securing a new source of income through a capital funds drive. . . .

Hartford Stage's Board undertook an in-depth assessment of the theater's requirements and determined that Hartford Stage must embark on a $6.1 million campaign to create a solid foundation to protect the theater as an important cultural resource. The $6.1 million is a realistic goal which presents an absolute need.

The Campaign for Hartford Stage is the only way in which the Hartford community can ensure the continued vitality and financial stability of a theater that has been at the heart of the region's cultural life for a quarter of a century.

The Goals of the Campaign for Hartford Stage

The Campaign will:

provide funds to restore the building and maintain the current level of artistic programming during the campaign period. Total funds required: $2,100,000.

The Campaign will:

establish an endowment to generate immediate annual income to bridge the projected gap between revenues and program expenses, sustaining the theater's outstanding level of artistic achievement. Total funds required: $4,000,000.

SOURCE: "The Campaign for Hartford Stage" (capital campaign materials). Sandra B. Wood-Holdt and Howard Sherman, Copyright © 1990 by the Hartford Stage Company. Reprinted with permission.

dance, theater, or opera company satisfy the current needs of the community? What needs are not being met? Can the arts organization fulfill any of these needs? In *Managing a Nonprofit Organization*, Thomas Wolf identifies three important steps in this process, which he calls *the case for support*[5]:

1. Identify the important problems or needs that the organization intends to address with the help of the contributions.
2. Demonstrate the organization's ability to address these needs.
3. Match the proposed areas of organizational activity with the funder's own philanthropic interests.

The obvious starting point for an organization attempting to identify problems or needs is to address the current programs and projects. For example, the fictitious theater company examined in Chapter 11 had established a home season, touring operation, education program, and a building fund (capital campaign) to meet the various needs of the community. To make its case for support, the theater company would offer proof of how it is uniquely qualified to meet the community's needs through its regular season, which enriches the cultural life of the community by presenting quality productions. The theater could argue that the touring operation provides a service to a wide geographical area and a diverse audience and that the education program offers a school or apprentice program to teach acting skills to young people. The building fund is targeted to provide a permanent home for the theater company so it may increase its effectiveness in presenting its season and its projects in the community.

After the theater company outlines what it is doing and how it is effectively addressing the community's needs, the important process of matching activities to funders takes place.

Accurate research about potential donors is critical to making the optimal match. For example, a foundation may focus on education, a corporation may support high-visibility activities such as touring, and individual donors may want to be associated with a new facility. A significant amount of time can be wasted if the wrong donor is approached. Even worse, a potential donor may be turned off to your organization because of an inappropriate approach. A closer examination of this matching process is presented later in this chapter when the various funding sources are discussed. (See "Campaign for Hartford Stage" for an example of a fund raising effort.)

Staff and Board Participation

One of the expectations of any fund raising campaign is that staff and board members will actively participate to help reach the goal. Potential donors may ask how much support staff and board members provide to the organization. Answers about the average contribution per staff or board member must be at hand. After all, why should donors give to an organization that does not have the support of its own people?

Many nonprofit organizations expect their board members to contribute a particular amount each year. This could range from a few hundred to many thousands of dollars. For some organizations, especially organizations made up of board members selected for their expertise and not for their wealth, there are expectations about contributing specific amounts of time to projects each year. In addition, nonprofit organizations are now asking staff members to make an annual donation. The amount is usually significantly smaller on average than the board member's gift, but the demonstration of strong internal support for the organization is important when talking to outside donors.

Data Management

A well-designed management information system to gather data about potential donors is critical if the organization is to organize its fund raising campaigns. Chapters 9, 11, and 12 provide examples of management information systems designed for the overall operation, the financial system, and marketing. In arts organizations, which usually have very limited staff resources, the data gathered about donors should be integrated with the sales and marketing systems. This integration can be achieved if the computer software is designed to capture and store information about sales and giving. Several companies specialize in fund raising software for nonprofit organizations. Any issue of the *NonProfit Times* or the *Chronicle of Philanthropy* will contain advertisements for such systems. A careful analysis of the software's capabilities is required, and the support available from the company after the system is purchased should be explored to ensure that the organization's data management needs will be met in the long term. For example, the fund raising financial record-keeping system must be able to track revenue through the entire accounting system. As we saw in Chapter 11, the balance sheet and account statements must reflect changes in the fund balances based on these donations. The donor tracking system must therefore be integrated with the accounting software used by the organization.

Data system needs Members of the development staff should be able to sit down at a computer terminal, enter the name of a subscriber, single ticket buyer, or member, and pull up a complete list of all transactions or donations made by that person. Staff members might also want to identify everyone who donated more than $50 and less than $250 in the last year. Donor tracking systems usually contain data fields about the estimated salary and giving potential for each subscriber or member. Staff members might also want to know who gave from a particular range of zip codes. The ability to cross-reference donors with sales of subscriptions or memberships is also important. For example, if some patrons only purchased single tickets to the musicals that the theater company performed, this information could be effectively incorporated in a fund raising letter. The letter would mention the individuals' fondness for musicals and suggest that they make a donation to support the production fund so more great shows could be produced for their enjoyment.

The donor data management system is also critical in developing confidential financial information. For example, if the business section of the local newspaper announces that one of your patrons just received a promotion, this information should find its way into the data file. A promotion probably means a larger paycheck. This information is noted so that the next time a solicitation is made, a higher gift amount is requested.

A word of caution about confidential information. The tendency to put large amounts of irrelevant personal data in a computer is directly proportional to the ease with which the data can be entered. A clear policy about what information may be kept in the donor file and who may have access to these data is important if the organization is to have any credibility in the community. Policies about the confidentiality of the data gathered by the development staff must also be enforced. Passwords or security codes to access the donor data will be meaningless if staff members sit around the lounge discussing how much someone gave to the operation. Any breach in security should be dealt with quickly and visibly.

Fund Raising Costs and Control

The annual campaign and the various capital fund drives contain a mix of activities designed to reach as many potential donors as the budget will permit. Development managers always seek ways to keep the costs of raising money as low as possible. The impact of these costs cannot be ignored. Potential donors want to know whether the organization is capable of using their gift efficiently. Although fund raising costs may vary with different types of campaigns, if they reach 20 percent of the total raised, it is time for the organization to reassess its methods. Organizations that can keep fund raising costs under 10 percent are viewed favorably by donors.

An effective budget control system must be in place before an organization undertakes any fund raising activity. In addition, legal requirements must be met when reporting income raised through donations on federal and state tax forms. Some states require special licenses before any fund raising may begin.

Direct and indirect costs Arts organizations usually have *direct fund raising costs* for salaries, wages, and benefits. These costs should be distributed across the budget if several fund raising activities are supervised by one staff. Refer back to Figure 11–4. Consultant fees would also be listed as a direct cost to the project. Other costs include supplies and services (paper, copying, printing, telephones), equipment (computers), and travel. *Indirect fund raising costs* reflect such items as a portion of the rent, lease, or mortgage, utilities, and the maintenance of general office equipment used for fund raising activities. For the purposes of budgeting, the financial manager must calculate the various costs of each area's use of the common resources and formulate a distribution that can be used to prepare fund raising budgets.

When applying for government grants, the organization can be reimbursed for indirect costs if these costs are reflected in the budget. For example, if a museum gets a grant for $1 million to run an educational program, 30 to 50 percent of the budget could be allocated for indirect costs. The organization could therefore expect that an additional $300,000 to $500,000 would be provided above the $1 million to support the costs of supporting the project. Grant applications to foundations and corporations normally show indirect costs as part of the overall project budget. Foundations and corporations may place restrictions on or refuse to support indirect costs. The application guidelines for these granting agencies normally outline the costs they consider to be legitimate.

Fund Raising Techniques and Tools

A successful fund raising campaign never ends. Most organizations must continually seek donations if they are to survive financially. As soon as the annual campaign has been completed for one fiscal year, it is time to get started on next year's fund drive. The overall goal remains the same each year: to establish a regular pattern of giving to the organization. Let's examine some of the specific details of the various ongoing campaigns that an organization must maintain. In addition to the occasional large *capital campaign* (usually conducted every three to five years), *annual campaigns* targeted to individuals, corporations, foundations, and government agencies require constant attention and fine-tuning.

Individual Donors

All organizations want to have a great many individual donors who make regular unrestricted gifts to the organization. An *unrestricted gift* carries no stipulation as to how the funds may be spent. Unrestricted gifts give the organization the flexibility to shift funds to fill the greatest need. *Restricted gifts*, on the other hand, are given on the assumption that the funds will be used for a specific project or program. The organization has a legal obligation to use restricted funds in the manner designated by the donor. An unrestricted gift might be added to the operating fund balance or be used to cover the expenses of a specific production or project. A restricted gift might be designated only for the

For Your Information

While corporations and foundations get all the attention and headlines, individual giving accounts for 89.9 percent of all giving for philanthropy in the United States, with corporate support at 4.4 percent and foundation support at 5.8 percent.

SOURCE: Milton Rhodes, "Forging Ahead," *Vantage Point*, Fall 1990, p. 3. Copyright © 1990 by the American Council for the arts. Excerpted with permission.

building fund endowment. Because solicitations to corporations, foundations, and the government often carry distribution restrictions, the more unrestricted gifts the organization can regularly gather, the better.

Because regular giving is the lifeblood of many organizations, it is fairly common for organizations to maintain a standing committee of board members and staff to coordinate the fund raising activity. Yearly funding goals and objectives are set, a detailed time table is created, specific details—such as who makes the calls, who signs the letters—are worked out, and assignments are distributed to the board and the staff. This is an example of all of the management theories coming together. The fund raising committee must plan, organize, and lead effectively if the organization is to remain strong.

The techniques for building a large base of individual donors include donor research, offering numerous funding options, personal contact, telephone solicitation, direct mail, and special events.

Donor research Many of the techniques used to develop an audience are used in donor research. The current subscribers, members, and single ticket buyers form the core of the donor base. This core group should be subjected to the most intense research, and as complete a donor file as possible should be compiled on each person.

The next level of research focuses on prospects. Vast amounts of data about prospective donors must be gathered and rated in terms of potential for further use. As Thomas Wolf says, "Only prospectors find gold."[6] The organization must commit personnel to go through lists of former subscribers, patrons of other arts organizations, country club members, and members of social or business organizations. They must also explore school phone directories (including college or university phone books), references given by current donors, and published social registers. Sources such as *Who's Who in America*, which publishes regional directories, may also be of use.

Funding options Fund raisers like to speak of gift giving as an "opportunity" or a chance to make an "investment in the future." Well-organized development managers design several choices for donors using a concept not unlike a menu. For example, gifts can be targeted for the current operating fund for those people who want their gifts to be put to use immediately. Others may want their gift to go to an endowment fund, which is invested, and only a portion of the interest is used to fund operations or special programs. A scholarship fund is a good choice for donors who want their gifts to have maximum longevity. Others may want to offer their funds in the form of a *deferred gift*, that is, a promise to provide funds, property, stocks, bonds, life insurance, property, or jewelry at some future date. Another form of deferred giving is a *bequest*, which is a gift that is distributed through the donor's will. Some donors specify that a portion of their life insurance will be donated to an organization.

As with any menu, it should be possible for the donor to combine several options. For example, an organization with annual donors who make bequests and give regularly to an endowed fund is in a fortunate position. Offering a selection of donor options is taking the principles of being a customer-driven, marketing-oriented organization to its logical conclusion. The donor programs must be designed to provide the maximum exchange satisfaction for donors who write that check or sign a document giving something to the organization.

Personal contacts Personal contact is the preferred method for securing a gift because fund raising is most effective when people ask other people for their support. In a typical capital campaign in which large sums of money are sought, the organization usually engages in a long courtship with the donor. Let's assume that a museum's donor research targets a local business executive for a gift to the building fund. On paper, this person looks like a good candidate, but without an introduction, the organization might get a firm refusal. If the research indicates that a current member of the board knows the executive, a valuable link exists. If there is no existing contact, an appointment should be made to get things rolling. The first meeting will probably be brief and informational. The board member who knows the executive should attend the meeting to help smooth out the initial communication. A staff member might accompany the board member to be introduced and answer questions, but his or her presence is merely functional. A package of information about the organization and a brief outline of how the current fund raising drive is doing would probably be enough for this first visit.

The key goal in making the first contact is for each party to learn more about the other. Follow-up meetings, invitations to various events sponsored by the organization, lunch at a local restaurant, informative notes, and updates by telephone would round out the process. When the fund raising committee feels that the time is right, the designated contact makes the request. The staff member should not actually ask for the gift. It is unethical for staff members to make such requests because they are paid by the organization. The board member or a volunteer should do the asking.

Telephone solicitations As shown in Figure 13–2, the typical funding pyramid includes many gifts valued at $500 or less. Personal contact is not an effective way to reach all of these donors. One of the most cost-effective alternatives is to make phone calls to as many prospects as possible.

Enthusiastic leadership is a must when motivating the board and other volunteers to ask for money over the telephone. This is especially true because the process of asking people for money can be very discouraging. Fund raisers always tell volunteers, "Don't take that 'no' personally!" However, it is only human nature to feel as if your request was rejected because you did not ask in the right way or you were not convincing enough.

Small organizations normally schedule a week or two each year for telephone fund raising campaigns. Banks of telephones, eager volunteers, and a little training with a well-written and flexible script can translate into thousands of dollars for an organization. Again, solid research can pay off for the organization. When a caller begins talking with a prospect, an information card or computer screen should help guide the interaction. Potential donors are initially very hesitant about getting a phone call from a stranger, so the first 20 seconds of the conversation is usually scripted to ask questions designed to get the prospect to respond. Assuming that the information about the potential donors is correct, it should be possible to establish what they thought of the last performance they saw or the last exhibit they attended. The key is engaging the person and building to the request in a timely manner. If the caller is able to connect with the prospect, the typical tactic

Figure 13–4

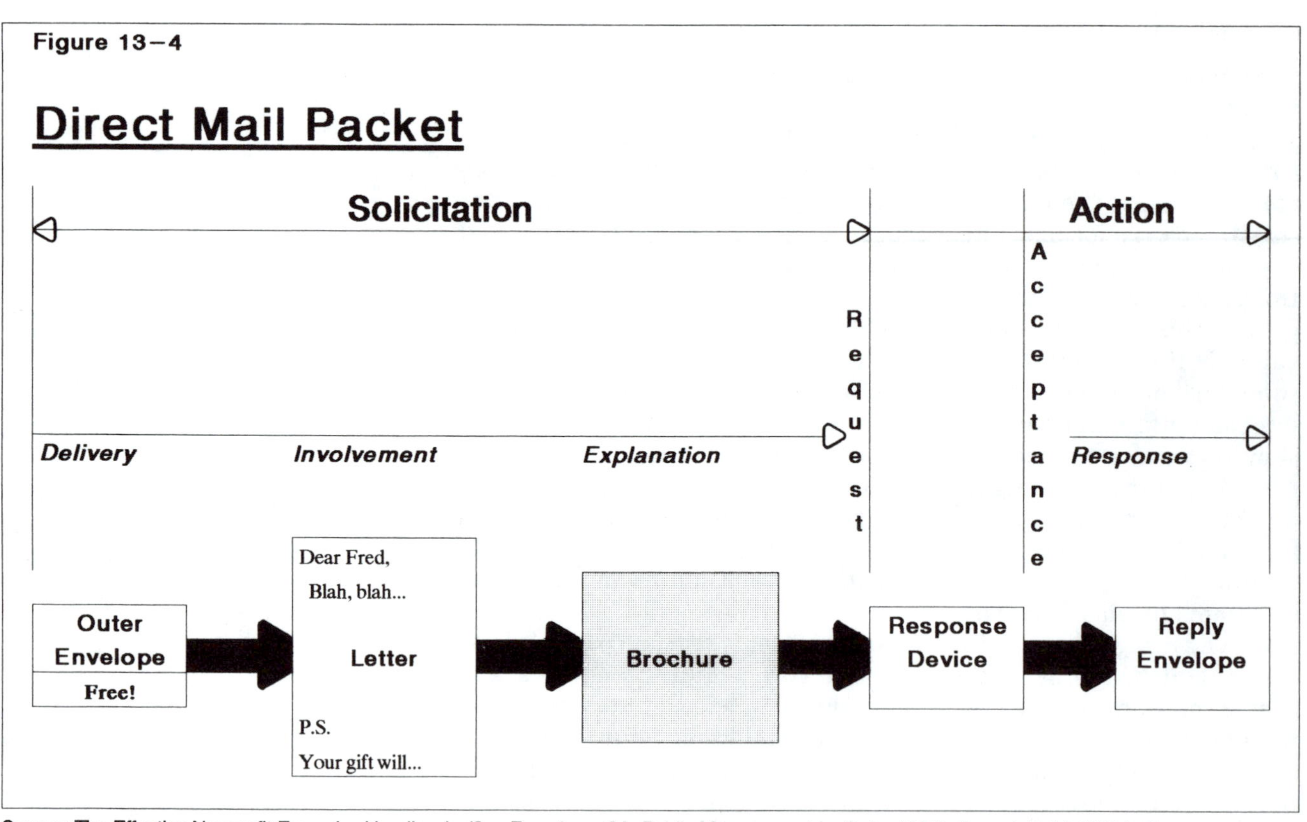

Source: The Effective Nonprofit Executive Handbook, (San Francisco, CA: Public Management Institute, 1982). Copyright (c) 1982 by the Public Management Instititue. Used with permission.

is to ask for a bit more than the donor research indicates. For example, if its possible that the potential donor might give $150, the caller might be scripted to suggest initially an "investment" of $500. Eventually, an amount with which the donor feels comfortable will be reached, and the "closing" can take place. The logistics of getting the gift information and payment method correct and thanking the donor concludes the process.

Advances in technology have permitted computers to dial automatically and play prerecorded sales messages, but many donors hang up as soon as they realize that someone has invaded their privacy with a request for money. However, the benefit for nonprofit organizations is a cost savings when they use this technology.

Direct mail Direct mail marketing and fund raising is a big business in the United States. Every day, high-speed computer printers merge-print millions of pieces of what most people call "junk mail" with names and addresses purchased from list brokers. Many people never open the handiwork of the direct mail marketer, but enough people respond to these offers to convince businesses that the cost is worth incurring.

Arts organizations have used the techniques of the direct mail marketer for years in attempting to build a subscriber or member base. The mailing of a solicitation for funds follows the basic direct mail principles and is traditionally part of the mix of options used by the development staff.

Figure 13–4 shows the usual steps involved in assembling a direct mail packet for a fund raising campaign. The outer envelope represents the critical first contact with potential donors. The envelope must

Figure 13–5

Response Table -- Fund Raising Campaign

	Total Solcitation	Number of Responses	Cost Per Unit	Response (%)	Average Gift	Total Raised	Campaign Cost	Net/Year or (loss)
Mail Year 1	10,000	150	$0.60	1.5%	$25.00	$3,750	$6,000	($2,250)
Phone Year 2	150	60	$1.00	40.0%	$75.00	$4,500	$150	$4,350
Phone Year 3	60	30	$1.00	50.0%	$125.00	$3,750	$60	$3,690
Phone Year 4	30	15	$1.00	50.0%	$125.00	$1,875	$30	$1,845
					Total	$13,875	$6,240	
							Net Raised	$7,635

communicate a short, strong, and clear message that something of interest is inside. A popular technique uses the word *Free* on the envelope with the assumption that people will be curious to see what the offer is. If the fund raiser can get potential donors to open the envelope, the combination of a well-written letter and an informative brochure will bring readers further into the solicitation. Because most people initially scan the text of the letter and brochure, the copy must be written and laid out in such a way as to get the message across in as few words as possible. The response device (the piece returned to the organization) and the reply envelope should provide a fast and easy way of completing the solicitation.

Tracking responses Direct mail is a long-term investment. Organizations that expect much more than a 1 or 2 percent response rate may be in for a surprise. For example, a 1.5 percent response rate for a 10,000-piece mailing would yield 150 donations. Let's say the average donation is $25. The first mailing thus yields $3750, as shown in Figure 13–5. First-class postage, letter, brochure, reply device, and reply envelope cost an average of 60¢ per unit. The first mailing therefore costs $6000 and raises $3750. However, if you view these 150 donors as a long-term investment, the $2250 loss will eventually be recouped. Suppose that the arts fund raiser tracks these donors as a target group that responded to a particular campaign. In the next year, a telephone solicitation of these 150 people yields a 40 percent response rate, and the average donation is $75. The organization has gained $4500 minus the cost of telephone solicitation. Let's say it costs $150 to get this next $4500. This leaves the organization ahead by $2100 on its total investment of $6150. In the third and fourth years, 50 percent of the remaining donors from the first year's campaign give an average of $125 each. The net yield now increases to $7635 in donations after subtracting $6240 in solicitation costs over four years. If carefully monitored and tracked, the organization should be able to create an overall system of periodic direct mail solicitations that yield a regular cash flow.

Note that this simplified example of a direct mail cost analysis is used to illustrate a point. Direct mail marketers have very comprehensive formulas for calculating campaign costs. For example, the data in Figure 13–5 does not take into account inflation, which is a cost to the campaign. In addition, the cost of raising money is fairly high in this example. Costs of 20¢ or less per dollar raised would be more appropriate.

Special events Arts organizations usually try to hold at least one event a year as a fund raiser. A group of volunteers coordinates and produces a costume ball, a silent or live auction, a raffle, or a benefit performance. The effort and time required to produce a major event can be overwhelming if the organization does not have the resources to make it happen. The costs of producing an event like a costume ball may run into the tens of thousands of dollars. Careful control of the budget is required, or the event may end up costing the organization more than it earns in donations. However, with a good planning committee and a realistic schedule, it is possible to earn thousands regularly for the organization and to provide a memorable experience for the donors.

These events can also provide visibility in the community for the organization. Raffles, for example, can be a way of involving the local business community in the arts by persuading business owners to donate goods and services. State and local governments may place restrictions on certain types of events, so it is always a good idea to consult with a lawyer before proceeding.

Corporate Giving

According to *Giving USA,* corporations donated $5.9 billion to charities in 1990. Although this is a substantial amount, it represents less than 5 percent of the $122.6 billion given by all funders to charities in 1990.[7] As the author of *"Corporations Tie Contributions to Their Profits"* points out, there is also a strong relationship between the economy and corporate giving. In addition, corporations undergo constant changes in ownership as they are bought, sold, and merged. Nonprofit organizations must adapt to the changing business environment if they intend to capitalize on the available funds. In the best of times, arts and cultural organizations are usually not at the top of the corporate funding priority list, but regular support can be found if fund raisers are willing to make the effort to track down the sources.

Corporate support is based in large part on the concept of *reciprocity*: What will the corporation gain by supporting a performance or an exhibition? A company may have motives for funding a specific event because of its public relations value, marketing potential, or benefits to its employees. The fund raiser's research must focus on trying to fit the organization into the corporation's donor strategy. The lack of a good *strategic fit,* as it is called, is the primary reason why support is not given to an organization. Arts organizations must remember that establishing a good strategic match is part of their marketing process. The packaging and emphasis of a proposal may need to be adjusted as the priorities of corporations change.

Corporate support is usually restricted to the immediate community because businesses are concerned about raising their profile in their immediate market. Larger corporations sometimes sponsor performances or major exhibits that have a highly visible national tour program. For the most part, a regional arts group has little chance of

In the News—The Realities of Corporate Funding

Corporations Tie Contributions to Their Profits: Bottom Line Comes First, Before Good Works

Claudia H. Deutsch

Executives in charge of corporate contributions used to worry exclusively about society, while their counterparts on the business side tended to profits.

Their philosophical paths never crossed.

Now, as both sides try to make do with slashed budgets, the separatism is ending. "We won't make a grant that doesn't fit the business strategy," said Cornelia Higginson, vice president in charge of overseas giving for American Express Co.

Neither will Paul M. Ostergard, director of corporate contributions for Citibank. "A professional contributions manager put the company's interest first," he said. "It's a classic case of doing well by doing good."

The timing is no coincidence. As the people who grew up influenced by the 1960s hippies and the 1980s yuppies move into positions of power, social conscience and capitalistic pragmatism have replaced social climbing as a prime mover behind corporate giving.

"Old-line CEOs used donations to get on symphony boards; today's CEOs use the money as a strategic tool," said Craig Smith, publisher of Corporate Philanthropy Report, a Seattle newsletter.

The recession has caused most companies to cut their marketing and contributions budgets at a time when nonprofit groups are also strapped for cash.

While the trend may hurt small, esoteric nonprofit groups, the larger savvier ones have used their new entree to marketing departments to get advertising and promotional advice. And they pander to self-interest—pointing out that sponsoring an avant-garde art show will position a company as innovative or that supporting child-care programs will help attract women as customers.

The overall result is that the wall between giving and selling has fallen. Smith said 55% of contributions managers have regular contact with marketing directors.

Nissan Motor Corp. U.S.A. asks line managers to review philanthropic projects. Citibank, American Express and GTE Northwest offer discounts or free tickets to customers at events to which they contribute.

When American Telephone & Telegraph Co. sponsors a play, its foundation often pays production costs, while marketing handles promotion.

International Business Machine Corp. authorizes local managers to make grants of up to $25,000—presumably to causes favored by local customers and employees. "If a customer is on the board of a worthy museum, why shouldn't that be the one we support?" asked Juan Sabater, director of corporate social policy.

Corporations have grown particularly adept at finding causes that tie in with their products. Since its purchase of Kraft and General Foods, Philip Morris Co. has sponsored hunger-relief and nutritional-education programs.

US West Inc., whose business strategy includes upgrading rural services and courting small businesses, announced a program this year to offer financial help and advice to small businesses in rural areas. "We always try to marry corporate and community interests," said Jane J. Pracan, executive director of the US West Foundation.

SOURCE: Claudia H. Deutsch, "Corporations Tie Contributions to Their Profits," *New York Times*.

attracting national corporate support unless there is an active branch of the corporation in the area.

A method once used to raise corporate support in some communities was to develop a United Way–type fund raising campaign for the arts. However, for many nonprofit organizations, the benefits of a regularly funding source were offset by a drop in overall corporate giving. Companies no longer had to give as many grants because they consolidated their giving and reduced their overall commitment of funds. Corporations saw an opportunity to continue to do good, but for less.

Potential problems The most problematic issue related to corporate support is the conditions (some direct, some implicit) that may be attached to a gift. For example, a performing arts group or museum may find its corporate support quickly withdrawn at the first sign of controversy. Once withdrawn, the chances of getting this support back may be very limited.

There may also be ethical considerations that the organization must take into account before applying for corporate support. For example, seeking funds from companies that produce products thought to be harmful to the environment or to people, or from companies that have holdings in politically repressive countries could be detrimental to the community perception of the arts organization.

Fund raising process Figure 13–6 shows the typical process used to seek a grant directly from a corporation or foundation. Corporations that make direct grants are listed in the *National Guide to Funding in Arts and Culture*. They number slightly more than three hundred.[8] Corporations generally establish foundations to distribute their gifts.

One of the most important steps in this process is the direct contact with an individual in the corporation or business. In this simplified model, if there is no existing contact, then a courtship process is undertaken. In many cases, the organization rewrites the proposal before asking for the gift because the initial meeting with the corporate contact made it clear that the original proposal did not address the company's current funding interests. In other cases, contact occurs before the proposal is written.

Whatever the situation, the overall process of corporate fund raising must be integrated within the master plan (see Figure 13–1). The funding manager for a nonprofit organization should read the business section of the daily paper and follow national trends in the business publications in an effort to stay in tune with the opportunities that may arise.

Foundations

A foundation is defined as a "nonprofit, nongovernmental organization with a principal fund or endowment of its own that maintains or aids charitable, educational, religious, or other activities serving the public good, primarily by making grants to other organizations."[9] The *National Guide to Funding in Arts and Culture* lists 3360 granting organizations. They are distributed as follows:[10]

Independent foundations	2168
Community foundations	129
Grant-making operating foundations	53
Company-sponsored foundations	707
Corportate giving programs	303

The *National Guide* is an excellent research source for information about types of grants, amounts granted, purposes, limitations, publications, and application procedures. Many other resource books are also available. Two are listed at the end of this chapter.

As with corporate fund raising, a good match between organization and foundation must exist. Foundations usually fund specific types of activities. For example, the Gap Foundation supports "employee matching gifts, capital campaigns, general purposes, operating budgets" in the San Francisco area. The Autry Foundation gives "primarily for cultural, educational, medical, and youth-related programs" in the Los Angeles area.[11] The fund raiser can investigate the kinds of grants given in the last few years to see if a match can be made.

As always, a clear, concise proposal and ability statement is the first step in the application process. Because many small foundations have little or no staffing, the application procedure may be as simple as a cover letter, a one-page proposal, and a budget. A large foundation

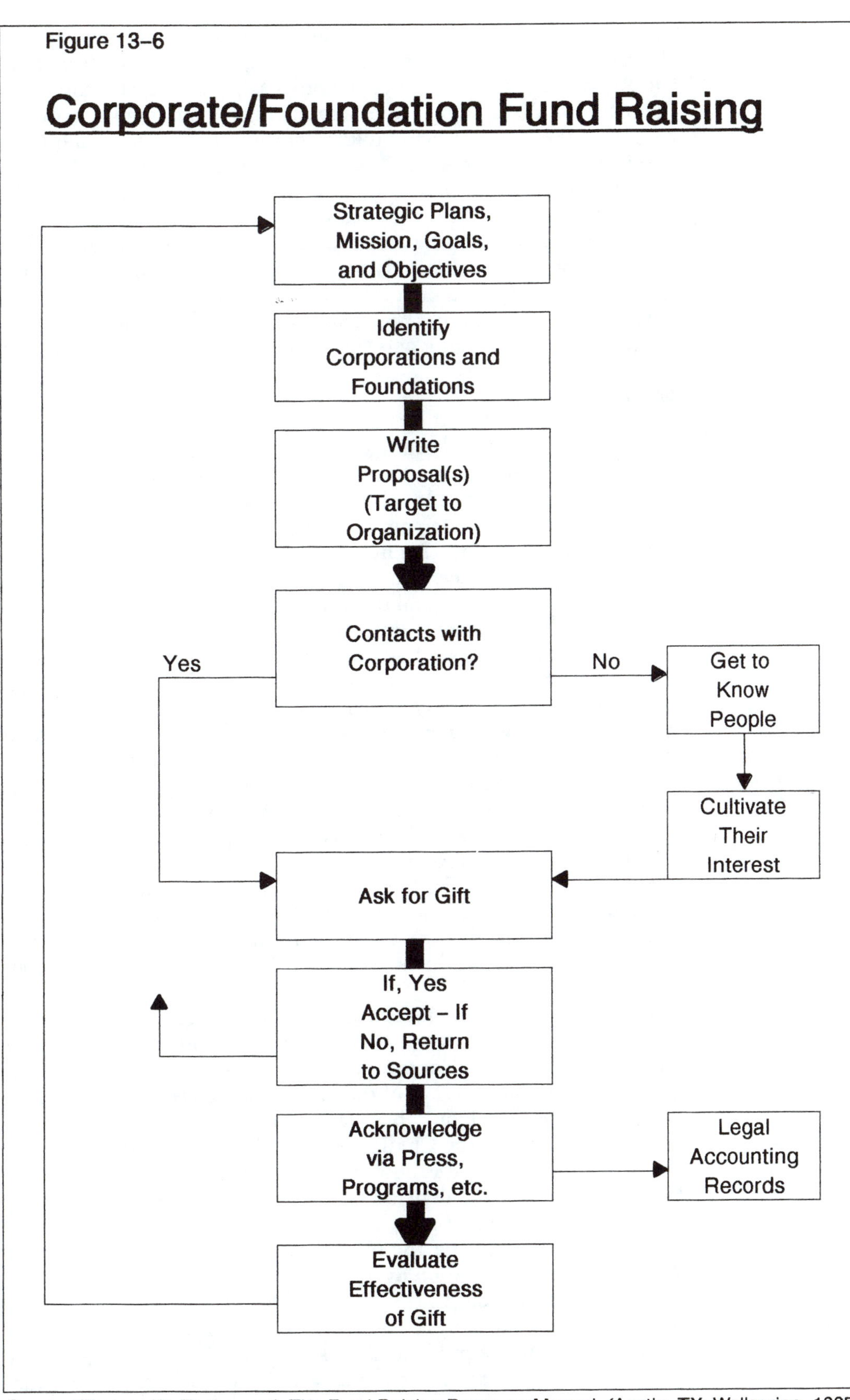

Source: Thomas W. Tenbrunsel, The Fund Raising Resource Manual, (Austin, TX: Wellspring, 1985).

may require a proposal of eight to ten pages, and a screening committee may review applications before referring them to a grants committee. Regardless of the length of the proposal, the applicant must state the problem, describe how the organization is qualified to solve the problem, explain the benefits to the community, and outline how the effectiveness of the project will be measured and evaluated.

The foundation application procedure is outlined in Figure 13–6. The key element again is personal contact with the foundation. Fund raising activity remains a person-to-person business, and without the proper introductions, the applying organization is an unknown entity. The greater the depth of involvement of the board in the community, the better the chances that the organization will be able to make itself a part of the grant-making network that exists in foundation funding.

Government Funding

It is possible to find funding for the arts at all levels of government in the United States. Local arts agencies usually have limited funds, but if an organization is trying to establish a positive record of effectively using grant money, the local level is a good starting point. For example, local agencies often provide funding for outreach programs into the schools. They may also help sponsor programming or subsidize ticket discounts to students or older audiences. The application procedures are usually simple, and the amount of time and money spent administering the support is minimal.

State arts councils usually have permanent staffs, standard application procedures, review panels, and standard evaluation and reporting procedures. They generally offer numerous types of grants, including grants for programming, new works, outreach touring, and individual artists. In many cases, the state agencies parallel the National Endowment for the Arts. Funding research is again required to achieve the best match between the arts organization and the granting agency.

The National Assembly of Local Arts Agencies and the National Assembly of State Arts Agencies, located in Washington, D.C., sponsor annual meetings and offer their members regular workshops on many areas of operation. Because NALAA and NASAA members form the core of agencies that distribute state and federal funds, an arts organization would be wise to cultivate a relationship with these organizations.

The National Endowment for the Arts As noted in Chapter 4, the National Endowment for the Arts, is the major source for funding and recognition in the arts community. The NEA publishes a guide that gives applicants an overview of the major grant areas. The thirteen program areas, the Office of Public Partnership (Arts in Education, local and state programs), the Challenge and Advancement Programs, and the Arts Administration Fellows Program form the core of the agency. Each program area and division of the NEA publishes detailed guidelines to help applicants through the process.

The 1992 budget for the NEA was approximately $176 million. When adjusted for inflation over the last ten years, the NEA now has less to give than it did in the late 70s. Despite the small size of the NEA's budget (compare the NEA's $176 million with the $1.4 billion given by corporations and foundations),[13] the NEA matching grants program helped stimulate an important partnership between arts organizations and donors. Organizations that received funds from the NEA also benefited from the recognition. Having received a grant added to the legitimacy of the enterprise and, though the NEA never intended it, created a "stamp of approval" for the arts group. Donors assumed that if a group or an artist had successfully passed through the grant review process, the work must be worthy of merit. As discussed in Chapter 4, the political turmoil that invaded the NEA's operation in the 80s and early 90s affected the agency's image. While the public's attention was focused on a limited number of controversial grants, thousands of grant requests were being reviewed, processed, and funded.

The peer review process The core of the NEA granting process is the peer review. A panel of experts are assembled at specific times during the year to review applications. Members of the committee are assigned specific grants to study in detail and to discuss at the review process. The time for each presentation is limited, and the competition for support is intense. Each year, the NEA receives about 18,000 grant applications.[14] The average grant application receives only a few minutes of discussion. Therefore, the proposal must be brief and to the point. The NEA staff reviews the details of the application, but the opening proposal, which is very limited in space, is what is most often read. If the reader's interest is not captured immediately or if the proposal raises more questions than it answers, the request will be pushed to the bottom of the stack. Consulting with the NEA staff may help when researching the kinds of key words, phrases, or concepts that are likely to catch the panelists' attention. At the same time, a brilliant proposal may fail because the organization has no track record, meaning that it has no previous history of having effectively used donated funds.

Applying for NEA funding is not very different from the corporate/foundation process shown in Figure 13–6. Establishing contacts within the agency and cultivating relationships with key people will help to establish the arts organization as a viable target for funding.

Other government sources Up to this point, the discussion of grants and the government has focused on the performing arts. The National Endowment for the Humanities also provides grants covering such areas as design, museums, research, music history, and interdisciplinary projects. In fact, thousands of grants are available from the federal government. Many of these grants have criteria that may make it difficult for an arts organization to qualify, but occasionally an opportunity arises that is worth pursuing.

The key to finding government support is research. One helpful source is the *Federal Register*, which is a very thoroughly indexed and cross-referenced publication that lists all federal grant programs. The *Catalog of Federal Domestic Assistance* contains information on government funding programs and is indexed by agency and subject area.[15]

Conclusion

Arts organizations have come to depend on funds from a mix of donors. Funding levels from individuals, corporations, foundations, and government agencies are subject to changing environments. For example, support for arts and culture changes as the economy improves or declines, as public attitudes about censorship shift, as government support for social services decline, as state arts budgets are slashed, and as companies disappear through mergers. Because the funding arena can be so volatile, arts groups are usually advised to avoid becoming dependent on any one source of funds.

The situation of too many nonprofit groups chasing too few donations by foundations and corporations probably will not improve in the next few years. In fact, as the demands for private support increase to cover the budget restrictions on government support for needed social and medical services, the actual amounts available to distribute to arts and culture groups may decline significantly.

Individual donors find themselves inundated with direct mail appeals and regular telephone solicitations from every conceivable cause. As more organizations learn the tricks of the fund raising trade, individual donors will be asked to support even more groups. Arts organizations may find a backlash from donors cutting into their major source of support. There is no doubt that the fund raising staffs and the board of directors for many organizations will have to reevaluate their fund raising strategies in the 90s. The trend may be toward more personal appeals in an attempt to form a tighter bond with donors.

Summary

Arts organizations in the United States depend heavily on the support of individuals, foundations, corporations, and the government to achieve their objectives. By tradition, U.S. government involvement in the arts has been minimal.

Giving, which is a person-to-person business, depends on the careful design and integration of the organization's strategic and operational plans. Auditing the organization's readiness to undertake a campaign includes analyzing its mission, objectives, resources, activities, and programs. The fund raising process requires a marketing orientation directed at donors to be effective.

The case for support is a key element in the organization's overall fund raising strategy. Board and staff support are needed to demonstrate to donors the commitment in the organization. An effective data gathering and management system is also needed. Careful control of the costs of raising money and disbursing funds are required for legal reasons.

Campaigns are designed to target funding groups, which include individuals, corporations, foundations, and government agencies. Donor research leads to designing different funding options to fit the needs of different funders. Gift programs are designed to accept current support, deferred giving, and bequests. Personal contact is the most effective way to solicit large gifts from a limited number of wealthy individuals. Telephone and direct mail campaigns are used to reach a wider audience. Corporations and foundations are usually approached based on the strategic fit between the donor's objectives and the organization's needs. Campaigns for government support usually involve meeting program requirements and criteria established by agency staff members.

Key Terms and Concepts

Funding pyramid
Fund raising audit
Case for support
Fund raising data management

Direct and indirect costs
Capital campaign
Annual campaign
Restricted and unrestricted gifts
Deferred gift
Bequest
Direct mail promotion
Reciprocity
Strategic fit
Foundation

Questions

1. What will be the impact on giving with the change in U.S. demographics over the next 20 years?

2. Using the excerpt from "Campaign for Hartford Stage," analyze how effectively the theater makes its case. What is the important problem Hartford intends to solve? Do you feel that Hartford has demonstrated its ability to solve the problem? How would you reword the document to restate the organization's needs and abilities?

3. Do you agree with the concept of expecting staff to donate regularly to the arts organization for which they work? Explain.

4. Do you think it is appropriate for arts organizations to gather personal data on potential donors for future use? What data would be inappropriate to keep? Why?

5. Have you ever been approached to make a gift to an arts organization? What techniques were used to solicit the donation? Did those techniques work? Explain.

6. Should arts organizations reject donations from corporations because of what the company manufactures or the politicians it supports? Explain.

7. If an arts organization unknowingly received donations from an individual later found guilty of defrauding people out of their money, should the organization return the gifts for redistribution back to the people who were defrauded? Defend your position.

Case Study

The following article illustrates the impact that the changing funding environment can have on an organization.

Studio Arena to Seek Funding Sources

Tom Buckam

Cuts in state and city support will force Studio Arena Theatre to seek funds from other sources, a Studio executive said Tuesday.

The regional theater in Buffalo Place expects to receive $112,000 less in 1991–92 than last year from the city and state, said Raymond Bonnard, producing director.

The bad news includes a 16 percent drop in city aid, to $157,250 from $187,000, and an especially sharp 63 percent reduction in the annual grant from the New York State Council on the Arts, to $46,515 from $128,250.

Since Erie County's contribution won't be known until December, half-way through the theater's fiscal year, "we've been put in a very tenuous situation, financially," Bonnard said.

Although the cuts may appear relatively small, measured against the theater's total budget of $3.2 million, their impact will be disproportionately large, Bonnard contended.

Because of annual increases in inflation and wage rates, the operation always needs more money than the year before—not the same amount or less—to meet expenses, he said.

"This business is extremely labor-intensive. Sixty percent of the budget goes to pay people," Bonnard said.

"When you're involved in a hand-crafted art form, trying to attract the best actors, directors and designers, they simply aren't going to come here for less money than we paid them last year," he said.

"We need that $112,000 just to keep up."

Studio Arena already supports itself about as well as can be expected, he said. Earned income, primarily ticket sales, accounts for 70 percent of the budget—more than any of the 20 other regional theaters of comparable size nationally.

"Yes, we could push that figure up to 75 to 80 percent by doing plays that sell more tickets and bring in more income. But the danger in that is that at some point you begin to redefine yourself and start turning into something other than your mission says you should be."

"Our mission is to do plays that have substance and ideas—that challenge the theatergoer. Plays that are entertaining but also illuminating."

The cutback from the state Arts Council caught the theater off guard, even though it was in line with the 66 percent across-the-board reduction in cultural funding proposed by Gov. Cuomo at the beginning of the protracted budget battle in Albany.

Three years ago, the state agency agreed to grant the Studio Arena $142,500 a year for two years. Last year the Studio, recognizing the state constraints, not only agreed to take less in the second year—$128,250—but to accept the same reduced level of funding this year, Bonnard said.

"The advantage of rolling over the grant into a third year, even if it was less money, was that you could at least lock it into your budget," he said.

But the agreement fell victim to budget cutting just the same.

Theater personnel find small consolation that regional theaters in Rochester and Syracuse were stung just as sharply.

SOURCE: Tom Buckam, "Studio Arena to Seek Funding Sources," *Buffalo Evening News*, July 17, 1991.

Questions

1. What percentage of the $3.2 million operating budget is made up of state and city funds before and after the cuts?

2. What are some of the funding alternatives that Studio Arena might investigate to make up for the projected shortages?

3. As a potential donor, what is your reaction to the idea that doing plays that sell more tickets is not a viable fund raising strategy for Studio Arena? Do you support this idea? Explain.

4. Based on this article, how well does it appear that Studio Arena

planned for changes in the political environment? What are some specific contingency plans that could have been made to prepare for the shortfall in local and state funding?

5. Based on Raymond Bonnard's summary of Studio Arena's mission, are you convinced of the need for support? Explain.

References

1. Melissa Mince, "History of Nonprofit Organizations: Summary," in *Nonprofit Corporations, Organizations and Associations*, 5th ed. Ed. Howard L. Oleck (Englewood Cliffs, NJ: Prentice-Hall, 1988), 41.
2. Neil Pendleton, *Fundraising* (Englewood Cliffs, NJ: Prentice-Hall, Spectrum Books, 1981), xi.
3. Loren Renz, "Foundation and Corporate Support for Arts and Culture," in *National Guide to Funding in Arts and Culture*, Eds. Stan Olson, Ruth Kovacs, and Suzanne Haile (New York Foundation Center, 1990), vii.
4. James Gregory Lord, *The Raising of Money* (Cleveland: Third Sector Press, 1986), 75.
5. Thomas Wolf, *Managing a Nonprofit Organization* (Englewood Cliffs, NJ: Prentice-Hall, 1990), 211.
6. Ibid., 225.
7. *The Chronicle of Philanthropy III*, (July 2, 1991), 1.
8. Renz, "Foundation and Corporate Support," vii.
9. Ibid.
10. Ibid.
11. Ibid., 26.
12. Ibid., 13.
13. Ibid., vii.
14. National Endowment for the Arts, *Guide to the National Endowment for the Arts*, (1990), 3.
15. Thomas W. Tenbrunsel, *The Fundraising Resource Manual* (Austin: Wellspring, 1985), 90.

Additional Resources

Porter, Robert, ed. *Guide to Corporate Giving in the Arts 4*. New York: American Council for the Arts, 1987.

Swaim, C. Richard, ed., *The Modern Muse: The Support and Condition of the Artist*. New York: American Council for the Arts, 1988.

14

Integrating Management Styles and Theories

Throughout this book, the stress has been on applying business theory and practice to managing an arts organization. In this chapter, we will summarize different styles of management and various strategies for integrating management systems into the operation of an organization. We will also review the specific functions of arts management and see how they can be applied to various management styles and systems. The goal of this chapter is to give the reader a model from which to work.

Management Styles

There are as many different ways to run an organization as there are people in this world. Everyone has a slightly different view of what techniques work best in managing an arts organization. In the interests of developing some practical approaches to management, let's examine three basic styles of management that can be used to lead an organization: rational, institutional, and organic approaches. Obviously, many other management styles may be applicable. Flexibility remains the foundation of any style of management. It is important to develop a repertory of responses from which to choose as operational situations change.

Before we focus on three management styles that can help keep an organization operating effectively, let's visit with an arts manager struggling with a dysfunctional work situation. There is always something to learn—even from bad examples.

The Irrational Arts Manager: A Model in Overextension

No one starts off in management with the goal of becoming an irrational manager. An irrational manager is the product of an organizational culture that thrives on dysfunction and chooses confusion as the standard operating mode. Organizations usually become dysfunctional either through evolutionary development or when an individual with a strong dysfunctional personality is allowed to take control of the management. Let's trace the development of an organization that creates an irrational manager and then briefly discuss the more serious problem of the dysfunctional manager.

Evolutionary dysfunction In Chapter 2, we saw that when an organization starts up, there is usually a small group of extraordinary people willing to spend 18 hours a day doing everything, including marketing, advertising, contracts, schedules, and budgets. Fayol's organizational esprit de corps is seen everywhere. Ambition, optimism, ceaseless energy, and a degree of ignorance about how impossible the job really is—these elements are all mixed together in a flurry of high-speed activity.

The volume of work increases as the number of productions, programs, or exhibits grows each year. Because everyone is so busy working, no one notices the gradual increase in the workload. Planning, if it is done at all, is never for more than a few days or a few weeks at a time. Little crises are put aside until they become big enough to disrupt operations. Before long, the problems multiply until the small staff spends all of its time solving one organization-threatening crisis after another. For example, one month there suddenly aren't enough funds in the bank to cover the payroll. "How could this happen?" everyone asks. No one is really certain because the payroll always managed to get done. Someone points out that it isn't a payroll problem; rather, the issue is cash flow. The investigation into the problem leads to the discovery that everyone was so busy last week dealing with a different crisis that no one deposited the box office receipts in the bank.

This example above may seem extreme, but unfortunately, it isn't. Overextended staff members who sometimes handle three or four major functional areas are often the norm in arts organizations. The corporate culture can be summed up as follows: "Because you love the arts, you will have the privilege of working long hours at low pay." Arts groups often thrive on having a work force addicted to the organization and the constant adrenaline-producing excitement associated with getting the show or exhibit finished minutes before it opens to the public.

The irrational manager is a product of this type of organizational system. The stress levels are high, so reason and logic are in short supply. Decisions are made and then quickly reversed because no one thought through the consequences. On any given day, no one knows what is really going on in the organization because of a lack of clear-headed thinking.

One obvious symptom of an organization suffering from an irrational management style is frequent staff turnover. High-energy people burn out quickly in a culture that requires them to sacrifice their private lives. Workaholic managers who drive their staffs to exhaustion assume that everyone is capable of matching their own work level. A newly hired employee is expected to adopt immediately the intense work ethic and, given the tendency of people to go along to get along, to adopt quietly the value system, no matter how unpleasant. The beginning operations-level staff person, with no point of comparison, accepts the required work level as the norm. A staff member without the power to effect a change in the work ethic usually opts to resign. The employee who resigns is immediately identified by the remaining staff as someone who "just didn't like to work hard," thus carrying on the cultural values of the organization.

The dysfunctional manager An organization can also become dysfunctional when the management team itself is dysfunctional. The cause of this problem is that the person in the role of manager is simply not suited for the job. In reality, some people were never intended to be managers. Their basic personality, for whatever reason, is inadequately formed. These individuals may have a whole range of character faults, including excessive defensiveness, aggressiveness, or passivity, verbal abusiveness, or being withdrawn. Unfortunately, the list could go on. Everyone has their problems, but the inescapable fact is that the managing process always reflects the personality of the manager. As character faults become more pronounced and, in some cases, are manifested in severe psychological problems, the work place becomes dysfunctional.

Unfortunately, a person hired into this situation is usually not aware that there is a problem until a few weeks have passed. After the first few explosions reveal the true personality of the manager and the character of the organization, the new employee has the option of adapting to this dysfunctional culture, trying to change it, or leaving.

How important is it that a manager be able to exhibit a positive personality profile and possess skills and expertise to help further the goals of the organization? It is central to the success or failure of the entire operation! No matter how beautifully crafted the mission statement or how detailed and comprehensive the strategic, marketing, or fund raising plans may be, if the individuals hired as managers cannot work with people in a way that promotes commitment, responsibility, and a sense of enjoyment about the work to be done, then the chances of ever achieving anything more than mediocrity are slim. It is important to remember that an organization is only as good as the people it employs. There is no escaping the fact that organizations become dysfunctional because of the people who work within them, not because of outside forces.

Let's now turn to three positive management approaches and contrast them with the dysfunctional and irrational manager. See Figure 14–1.

The Rational Manager: Changing the Culture

Applying a rational style of management to an irrational or dysfunctional situation takes persistence on the part of the manager. The first step in the process is to identify the steps that will be most effective in accelerating change where it is needed the most. Some parts of the organization will be impossible to change quickly, and others may be ready and willing to assist with making things different. There are no rules or guidelines that apply universally when trying to change a culture that has grown self-destructive. However, one obvious point to keep in mind is that changes are usually a great deal easier to instigate by moving with the flow of the organization rather than against it. This simply means that changing the attitudes and values of people by cooperation rather than coercion will greatly accelerate the acceptance of the rational manager's point of view.

One strategy to pursue is to enlist the support of the other senior staff and the board to undertake an organizational audit modeled after the marketing audit that was outlined in Chapter 12. The objective is to make board and staff members more aware of how the organization behaves and where the values and beliefs need to be changed to make the operation more effective and humane. Making sweeping changes to an organization is a daunting task, especially when it is dysfunctional. It can be done, but it may take longer than anticipated. After all, it usually takes years for an organization to develop a culture of elaborate values and beliefs.

Cultures in conflict were illustrated in numerous examples in this book. For instance, bringing in new artistic leadership is an opportunity for change, but it can also lead to counterproductive disruption. Rational managers understand that rapid change in any organizational system leads to a great deal of psychological stress on everyone. Being able to gauge how fast change can be effected is part of the art of managing.

Figure 14–1

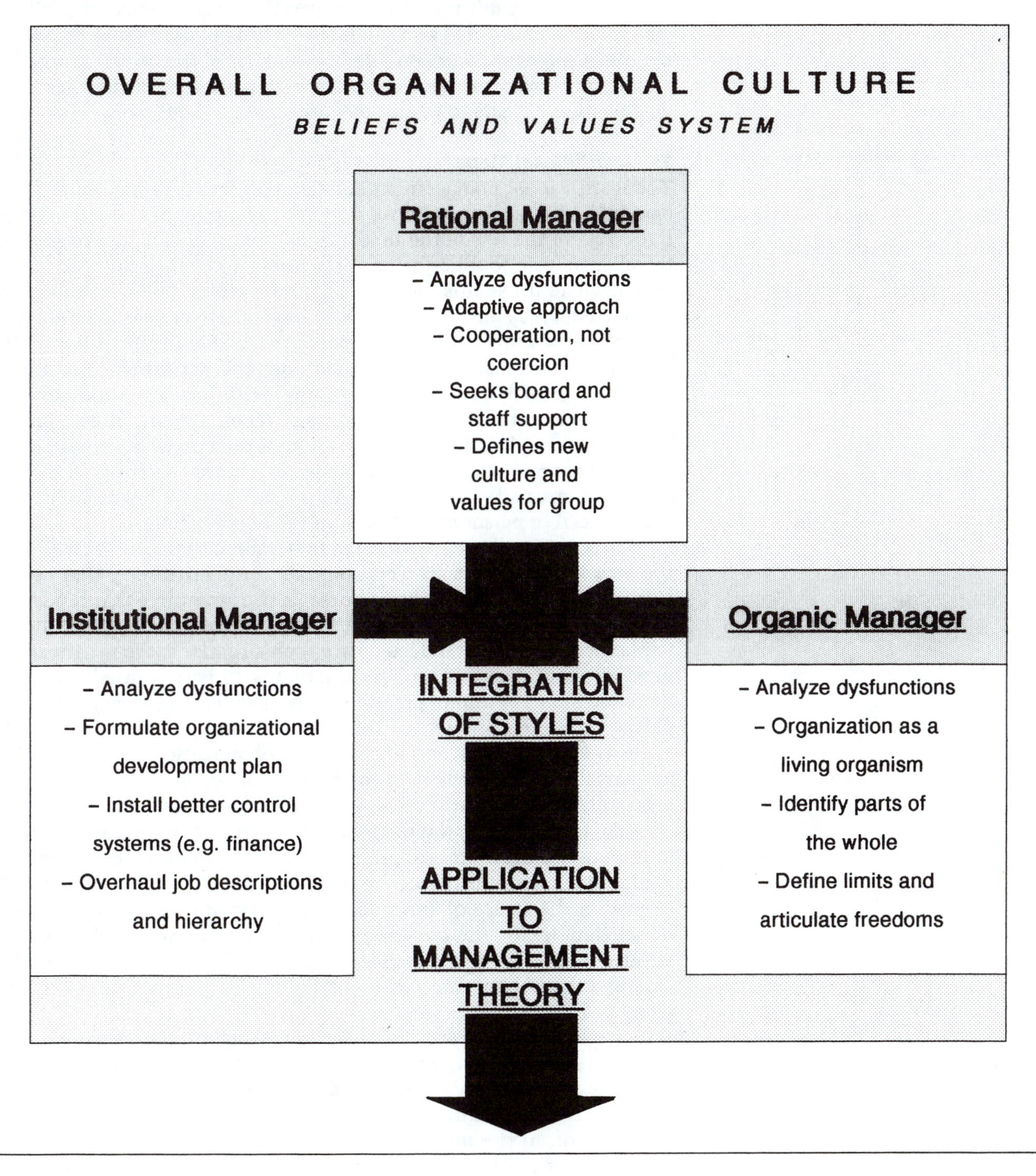
MANAGEMENT STYLES
OVERALL ORGANIZATIONAL CULTURE
BELIEFS AND VALUES SYSTEM
Rational Manager
– Analyze dysfunctions
– Adaptive approach
– Cooperation, not coercion
– Seeks board and staff support
– Defines new culture and values for group
Institutional Manager
– Analyze dysfunctions
– Formulate organizational development plan
– Install better control systems (e.g. finance)
– Overhaul job descriptions and hierarchy
Organic Manager
– Analyze dysfunctions
– Organization as a living organism
– Identify parts of the whole
– Define limits and articulate freedoms
INTEGRATION OF STYLES
APPLICATION TO MANAGEMENT THEORY

Rational managers believe that organizations and people learn from their experiences and mature over time. A creative artist and a manager (perhaps the same person), working in cooperation with a board of directors, can use the functions of management to chart a course for the organization that replaces the state of constant crisis and anxiety with controlled growth. Granted, it is not easy to set aside the time for planning and organizing in the midst of the unremitting press of daily business, but once a system is in place, the difficulties every organization expects to go through are smoothed out a bit.

The Institutional Manager

The institutional management style emphasizes organizational development techniques and control systems to help reach the stated objectives. Part of the rise of the institutional manager can be attributed to the change from founder-driven arts groups to board-driven arts institutions. The museums, performing arts centers, opera, theater, and dance companies that started off with one or two people 30 or 40 years ago may now employ 50 to 100 people and have multimillion-dollar budgets and extensive office, production, and performance facilities. With this growth and increasing complexity has come the development of the professional arts manager. When a board hires a professional manager, it is usually expected that the organization will adopt a more corporate structure. The change to a more "business-like" style of operation usually follows a management crisis brought about by a big financial or personnel problem in the organization.

Not everyone greets what could be called the *managerization of the arts* with enthusiasm. For example, some critics say that adding layers of management produces bureaucratic structures that hinder the accomplishment of objectives and add to the operating costs. Although unplanned growth can indeed create such situations, the reality is that most arts organizations tend to function with very limited resources; therefore, it is a common practice to give people two, three, or four job titles—any one of which could be a full-time job. What is seen by some as too many managers is often the organization's way of finding a reasonable balance between the number of people required to do the tasks and the organization's stated mission, goals, and objectives.

At another level, the role of the institutional manager has grown in response to the increased pressure to produce a balanced budget and the funding community's desire to see its money used responsibly. To cope with the issues of fiscal accountability and increasing organizational complexity, the two- and three-headed management structures are also finding their way into larger arts organizations. (See "Kennedy Center Adopts New Management Plan" in Chapter 6 and the case study in Chapter 8.)

Regardless of debate about the increasing number of middle- and upper-level managers in arts organizations, the fact remains that what was once an idea in someone's head is now a multimillion-dollar institution in the community. The time and money given by the board of directors, often the most influential people in the community, frequently carry subtle and not so subtle restrictions that can redirect the organization to a safer or less controversial path. A vigilant management and artistic team can work with the board to keep the art alive and challenging. However, given the economic and political pressures, it is not hard to see why many arts institutions begin to engage in forms of self-censorship. If the community support for the organization is strong enough, the artistic reasons for performing a particular play or mounting a particular exhibit should outweigh the financial or political pressure placed on the organization.

The Organic Manager

The organic management style recognizes that a changing dynamic exists in organizations and that, like a living organism, the group will grow and change over time. In some ways, the organic manager functions like a doctor who practices preventive medicine. The organic manager works with a dynamic system and focuses on spotting problems that could affect the health of the organization. Intervention is designed to treat the symptoms and the causes. In this case, the tools used to practice this preventive medicine are the theories and practices of management, economics, marketing, and so forth, which have proved to work given specific circumstances.

Organic managers realize that they are working with very distinct groups, and that these groups have flexible boundaries of skills and interests that overlap. The ability to give each group a sense of its own importance in the whole, while at the same time promoting communication and understanding, is the most important job facing the organization's leadership.

The organic manager also recognizes that there is an element of chaos in any organizational system. However, chaos doesn't necessarily mean that the organization is out of control. In the context of an arts enterprise, chaos is recognized as an element of creative unpredictability. No one is ever sure that a production will really work as planned. Giving artists the freedom to experiment carries with it risks that an organic manager recognizes. The organic manager makes allowances for the different levels of structure required for the various parts of the organization. The accounting department has rules and regulations restricting what can be done, but directors, choreographers, and designers are given more freedom to explore alternative solutions. Recognizing the differences in the way subunits need to operate does not mean abdicating control. The latitude given to creative artists still fits within the overall control system that everyone agrees is necessary.

Management Theories

Having established a management style, which may include a combination of rational, institutional, and organic techniques, the manager can turn to adapting various management theories to arrive at an overall operating approach to the organization.

Scientific Management

Many of the fiscal and production aspects of an arts organization can benefit from the application of quantitative procedures and ongoing statistical analysis borrowed from the scientific theories of management. There are potential gains in productivity if constant monitoring of routine procedures is a part of the organization's culture. Is there a more effective way to go about the process of constructing, storing, or rigging scenery? Can money be saved if a rehearsal sequence is altered? Is the method used to enter sales data producing the timely information management needs to quickly spot financial problems? Is the procedure for

FIGURE 14–2

INTEGRATION OF MANAGEMENT STYLES WITH THEORY

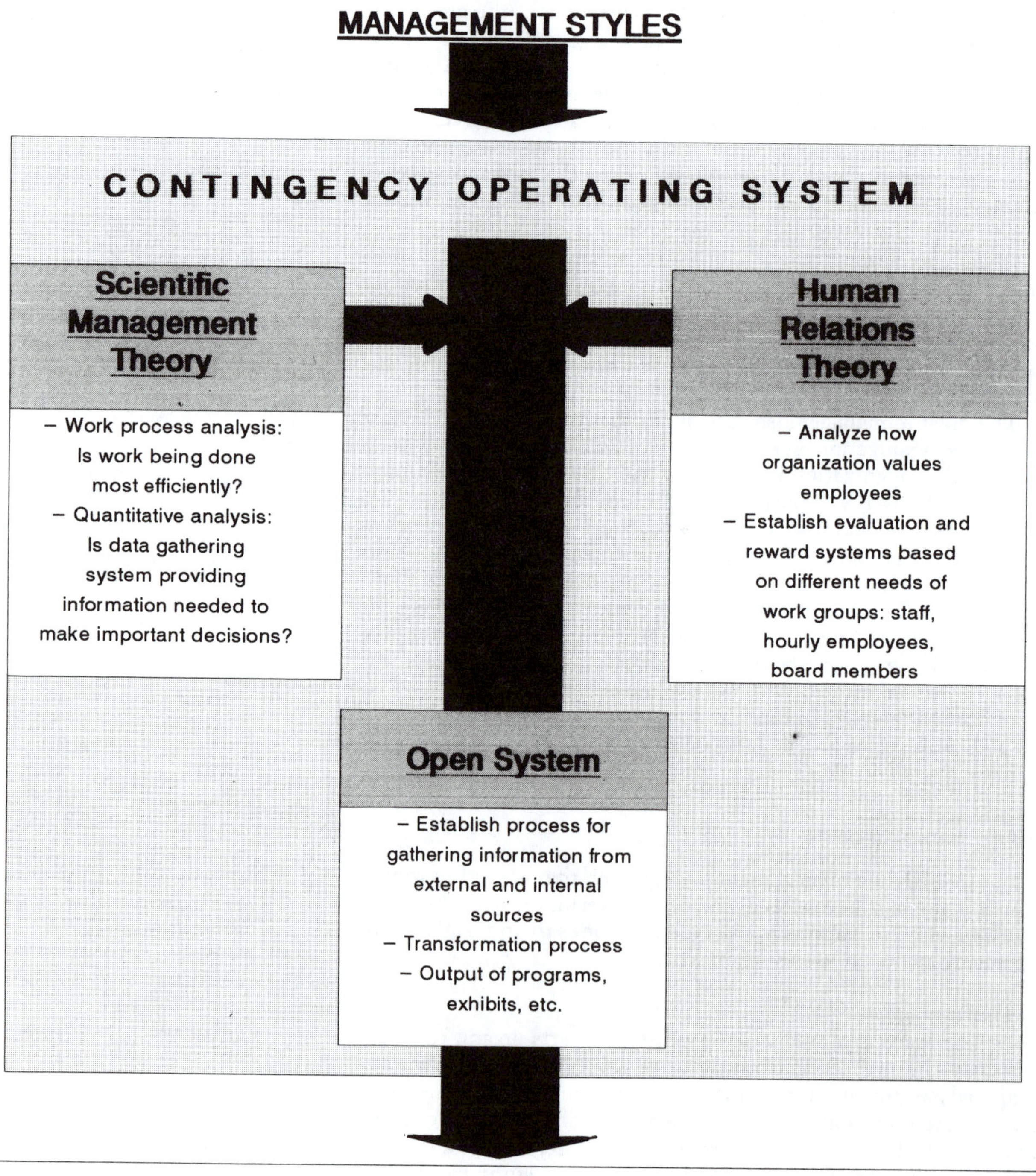

MANAGEMENT STYLES

CONTINGENCY OPERATING SYSTEM

Scientific Management Theory

– Work process analysis: Is work being done most efficiently?
– Quantitative analysis: Is data gathering system providing information needed to make important decisions?

Human Relations Theory

– Analyze how organization values employees
– Establish evaluation and reward systems based on different needs of work groups: staff, hourly employees, board members

Open System

– Establish process for gathering information from external and internal sources
– Transformation process
– Output of programs, exhibits, etc.

processing an order organized so as to reduce the number of steps required? Nearly all of the tracking of responses to mailings, donation requests, marketing, and sales campaigns relies heavily on techniques related to scientific management. As we saw in Chapters 9 and 11, an organization's MIS and FMIS require the vigorous application of quantitative systems if the organization is to stay informed about its fiscal health.

One clear signal that it is appropriate to undertake a more scientific approach to aspects of the operation is when you hear the phrase, "But we have always done it that way around here." There may be good reasons why certain procedures are accomplished in specific ways, but nearly everything that is done routinely can usually be done more efficiently if given some thought.

Human Relations

As we saw in Chapters 3 and 8, the human relations approach to management grew out of McGregor's Theory X and Theory Y, Maslow's hierarchy of needs, behaviorist theory, and other psychological approaches to the work place. Because the product of a performing arts organization is the work of people, there is a natural fit between human resource management and an arts organization. The arts manager must realize that each employee group has its own set of behaviors and expectations about the work of the organization. For example, the stagehands working on a show will have a very different perspective about their job and their place in the organization than that of fund raising assistants. Both support jobs are needed to make the show work, but employees in each category need different types of recognition and rewards for their contributions to the organization.

Developing performance standards and an appraisal system that recognizes the similarities and differences in employee groups while keeping employees focused on the defined objectives requires a significant commitment of the manager's time. Given the dynamics of an arts organization, once-a-year job reviews will simply not monitor adequately the work output of employees. Daily, weekly, or monthly reviews of employee performance may be more appropriate, depending on the type of work being done.

Attention must also be paid to how effectively the organization's communication and management information systems are working. Information is often equated with power. Those who have the information have power over those who do not. However, this is a very destructive approach to information management. As we saw in Chapter 6, all organizations have formal and informal structures and communication systems. If any employee group is excluded from the communication system, the risk of rumors and the harmful distortion of information increases in great leaps. No employee communication system is perfect, but it makes little sense to establish communication approaches guaranteed to alienate people.

The Open System

The open system and the ability to adjust to changing circumstances were stressed in many chapters in this text. Input from clients, audiences, staff, and so on, is combined with input from the external environments—the economic, political and legal, cultural and social, demographic, technological, and educational environments—to produce an organization that constantly changes and adjusts to the world around it. The open system approach to management does not mean that the organization's mission undergoes constant change. Rather, the open system allows the organization to capitalize on opportunities that support its mission while minimizing the impact of threats to the enterprise. For example, new video technology may provide additional cash flow opportunities for the arts by opening up new markets for distribution of the product and by creating new viewing audiences. At the same time, an adverse tax ruling by the IRS or a proposed Federal law that affects labor practices can be addressed through active participation in the political system.

The Contingency System: An Integrating Approach

The approaches and theories reviewed thus far are all directed at finding a way to integrate the various styles and theories of management into a workable system. Combinations of rational, institutional, and organic management techniques applied to theories of scientific management, human relations, and the open system can help a manager achieve results when carrying out the functions of management.

One model that meshes these styles and theories is shown in Figure 14–2. The integration of these styles and theories into what is called a *contingency system* holds much promise. The important point to remember with this system is that, depending on the circumstances and the nature of the problem, the manager can pick and choose options, combining some portion of each approach. Sometimes, it is a human resource problem; at other times, it is a quantitative problem; and on other occasions, the problem relates back to a change in one or more of the environments or input groups. Each individual will feel comfortable with different applications of a contingency system; however, the central point is that a manager must actively choose the particular combination of styles and theories that will best solve the organization's problems.

Let's review the functional areas of management from the perspective of applying the contingency system of management.

The Management Functions

The center of the arts organization's operational system is found in the functions listed in this section. The goal of the entire organization should be to take the contingency system, which integrates management styles and theories, and apply it to the operational areas in ways that achieve the organization's stated goals and objectives.

Planning and Development

Looking toward the future is a major responsibility of a manager in any organization. As we saw in Chapter 5, the ability to plan requires no special genius. The key ingredients are self-discipline and an established process. Of course, the underlying culture of the organization must value looking ahead and should stress the involvement of the board and all levels of staff in the process. Every action that an organization takes must relate to the overall master plan. If it doesn't, human and other resources will be misdirected and ultimately wasted.

Types of plans All managers, including overextended managers, engage in short-term and crisis planning. Problems such as cash flow difficulties, a show that might have to be canceled, a tour cut short because presenters backed out at the last minute, a work of art damaged at an exhibit, and the death of a member of the cast are part of the business. Managers with a methodology in place for problem solving quickly identify the source of the difficulty, generate alternatives, and implement and evaluate decisions before a crisis develops that threatens the operation of the organization. Written outlines and procedures to follow when a crisis does strike can help the organization keep its balance.

Intermediate-range planning (one to five years) is integral to the program development and fund raising activities of an arts group. Launching any fund raising campaign requires coordinated planning to organize resources and people in cost-effective ways. The case must be strong, and the reasons for giving must be clear to potential donors. The need for the funds must relate directly to the organization's strategic planning.

Long-range strategic planning and development (five to ten years), although always subject to revision, are also important processes in helping to shape the organization's future. Long-term planning is a serious business, but the enjoyable and creative side of the process should not be overlooked. "What if" discussions among the board, the artists, and the staff can lead to new ideas and directions. In fact, planning is a break from the routine of day-to-day work, and it should be a strong selling point for potential board members. After all, would you rather be on a board of directors that was actively engaged in charting the future of the organization, or would you prefer to sit through yet another report detailing fund balance transfers?

Marketing and Public Relations

Arts organizations are businesses that must function in the highly competitive entertainment industry. Effective marketing and positive public relations can help any enterprise target and inform people of goods and services designed to meet their wants and needs. The key to success, especially given the very limited resources that most arts groups have to work with, lies with making sure that you are talking to the right people.

The universe of goods and services seems to be expanding as thousands of new products come on the market every month, making the task of keeping an arts organization visible even more exacting. Therefore, expanded press and media relations must be a central part of the organization's strategic plan if the organization is to command any attention.

As we saw in Chapter 12, marketing is an organizational orientation that places the consumer at the beginning of the process. To be truly customer-driven is seen as an ideal in establishing a long-term product–consumer relationship. However, many arts organizations equate "customer-driven" with lower artistic standards and pandering to the public. They therefore engage in what is more accurately described as a "selling orientation" toward the public. The selling approach assumes that if consumers are made aware of the product, they will buy it. In more extreme cases, organizations adopt a product orientation, which assumes that because the product is so inherently good, people will want to buy it.

Marketing strategies In Chapter 12, it was pointed out that the adoption of a marketing orientation for an arts organization requires a careful analysis of the four P's (product, price, promotion, and place) and, at the same time, an understanding of the limits of marketing. For example, no marketing campaign will suddenly create an arts audience by changing the well-established behavior of masses of people overnight. As we have seen, the audience for high-culture events is a by-product of the education system, especially at the college and university level. Arts marketers interested in building a long-term purchase and donation relationship with consumers would assemble demographic and psychographic profiles of the community, and a distribution of neighborhoods by spending type (for example, money and furs, pools and patios) in an effort to piece together the most effective campaign that the available resources permit. The objective is to get inside the consumer's head to find out what combination of the four P's will lead to the exchange process: money in exchange for the arts experience.

Of course, marketing strategies need not be targeted to only a limited segment of the population. Marketing to reach a diverse audience is important and will become critical as the demographic composition of America continues to change. The key to reaching people lies with how and what you communicate about the product to the public. Reaching new audiences, especially minorities, depends on the mix of programs and community outreach, the right price, and access to the facility.

It must be stressed again that all of the innovative marketing strategies in the world cannot create an arts consumer overnight. Marketing is a long-term investment of the organization's resources. Results should be measured over three- to five-year periods of time, not just one season.

Finally, it is also important to remember that marketing and fund raising share the similar goal of establishing an exchange relationship with audiences and donors. The progression from single ticket buyer to subscriber or member, and finally to long-term donor dictates an integrated plan under the general heading of development.

Personnel Management: Staff, Labor, and Board Relations

Even when the manager adopts a human resource management strategy that takes into account the various attitudes and values of the different work groups within the organization, being able to keep employees happy and productive is still an enormous challenge. Establishing working conditions that support the creative process and encourage artists, the board, and staff members is one of the manager's primary objectives. This would seem to be self-evident, but judging by the stories of performers, designers, staff, and technicians about the abuses of management—which also includes directors, conductors, and choreographers—it is a miracle that there isn't more strife in arts organizations.

Unions The abusive past practices of management were a primary reason for the establishment of unions in the arts. Although managers and unions may never see eye-to-eye about how the organization operates, the fact remains that it is management's job to establish the criteria for performance. For the arts manager, defining acceptable practices is at the heart of the relationship with the union because, when both sides sit down to negotiate a contract, the odds are very high that labor and management will not agree about levels of compensation, benefits, and work rules. The arts manager who does not understand how the backstage really works in a theater will be placed at a huge disadvantage when it comes time to evaluate the rules written into a contract. A lack of appreciation for the work environment could cost the organization thousands of dollars every time a show is performed. In addition, an arts manager must remember that a union is also a complex organization and is therefore subject to the same external and internal forces that shape the arts group. Understanding the perspective of union members and the union organization can make the arts manager's job a little easier.

Board of directors Another group that is part of the overall mix of human resources in the arts is the board of directors. Managing board relations is just as important to senior- and middle-level staff members as is a successful contract negotiation with a union. The board, which could be made up of as few as eight or more than 50 people, has its power blocs, hard workers, and deadwood, just like any other organization. A powerful finance committee, for example, could prevent a manager from implementing new programs by not approving a budget. A board personnel committee could hold up a key appointment or lobby for a candidate of the board's choosing. Approval of a season might be held up if the board has doubts about title selection.

For the arts manager, a clear picture of the scope of the board's power and responsibility is of primary importance if the organization is to function effectively. In numerous case studies in this book, we have seen board actions both strengthen and undermine arts groups. There never seems to be a shortage of stories about the communication gap between artistic directors and the board.

Personnel management Meanwhile, as organizations mature, issues of salaries, staff training, and renewal will become more important. Finding new challenges, revising job descriptions, reorganizing departments, and combining or separating jobs should all become part of the overall operation of the personnel area.

When adding up all of the work required to manage the personnel functions of an arts organization, artistic directors or museum directors might wonder when they will have time to direct a production or engage in scholarly research. In reality, there is no time. Upper-level managers, like artistic directors who involve themselves in production, find that, more often than not, one or more functional areas are left unattended. What strategy can an artistic director use to resolve this conflict? One obvious approach has been to split the job into different positions based on operations and product development. This explains why more organizations are creating two or three upper-level management positions to supervise the fiscal, planning, and operational aspects of the arts enterprise. The tradeoff for some artistic directors is that they must share their power with others. At some future date, should conflict arise with the board, artistic directors might find one of their operational peers undermining their credibility and forming a power bloc against them. The case study at the end of this chapter demonstrates what can happen when the board and the artistic director move in different directions.

Fiscal Management

The area of fiscal planning and control is also at the core of the arts organization. A large portion of a manager's time will probably be spent on this area, often at the expense of such equally important areas as planning, programming, and staff development. Financial management is at the center of so much attention because expectations about the amount of money that can be generated through sales and donations are often unrealistic. At the same time, prices, labor, and operating costs continue to escalate higher than anticipated. The combination of these two elements establishes a perpetual deficit hanging over many organizations.

In *The Quiet Crisis in the Arts*, McDaniel and Thorn note that arts organizations are in a state of constant debt crisis, which produces an unbelievable amount of personal stress on the staff.[1] The cycle of overspending followed by painful budget cutting takes its toll on people. When artistic directors speak of maintaining quality even in the face of budget cuts of millions of dollars, people know that it is simply untrue. If you cut the budget, you will reduce quality. The expectation that the performers, designers, technicians, museum preparators, and others will somehow be able to create the same quality product with fewer rehearsals, less money for resources, and so forth, is total nonsense. The assumption seems to be, "If we *say* often enough that we aren't going to lose quality by cutting budgets, then maybe everyone will begin to believe it." Managing and leading by fiscal self-delusion is hardly the most effective way to build board and staff confidence. As a matter of fact, a board member asked to raise yet another million dollars might ask, "Well, if we can do the same quality with a million dollar budget cut, why did we go a million dollars in debt in the first place?" The answer usually comes back, "Because we are dedicated to pursuing excellence in the arts." The more accurate answer would be, "Because we didn't know how to effectively manage the human and production resources we had available to us." Needless to say, the latter reason usually isn't a discussion item for staff and board meetings.

Strategies There are a few fiscal management strategies that are likely to keep the organization solvent and the board confident. They include the following:

- realistic budget planning procedures for revenue first, then expenses
- organizational attitudes and values that stress that budgets are not to be exceeded
- very tight control and oversight systems for expenditures
- a clear picture of the cash flow needs of the organization
- a system for accounts payable activity that pays no bill before its time except when there is a discount for early payment
- a very aggressive "asset management program" that involves investment in a wide range of financial instruments

Government Relations

As we learned in Chapter 4, government relations extend from the federal to the local level. A manager's involvement in the political arena is usually fairly limited. However, as we saw in the case study on the Los Angeles theater district in Chapter 4, without the support of the people who make the laws and sit on the appropriations committees, arts organizations will suffer. Arts groups must earn the support of elected officials at all levels of government. Support is not given just because the arts organizations and artists think their enterprise is nobler than other agencies established for a public good.

The first step for successfully interacting with the government system is education. The arts manager must learn how the various local, state, and federal systems work. The second step involves learning about the power brokers and the issues close to them and their constituents. The third step requires the arts manager to visit the various representatives in an effort to become visible. The fourth step involves making the new-found visibility mean something by updating the representatives about the organization's important activities and the positive impact of the arts on the community.

For a weary arts manager trying to cover all of the management functions described in this section, the political system, with its complex subculture, is not always the highest priority. If the arts manager is too busy to have lunch with the local director of cultural affairs or to go to a candidate's fund raising dinner, there should be no surprise when politicians don't spring to the rescue when the arts group has a budget crisis or some other problem.

Conclusion

The goal of developing an integrated approach to management styles and theories is to help the organization succeed at whatever it attempts. As we have seen, managing is an intensely personal process. The theories may provide the overall structure, but the effectiveness of any management system ultimately depends on the people who are doing the managing. The ability to establish an overall work environment where people can express a point of view without fearing for their jobs, or where they can make suggestions that will be heard is often overlooked when it comes to designing an organization. Of course, people who work in any organization want to earn enough money to live comfortably, and they want to have the support of health benefits should they have an accident or become ill, but on a day-to-day basis, people also have an intense desire to believe that their work is making a worthwhile contribution and that their effort is being recognized in some way. Therefore, a manager must always remember that it is the people who are important to the organization, not just the product. Treating people with respect and recognizing their daily contribution to the enterprise are key ingredients in successful organizations.

To integrate management styles and theories, managers must also know their own strengths and weaknesses. Undertaking a personal inventory helps a manager to see more clearly the things the organization does well and to identify the areas that need improvement. The advantage of using this approach is that it keeps the manager and the organization renewed. It is also important for the manager to reap a personal satisfaction when it comes time to evaluate how well things are working. Success is a great motivating force, especially when it is widely recognized throughout an organization.

Questions

1. Based on your own experiences, can you cite examples of situations in which dysfunctional, rational, institutional, or organic management styles were used to solve a problem effectively?

2. With proven approaches to managing, such as scientific and human relations management, and organizational models, such as the open system, to help guide operations, why do so many arts groups have trouble with their management structures?

3. Use any of the case studies in this text to provide examples of the effective use of the following functional areas:

- planning and development
- marketing and public relations
- personnel management
- fiscal management
- government relations

Case Study

Spoleto Feud Ends; Menotti Wins

Allan Kozinn

Gian Carlo Menotti, the 80-year-old founder and artistic director of the Spoleto Festival U.S.A., has won his 17 month battle with the festival's board.

At a tense and sometimes fractious three-and-a-half-hour board meeting at the Drake Hotel in Manhattan, 10 of the board's 46 members resigned, and 9 others either decided not to stand for re-election or were not nominated.

Among those who resigned were the festival's chairman, Ross A. Markwardt, and its president, Edgar F. Daniels, the two officers whose resignations Mr. Menotti demanded in a them-or-me showdown this May. Nigel Redden, the general manager, whose resignation was also demanded by Mr. Menotti, left his post last month but is staying on as a consultant.

The Spoleto Festival U.S.A., which is based in Charleston, S.C., is widely regarded as one of the most rewarding and expansive summer arts festivals in the United States. During its 15-year history, Mr. Menotti has had frequent clashes with the festival board, but since May 1990 his relations with the board have been particularly hostile, with disagreements over everything from budgets to programming.

Soft Voices Turn Loud

At the meeting, the 19 board members who left the organization were replaced by 14 new members and a new slate of officers. Other candidates are to be considered later. Although Mr. Menotti did not attend the meeting, his principal ally in the dispute, Joseph P. Riley Jr., the Mayor of Charleston, was on hand.

The public parts of the meeting unfolded beneath a placid veneer of Southern geniality. But after 90 minutes, the new chairman, Theodore S. Stern, moved to shift into executive session before new board members were proposed. With the exception of Mr. Riley, everyone who was not on the board was ejected from the room for an hour and 45 minutes. But raised voices could frequently be heard, and board members who intermittently left the room spoke critically of the proceedings.

"They are going over old history, rehashing every slight and every detail of every previous administration," said an exasperated Mr. Markwardt a few minutes before he returned to the meeting to resign. "They are debating the by-laws, and people are asking questions like,

'What does the board of directors do?'" Another board member who resigned, Bessie Hanahan, described the meeting as "amateur hour."

The dispute between Mr. Menotti and the board started in May 1990, when the composer objected to an exhibition of Conceptual Art proposed by Mr. Redden. The board voted to have the show, "Places with a Past," which proved to be one of the successes of the 1991 festival. In October 1990, Mr. Menotti threatened to resign unless Mr. Markwardt and Mr. Redden guaranteed him complete artistic control. At the same time he also objected to a retirement contract the board had offered him under which he would become director emeritus and would continue to direct productions of his own operas.

When the festival opened this May, Mr. Menotti arrived at a board meeting demanding the resignations of Mr. Markwardt and Mr. Redden. Adding Mr. Daniels to the list, Mr. Menotti insisted that if they did not leave, he would. At the time, the three maintained that they were answerable only to the board, which repeatedly assured them of its support. In July, 33 members of the board signed a letter saying they opposed a change in the festival's organization.

What changed, several members of the board said, was that Mayor Riley told them he would not allow the festival to use city-owned concert halls and other facilities unless Mr. Menotti remained with the festival. And Mr. Menotti refused to alter his resignation demand. Several of the board members expressed bitterness at the Mayor's position.

"The Mayor had told us that he regarded his role as primarily that of a mediator," Mr. Markwardt said. "But he was unwilling to peruse any compromise option that we approached him with. We raised the issue of whether the Mayor could actually stop the festival, but the board elected not to pursue it, and we absolutely did not consider relocating to another city. Our primary goal was for something to emerge from the ashes."

Mrs. Hanahan, one of the board members who resigned, described the Mayor's dealings with the board as heavy-handed and unethical. And Charlotte F. Sloan, in her resignation letter wrote: "It distresses and outrages me that Mayor Riley, by his ill timed, inappropriate, thoughtless and autocratic action, has effectively emasculated this hard-working group."

Mayor Riley would not respond to characterizations of himself as an autocrat. When asked whether he had actually denied the use of Charleston's facilities to the board he said:

"Spoleto is the most important development for my city in the 20th century. It brings under the direction of the greatest living opera composer the most comprehensive arts festival in the world, a human-scaled festival that has greatly enriched my city.

"My duty as Mayor was to make sure that this marvelous festival continued at the level of quality and renown that it enjoyed, and it would not have continued at that level if Gian Carlo Menotti had been removed from his creation. I was unalterably opposed to that, and I made it clear the city, as an important part of the festival, would be opposed to continuing without Maestro Menotti."

Both Mr. Riley and his father, Joseph P. Riley Sr., were among the 14 new board members.

Some Costs of the Dispute

Mr. Menotti's victory will not be without cost to the festival. Several of those who resigned said they would not contribute money to the festival again. The loss is projected in the festival's $5 million 1992 budget to be more than $300,000. There is also a $40,000 loss for tickets to a special opening party bought by AT&T, of which Mr. Markwardt is a vice president.

During the meeting's financial discussions, festival officials said the battle itself cost some contributions from private and corporate donors. In outlining the reasons the festival's 1991 fund drive fell just below its goal, Connie Baldwin, the acting director of development, cited "the softening economy, the gulf war and the Spoleto conflict, which was played out in the media."

The impasse between Mr. Menotti and the board has also put the festival behind in its budget planning because he has refused to submit his programming plans. "I have asked Maestro Menotti to at least give us the number of events he would like to have in each category, and he has been unable to do that" said Mr. Redden, who despite his resignation was still required to present the 1992 financial prospectus. Mr. Redden said this has hindered raising money for specific projects. "Special fundraising," he explained, "requires knowing what the program is before we ask for money." He added that the festival usually issued its first brochures in October and counted on the money from early ticket sales.

Mr. Markwardt, speaking after he left the executive session, said that during his year as chairman he "felt that it was more important to shape the future than to preserve the past," adding, "But we ran afoul of the preservationists."

Mr. Stern, on the other hand, had been the festival's first chairman, and his return to that post may be seen as an attempt to return to the festival's comparatively untroubled youth.

"I think that the role of officers and directors, both artistic and administrative, is to bring out harmony and mutual respect," Mr. Stern said after he was elected. "It is not unusual to sense discord in any festival. Dissension is healthy. It provides vigor. My purpose is singular: to assure the further success, achievement and acceptability of the 1992 Spoleto presentation. This cannot be accomplished by one person or group, but it can be achieved only by the energetic, cooperative efforts of all concerned."

SOURCE: Allan Kozinn, "Spoleto Feud Ends; Menotti Wins," *New York Times*, September 17, 1991.

Questions

1. Based on the comments by board members about the executive session, how much did the board have to do with the operation of the festival?

2. Part of the struggle between Menotti and Markwardt, Redden, and Daniels appears to be over the organizational structure of the festival. Why do you suppose the board made the latter three men answerable to the board and not to the artistic director? If Mr. Menotti was the artistic director, what message do you suppose the 33 board members were trying to send when they signed a letter saying they supported the current organizational approach?

3. Do you agree with the position taken by Mayor Riley that the

festival couldn't continue without Menotti? Two of the resigning board members called the Mayor's threat to withdraw city resources "unethical" and "autocratic." In your opinion, what was unethical about the Mayor's action?

4. The article indicates that Menotti has not provided information about programs for the next season. What are some of the other problems that are likely to occur because the organization's entire planning cycle has been dislocated by the power struggles of the festival?

5. Markwardt is quoted as saying, "We ran afoul of the preservationists." He and others on the board clearly had plans to offer Menotti a retirement package and forge ahead with new programs. What other strategies could Markwardt and others have taken to achieve what the incoming chair calls "harmony and mutual respect?"

Reference

1. Nello McDaniel and George Thorn, *The Quiet Crisis in the Arts.* New York: FEDAPT, 1991.

15 Final Thoughts

In the previous chapter, we discussed the integration of theory and practice with specific styles of management. It seems appropriate to end this text with a brief discussion of the future. When it comes to actually making an organization work, the pressing demands of the day-to-day routine seem to devour every available minute. There never seems to be enough hours in the day to put aside time for the future.

As we have seen throughout this book, middle- and upper-level managers must always look beyond the immediate future. The planning process can be viewed in much the same way that an artist views a rehearsal: as an opportunity to refine and perfect the work before an actual performance. Most performing artists always find room for improvement in their work, even after what many people would call a brilliant performance. The arts manager must adopt this attitude about planning: It never ends, and there is always room for improvement. If, as the saying goes, the way to Carnegie Hall is through practice, then the way to become a first-rate arts organization is through the equivalent of constant planning.

This last chapter focuses on the future. There are no right answers when venturing into this territory—except in hindsight. Keeping that in mind, let's look at current trends and possible future directions that point toward growth. We'll examine some of the key economic realities facing arts groups.

Current Trends and Future Directions

There are optimistic and pessimistic views on nearly every aspect of an arts organization's operation. The most troubling area is the fiscal health of the arts. Although there appears to be a great deal of grassroots support for the arts in most communities, competition for the entertainment dollar continues to increase each year. About the only thing that it is safe to say about today's arts organizations is that they will be more expensive to operate tomorrow.

An Uncertain Future

The pessimist, as we all know, describes the cup filled to 50 percent of capacity as half empty, and the optimist calls it half filled. For an arts manager facing drastic cutbacks in funding support from financially strapped states and cities, declining corporate gifts as profits fall and layoffs mount, increasing insurance costs, spiraling production and labor costs, an aging audience, and a decline in income levels for individuals, it is hard to see a rosy future.

The statistics clearly indicate that since the 1960s, thousands of new arts institutions have been created all across the United States. Regional theater, opera, and dance companies, music groups of all sizes, performing arts complexes, university arts programs, and numerous museums now exist in communities with no past history of arts patronage. Millions of people subscribe, have memberships, or regularly buy tickets to events that employ tens of thousands of artists and support personnel. By every account, the cup looks more than half filled. Why, then, is there pessimism?

The efforts of the last 25 years—which have produced significant funding increases from government, foundations, and corporations—are only closing the income gap enough to delay the inevitable financial collapse of many nonprofit organizations. In many cases, it appears that the increased funding and ticket sales associated with the baby boom audience helped fuel growth beyond the long-term abilities of private and public support. Therefore, the condition seems to be one of too many groups seeking too few funds, and as the market system theory predicts, some organizations will have to drop out of the market to restore an equilibrium.

Modest to No Growth

For the larger, well-established arts organizations, rapid growth is a thing of the past in some parts of the United States. Arts groups in the South and West will have the opportunity to capture new audiences as those regions grow. In the Northeast and Central regions of the country, the population loss slowed in the early 1990s, but it is unrealistic to think that the millions who moved away will be back any time soon. The continuing shift to a global economy and the need to keep labor costs as low as possible may further accelerate the exodus from many of the older cities in the East and Midwest.

Arts Facilities and Social Engineering

The process of creating or renovating performing arts centers in downtown areas in large cities, despite population losses, has helped lead to an economic revival in some communities and has provided homes for many of the newly created performing arts companies and visual arts organizations. The concept of arts centers surrounded by apartments, condominiums, and shopping malls is appealing because it can lead to a critical mass of economic development and expansion. For example, if enough people shop and use the services of the stores and the arts groups, a self-sustaining microeconomic system could be developed within the community. In reality, however, the loans used to start up these new buildings and businesses must be paid, and operating costs continue to rise. The cash flow required to keep these businesses alive may prove insufficient unless enough people can be enticed to resettle in central city areas. Noontime shoppers will not generate enough economic development to support the entire system. Meanwhile, the arts group may be left with a lot of new, expensive overhead in the form of a building that absorbs hard-earned dollars.

The development of the arts complex and the commerce to support it is, in many ways, an attempt at a type of social engineering. However, if the rest of the society and its leaders are not capable of creating a system of government and control that solves fundamental problems like poverty, racism, and crime, we shouldn't be surprised to find that these arts centers turn into ghost towns on the evenings when no performances are scheduled. When coupled with a constant barrage of stories in the media about the murder and mayhem in the streets, it is little wonder that the vast majority of the people who attend the arts live far away from the cities.

Until other serious social problems are solved, the urban arts center will require heavy subsidies in combination with substantial private giving if it is to survive. It remains to be seen how cities strapped for resources will be able to afford politically to support these expensive operations as city services continue to decline. The search for funds to create an operating endowment will probably be a high priority for several major arts organizations during the remainder of this decade.

Conclusion

Trying to predict glory or gloom for the visual and performing arts in the United States is a risky business. The variables seem almost infinite. The external environments, the audiences, and the always changing interests of artists make for a dynamic mix. Looking to the past will not prove to be a useful method for predicting the future of the arts because there are no similar combinations of circumstances to use as guides. There are far more arts organizations of all sizes distributed across the United States today than ever before. Although many of these organizations spring into existence overnight and go out of business just as fast, a substantial number have been established in the last 20 years. Why? One answer lies with the greater number of people who have attained some knowledge in or experience with the arts through their education. If the number of people earning college degrees continues to increase faster than the growth in population, the established arts organizations and many new groups may flourish in the next 20 years. At the same time, rising education costs and dwindling financial support for students may undermine this trend. It is hard to believe that with the United States facing increasing global competition, the political system will fail to find the resources to educate its people to the level required to stay competitive.

The Partnership: Artist and Manager

Running an arts organization never was and never will be easy. Although arts organizations can benefit from applying management theory and practice to become better organized and more efficient, there is still no single element that guarantees survival. Getting better at doing well is an ambitious goal that any arts manager should be proud to work toward. To achieve that goal, an arts manager should be prepared to borrow any techniques that work from business, government, educational institutions, or other nonprofit organizations. Successfully integrating these different approaches to effectively manage an organization requires as much creativity as any visual or performing artist. Forging a successful partnership of manager and artist is therefore predicated on each party recognizing the other's creative contribution to one goal: creating a world in which life is enriched by the accomplishments of both parties.

Questions

1. What are your predictions about how the arts will fare in the United States during the next ten to 20 years? Will there be growth or shrinkage? Will demographic shifts and changing cultural values change attendance patterns at arts events? Will the political system help or hinder the arts?

2. What specific strategies should arts groups adopt today to be ready for the year 2000?

Index

Abady, Josephine, 75
Abruzzo, James, 2, 26, 96
Aburdene, Patricia, 42, 52, 63
acceptance theory, 135
accountability, 25–26, 78, 156
accounting system, 215–218; and donor tracking, 263–264
accounts payable activity, 292
accrual-based accounting, 217
Actors Equity Association, 123, 127–128
Actor's Workshop (San Francisco), 19
Addictive Organization, The (Schaef and Fassel), 105
ad hoc committees, 144
administrators, 8
advertising, 58, 183, 230, 245
affirmative action programs, 96, 117, 128
AFL-CIO, 55
African-Americans, 56, 65, 66, 245. *See also* minority groups; black theater, 83–85, 129–130; nontraditional casting, 127–130
Age Discrimination Act of 1967 (amended 1973), 117
agency organization, 95
age of audience, 193, 239
aging population, 57, 58
AIDS, 2, 57, 62
Alley Theatre (Houston), 19
American Ballet Theatre, 20
American Dance Ensemble, 21–22
American Express Company, 271
American Federation of Arts, 30
American Federation of Musicians, 58, 123
American Federation of Television and Radio Artists, 123, 124
American Guild of Musical Artists, 123
American Guild of Variety Artists, 123
American Marketing Association, 231
American Repertory Theatre (Cambridge, Mass.), 103
American Stage (Berkeley, Calif.), 129
Americans with Disabilities Act of 1990, 118
American Telephone & Telegraph Company, 271, 296
ancient times, 16–17
Andreasen, Alan, 230, 234–235
Annenberg Foundation, 255
annual campaigns, 265
apprentice system, 122
archon eponymous, 16
Arena Stage (Washington, DC), 19
Armstrong, Thomas N., III, 123
Art Institute of Chicago, 242
artist-manager, 15. *See also* arts manager
art museums: attendance, 51, 54, 236; case studies, 29–31; Great Britain, 18; manager profile and training, 21; number in United States, 20; and product orientation, 234–235
Artrain (Michigan), 31
arts agencies, local and state, 26, 274
arts calendars, 60
arts councils: Great Britain, 18; states, 26, 274
ArtSEARCH, 21, 115, 118
arts institutions: economic impact of, 1–2, 188–189; economic problems, 2, 52–54, 63, 190–194, 255, 298–300; funding, 2, 4, 15, 18, 19, 20, 25, 54, 57, 298; growth, 19–20, 49–50, 51, 57, 192
arts manager, 15, 16–23, 25–27, 34–35, 133–150; and artists, 300; and communication, 148–150; delegation of work, 100–101; and government relations, 292–293; and group dynamics, 143–148; higher education of, 26–27; historical role, 16–20; and leadership, 136–138; and management styles, 280–285; and managing change, 49; and motivation, 138–143; and the NEA, 25–26; and on-the-job training, 21, 34–35; and

personnel, 290–291; and power, 133–136; profile, 20–23
Asian Americans, 65, 66
asset management program, 292
assets, defined, 218
Association of Performing Arts Presenters, 26, 238, 239
audiences: demographics, 2, 56–57, 192–193, 238, 239, 290; feedback, 233, 234, 239; as information source, 59–60
auditions, 119–120
Autry Foundation, 272
average fixed costs, 189
average total costs, 189, 190
average variable costs, 189

Babbage, Charles, 36–37
baby boom, 19, 26, 49, 56, 57, 299
Bachman's inevitability theory, 77
balance sheet, 218–219, 220
Baldwin, Connie, 296
ballet: auditions, 119–120; born in France, 18
Ballet Comique de la Reine, 18
Ballet West (Utah), 20
Baraka, Amiri, 129
Barnard, Chester, 135
Barter Theater (Virginia), 19
base-line budgets, 211–212
Baumol, William J., 190–194, 239
Beal, Graham W. J., 30
Bednarek, David I., 236
behavioral decision theory, 80
behavioral modification approach, 142–143
behaviorist theory, 39, 41, 287; and leadership, 136, 137
behavior norms, 145
Belasco Theater (New York City), 198
Bennett, John, 55
Bennis, Warren, 138
bequests, 266
Berkowitz, Roger M., 30
bet-your-company culture, 103
Biddle, Livingston, 25
Bill, Mary, 75
Bing, Rudolf, 104
birth rate, decline, 57, 58
blacks. *See* African-Americans
board of directors, 60, 224, 263, 284, 289, 291; Spoleto Festival, 294–296
board of trustees, 110
Bonnard, Raymond, 277–278
bookkeeping, 215
Boston Symphony Orchestra, 19
bottom-up planning, 72–73
Bowen, William G., 190–194, 239
broadcast media, 1, 19, 55–56, 61
Broadway Alliance: A New Plan for Play Productions, 197–199
Brockett, Oscar, 17
Brown, J. Carter, 30
Buckam, Tom, 277–278
budgets, 71, 168–171, 206–213, 264, 292
bureaucracy, 89–90
Bureaucracy: What Government Agencies Do and Why They Do It (Wilson), 89–90
Bush, Vannevar, 39–40
Bushnell, William, 66–67
business environment, 103, 166, 270
business schools, 30

cable television, 47, 57, 63
Cannon, Martin, 108–109
Cannon Devane Associates, 108–109
capital campaigns, 262, 265
career management systems, 125–126
case studies, use of, 41
cash-based accounting, 217
cash flow statement, 210–211, 214
casting, multiracial, 127–130
Catalog of Federal Domestic Assistance, 275
Catholic Church, 17, 35
"Cat on a Hot Tin Roof," 127–128
censorship, 18, 25, 62; self-censorship, 62, 284–285
census data, 20–21, 62
centralization-decentralization, 101, 156
Central Opera Service, 20
chain of command, 98–100
chamber music associations, 51
charitable organizations, 205, 254
Chicago Lyric Opera, 20
child labor, 36, 37
children's concerts, 251
choregoi, 16
Chronicle of Philanthropy, 263
Church, the, 17, 35
Citibank, 271
Civil Rights Act of 1964
Claritas Corporation, 238–239
classical decision theory, 80
classical music, 58, 250–252
Cleveland Ballet, 218
Cleveland Museum of Art, 250–251
Cleveland Orchestra, 104
Cleveland Playhouse, 75
Clustering of America, The (Weiss), 238
coaching of employees, 122
coercive power, 134
colleges and universities, 19, 21, 26–27, 101, 234
Colling, Glenn, 58

color-blind casting, 127–130
Columbia Pictures Entertainment, 71
command groups, 144
committees, 144
committee-style management, 138
communication, 22, 59–60, 93, 94, 106, 148–150, 287
Communist infiltration, fear of, 54
community arts agencies, 21, 26
community fund raising, 261–263, 270–271
compact discs, 57–58
comparable worth, 117, 141
compensation, 22–23, 116
competition among nonprofit groups, 54, 63, 242–245, 298–299
complementary good, defined, 180–181
computer-aided design (CAD), 38
computer-assisted manufacturing (CAM), 38–39
computer-integrated manufacturing (CIM), 39
computer models, 38–39
computers, 36
computer skills, 22; and fund raising software, 263–264; and management information system, 163–165; and marketing, 233, 239, 246; and telephone solicitations, 268
confidentiality of financial information, 264
consultants, 61, 75, 242, 246, 254
consumer feedback, 233, 234, 239
content analysis, 51
content theories of motivation, 139–140
contingency approach, 42, 136, 288
contingency planning, 73
control system, 11–12, 155–162
cooperative management, 63
coordination, 98–101
corporate culture, 9, 102–107, 125, 159
Corporate Cultures: The Rites and Rituals of Corporate Life (Deal and Kennedy), 102
Corporate Philanthropy Report (Seattle), 271
corporate support, 254, 265, 270–272
Cousin, Jack, 126
Crawford, Robert W., 81–82
Crayton, Reyno, 84, 85
credit-rating companies, 239
crisis planning, 73, 288–289
critical-incident appraisal, 161
critical path method (CPM), 38
crossover projects, 250–252
cultural and social environment, 55–56
cultural boom, 20, 49–50, 51, 57
cultural network, 106
Cuomo, Mario, 278
customer orientation: 233, 235–236, 259, 289

Dafne, 18
dance companies, 19, 20, 120
Daniels, Edgar F., 294, 295
Dark Ages, 17
Dartmouth College, 38
data bases, 162–163, 238–239, 263–264
Davidson, Gordon, 66
Davis, Clinton Turner, 129
Deal, Terrence, 102, 103, 104, 106
debt crisis, 2, 52–53, 292
decision making, 41, 78–80, 81
deferred gift, 266
delegation of work, 100–101
demand curve, 180–183
demographics, 2, 56–57, 192–193, 238, 239, 290
departmentalization, 96–98
detailed budget, 207, 209
Deutsch, Claudia H., 271
differentiation strategy, 242
DiMaggio, Paul, 21, 35
direct fund raising costs, 265
direct grants, 272
direct mail marketing, 268–269
disabled people, rights of, 117, 118
discretionary income, 240, 255
diseconomy of scale, 190
distributed leadership, 148
divisional structure, 96
division of labor, 8
domini, 16
donors, individual, 265–270. *See also* corporate support; foundations
Downs, Patty, 188
Drape, Joe, 188
Drucker, Peter, 42
dysfunctional groups, 145–146, 147
dysfunctional manager, 280–282

economic environment, 176–196; application of economic theories, 188–190; economic problems, 52–54, 63, 190–194, 255, 298–300; introduction to economics, 176–179; macroeconomics, 179; microeconomics, 180–188
economies of scale, defined, 190
education, 56, 58–59, 239, 290, 300
Effective Nonprofit Executive Handbook, The, 73–74
Eisenberg, Alan, 199
Elder, Lonne, 128

elderly, the, 57, 58
electronic synthesizer, 58
empirical school, 41
employee benefits, 116
employees, 60, 95; career-management system, 125–126; and fund raising, 263, 267; job description, 115–116; orientation and training, 121–122; performance appraisal system, 160–161, 287; personnel management, 39–41, 291; recruitment, 118–119; replacement and firing, 123; selection process, 119–121; turnover, 281
employment-at-will agreements, 123
employment discrimination, 117
employment statistics, 2
endowment fund, 266
entertainment industry, 1–3
environments, external, 5–6, 9, 47–48, 53–59, 75, 96; and fund raising, 254–255
Equal Employment Opportunity Act of 1972, 117
Equal Employment Opportunity Commission, 117
Equal Pay Act of 1963, 117
equity theory of motivation, 140–141
ethical considerations, 239–240, 271–272
Europe, 18, 20
Evans, Don, 84
Evett, Marianne, 75, 83–85
Evolution of Management, The (Wren), 36–37, 41–42
exchange process, 231–233, 259, 290
expectancy theory, 141–142
expectations, 182, 186
expense, defined, 217
expense centers, 168
expert power, 134

factors of production, 183–184
family, redefined, 56
Fassel, Diane, 105
(Henri) Fayol's Fourteen Principals, 39, 40, 98–99, 280
featherbedding, 125
federal data bases, 239
federal grant programs, 275
Federal Register, 275
Federal Reserve Bank, 179
Federal Theatre Project, 54
feedback, audience/consumer, 233, 234, 239
Feuer, Cy, 198
fiduciary responsibility, 224
film industry, 1, 19, 234
Financial Accounting Standards Board, 215
financial management, 201–226, 291–292; budgeting and financial planning, 206–213; legal status and financial statements, 203–206; of profit/nonprofit organizations, 202–203
financial management information system (FMIS), 201, 203, 204, 206, 213–225; accounting, 215–218; financial statements, 218–224; investment, 224; and marketing data system, 246
Financial Management Strategies for Arts Organizations (Turk and Gello), 224
financial manager, 202–203, 265
financial risk, 224
financial statements, 218–224; and legal status, 203–206
Finn, Robert, 250–252
fiscal year, 206
Fitzhugh, Lynne, 58, 239
fixed assets, 224
fixed budget, 170
fixed costs, 189, 212–213, 215
flexible budget, 170
FMIS. *See* financial management information system (FMIS)
Ford Foundation, 20
Ford-Taylor, Margaret, 84–85
formal group, defined, 143
formal leadership, 133
form utility, 232
foundation application procedure, 272–273
foundations, 272–274; donation statistics, 254, 265
four P's of the marketing mix, 237, 289
four utilities of the exchange process, 232
France, 18
free-form narrative employee evaluation, 161
Fuller, Charles, 129
functional authority, 101
functional departments, 96
functional manager, 7
functional satisfaction, 231
fund balance, 206, 218
fund-based accounting, 217
funding options, 266
funding pyramid, 258–259
fund raisers (special events), 109–110, 270
fund raising, 254–276; and commu-

nity needs, 261–263; corporate and foundation support, 270–272; costs and control, 264–265; data management, 263–264; government support, 274–275; individual donors, 265–270; management, 259–261; and marketing, 229–230, 259, 290; planning, 255–259; techniques and tools, 265–275
fund raising audit, 259, 260–261
fund raising committee, 266
fund raising statistics, 254, 265

Gallo, Robert P., 217, 223, 224
Gap Foundation, 272
Gelb, Hal, 127–130
generally accepted accounting principles (GAAP), 215
general manager, 7–8
Germany, 18
gifts, restricted/unrestricted, 265–266
gift shops, 232, 246
Giving USA, 270
global marketing, 234
goals, defined, 11, 70
goods and services, 177; normal and inferior, 181–182
Gordone, Charles, 129, 130
Gorky, Maxim, 237
government bureaucracies, 89–90
government hiring regulations, 117–118
government relations, 292–293
government role in the economy, 178, 179
government subsidization of the arts, 15, 20, 54, 254, 255, 274, 275. *See also* National Endowment for the Arts;
criticism of, 4, 24–25; in Europe, 18, 19, 20
Graham, Katherine, 35
"Grand Hotel," 58
Great Britain, 18, 19, 20
Great Depression, 54
Great Lakes Theater Festival, 75
Greece, ancient, 16
gross domestic product (GDP), 179
group dynamics, 143–148
groupthink, 145–146
growth strategy, 76–77, 242–245, 256
Gruszewski, Jeff, 84
GTE Northwest, 271
Guggenheim Museum (New York City), 29–30, 31
Guthrie Theatre, 5, 74

halo effect, 149
Hanahan, Bessie, 295
Harakal, Eileen, 242
Hartford Stage Company (Hartford, Conn.), 262
Harvard University, 30, 38
Hawthorne effect, 39–40
head-hunter services, 118–119
health care costs, 57
Helms, Jesse, 52
hero-leader, 104
Herzberg, Frederick, 140
"hidden hierarchy," 106
hierarchy of authority, 9, 92
hierarchy of needs (Maslow), 40–41, 139–140, 238, 287
Higginson, Cornelia, 271
high-definition television (HDTV), 58
Hispanics, 56, 65, 66, 129
history museums, 20
Hogan, Bill, 242
home entertainment centers, 2, 58
horizontal coordination, 98, 101
Hoving, Thomas, 30–31
human behavior school, 41
human relations management, 39–41, 290–291
human resources planning, 112–115
Human Side of Enterprise, The (McGregor), 133
Huston, Don, 124

Iaccoca, Lee, 104
Immigration Act of 1990, 55
income, individual, 181–182
income effect, 188
income gap, 193–194, 225
income level of arts consumer, 56, 239
incorporation, 203–205
indirect fund raising costs, 265
individual donors, 265–270
Industrial Revolution, 36–37, 233
inflation, 53, 193, 194
informal group, defined, 144
informal leadership, 133
information sources, 59–62
inner cities and the arts, 65, 66, 299–300
input standards, 157–158
In Search of Excellence (Peters), 151
institutional management style, 284–285
insurance, 57, 224, 298
interest groups, 144
intermediate-range planning, 70, 289
Internal Revenue Service Code, 205, 206
International Alliance of Theatrical Stage Employees and Motion Picture Machine Operators, 123

International Business Machine Corporation (IBM), 271
interracial cooperation, 84
interview process, 120–121
investment, short term/long term, 224
irrational arts manager, 280–282
Italy, 18

Janis, Irving, 145–146
Japan, 38
Jelliffe, Russell and Rowena, 84
job application process, 120
job description, 115–116
job enrichment, 140
job rotation, 122
job screening process, 120
Jobs in Arts and Media Management (Langley and Abruzzo), 2, 26, 96
jobs matrix, 116–117
job titles and job listings, 21–23
Joffery, Robert, 104
Joffery Ballet, 104
Jones, James Earl, 127–128
Journal of Arts Management and Law, 239
Jujamcyn Theatres, 23, 197–199
"just-in-time inventory" (*Kanban*), 38

Kalhorn, Joan, 124
Karamu House (Cleveland), 83–85
Kennedy, Allen, 102, 103, 104, 106
Kennedy Center for the Performing Arts (Washington, D. C.), 108–110
Kirstein, George G., 254
Koontz, Harold, 41–42
Kotler, Philip, 230, 234–235
Kotter, John R., 135–136
Kozinn, Allan, 294
Kramer, Hilton, 30, 31
Krauss, Marvin, 58
Krens, Thomas, 29–30, 31

labor-management relations, 124–125, 290–291
Landesman, Rocco, 23, 198–199
Lane, Jack, 29, 31
Lane, Louis, 251
Langan, E. Timothy, 237
Langley, Stephen, 2, 26, 96, 202
Lavine, Steven D., 65–66
law of diminishing returns, 189–190
lawsuits, 114, 121, 123
leadership, 132–143; and communication, 148–150; formal/informal, 133; and group dynamics, 143–148; and motivation, 138–143; and power, 133–136; trait, behavior, and contingency approaches, 136–138
leading function of management, 10, 11
League of American Theaters and Producers, 197, 199
legal environment: and employment policies, 117–118; and fund raising, 264–265, 270; and incorporation, 203–205; and political environment, 54–55; and tax exemption, 205–206
legitimate power, 134
Leibowitz, Leonard, 58
leisure time, 2, 56
Lev, Michael, 71
Lewin, Daniel S., 66
Lewis, William, 84
liabilities, defined, 218
Libin, Paul, 198
Lieberman, Barry, 126
line manager, 7
liquid assets, defined, 223
living standard, 56
Livingston, Sandra, 124
Llewellyn, John, 124
lobbying, 54, 55, 206
local arts agencies, 26, 274
long-term budget, 170
long-term investment, 224
long-term plans, 70, 289
Lord, James Gregory, 256, 259
Los Angeles Theater Center, 65–67
Lyceum Theater (New York City), 198

McCallum, Daniel Craig, 37
McDaniel, Nello, 52–53, 292
(Douglas) McGregor's Theory X and Theory Y, 41, 133, 159, 287
macho culture, 103
macroeconomics, 177, 179
mailing lists, 237, 238–239, 268–269
maintenance activities, 146–148
maintenance factors of motivation, 140
Malone, Mike, 84
management, committee-style, 138
management, cooperative, 63
management, defined, 87
management by exception, 158–160
management by objectives, 160
Management for Productivity (Schermerhorn), 77
management functions, 10–12, 288–293
management information systems, 162–168, 246, 263–264
management levels, 6–7, 92–93
management process, 41
management styles, 280–285

management theory, 34, 35–44, 98–99, 285–288
management theory jungle, 41–42
manager: defined, 4, 87; functional areas, 12; types of, 6, 7–8
managerization of the arts, 284
Managers of the Arts (DiMaggio), 21, 35
Managing a Nonprofit Organization (Wolf), 201, 207, 262
managing change, 49
Mandatory Retirement Act; Employment Retirement Income Security Act of 1974, 117
marginal costs, 189
market, defined, 180
market demand, 180–183
market equilibrium, 186–188
marketing, 229–249, 289–290; data system, 246–247; defined, 230–231; ethics, 239–240; exchange process and utilities, 231–233; and fund raising, 229–230, 259, 290; modern marketing, 233–234; needs and wants, 231; product, sales, and customer orientation, 234–236; strategic plans, 240–247, 289–290; targeting, 237–239
marketing audit, 242, 243–245
marketing mix (the four P's), 237, 289–290
market research, 238–239
market segments, 237
market supply, 183–186
market system, 178
Market the Arts!, 239, 246
Markwardt, Ross A., 294–295, 296
(Abraham H.) Maslow's hierarchy of needs, 40–41, 139–140, 238, 287
mass media audience, 1, 2, 19, 55–56
maternity rights, 117, 118
mathematical school, 42
matrix of jobs, 116–117
matrix organizational structure, 96–98, 99, 101
Mayo, Elton, 40
mechanistic organization, 89
media coverage, 61, 289
Megatrends 2000 (Naisbitt and Aburdene), 42, 52, 63
Menotti, Gian Carlo, 294–296
Metropolitan Museum of Art (New York City), 30–31
Metropolitan Opera (New York City), 19, 47, 91–92, 104, 232
microeconomics, 177, 180–188
Middle Ages, 17
Milwaukee Art Museum, 75, 236
Minneapolis Institute of Art, 31
minority groups, 89, 96, 290; nontraditional casting, 127–130
mission analysis, 74–76
mission statements, 11, 24, 73–74, 109, 234, 246
Mitchell, Arnold, 238, 239
modeling of employees, 122
modern dance, 20
monopolies, 19
Morris, Milton, 84
motivational factors of motivation, 140
Motivation and Personality (Maslow), 139
motivation theories, 138–143
Motivation to Work, The (Herzberg and Syndermann), 140
Multimedia, Inc., 124
multiple-manager leadership model, 138
multiplier effect, 188
musicians, 58

Nahat, Dennis, 218
Naisbitt, John, 42, 52, 63
"narrow-casting," 57
National Assembly of Local Arts Agencies, 274
National Assembly of State Arts Agencies, 274
National Association of Broadcast Employees and Technicians, 124
National Council on the Arts, 24, 25
National Endowment for the Arts, 20, 23–26, 274–275; art manager profile, 21; arts consumer profile, 239; arts organizations survey, 1–2; reauthorization battle, 18, 25, 52, 61, 178–179
National Endowment for the Humanities, 275
National Gallery (Washington, D. C.), 30
National Guide to Funding in Arts and Culture, 272
National Labor Relations Board, 124–125
NEA. *See* National Endowment for the Arts
Nederlander Organization, 197–199
needs: of the community and fund raising, 261–262; defined, 231; hierarchy of (Maslow), 40–41, 139–140, 238, 287
Negro Ensemble Company, 127, 130
net earnings, 206
New York City Ballet, 20
New York Philharmonic Orchestra, 19

New York Shakespeare Festival, 129
New York State Council on the Arts, 277, 278
niche strategy, 242
Nine American Lifestyles (Mitchell), 238
Nissan Motor Corp. U.S.A., 271
Nodal, Adolfo V., 66, 67
noise, 149
Nonprofit Organizations (Singer), 206
Nonprofit Times, The, 263
nontraditional casting, 127–130
Non-Traditional Casting Project, 128, 129
No Quick Fix (Planning) (Crawford), 81–82
Northwestern University, 26
numerical rating scale, 161

Oberlin College Theater and Dance Program, 115
objectives, defined, 11, 70
obscenity issue, 25, 61
Occupational Safety and Health Administration, 54
occupations of arts consumers, 239
Off Broadway theaters, 19, 197, 198
Ohio Chamber Orchestra, 218
Old and Kahn's law, 144
O'Leary, Daniel E., 31
on-the-job training: arts manager, 21, 34–35; employees, 122
open system model, 5–6, 21, 42, 49, 88–89, 91, 287–288
opera, 18, 19–20, 47, 51, 120
operant conditioning, 142, 143
operating budget, 168, 170, 189
operating costs, 63
operational plans, 70–71
operational redundancy, 63
operations research (OR), 38
opportunity costs, 177–178
organic management style, 285
organic organization, 89
organization, 5–6, 8–9; size and growth, 95, 101–102
organizational audit, 282
organizational chart, 90–93, 97, 114, 116–117, 144
organizational culture. *See* corporate culture
organizational management (1916 to present), 39
Organizational Psychology (Schein), 146
organizational structure, 6, 8; and coordination, 98–101; departmentalization, 96–98; formal, 90–93; informal, 93–94; and strategy, 94–95
organizing function of management, 10, 11, 87–94
organizing process, defined, 6
Ostergard, Paul M., 271
O'Toole, James, 42
Ottremba, Geraldine, 110
Our Government and the Arts (Biddle), 25
output standards, 156–157
outreach projects, 250–252, 274, 290
overhead, 189
over-qualified job applicants, 121
Owen, Robert, 36

Page, Ruth, 20
Pall, Ellen, 23
Palmer, Jane, 84
paper trail, 218
Papp, Joseph, 129
participative management, 41
payable, defined, 217
payroll department, 101
peer review system (NEA), 24, 25, 275
Pennsylvania Ballet, 217, 255
pension rights, 117
perception, 149
performance appraisal systems, 160–161, 287
performance costs and CPI, 194
performance norms, 145
performing artists, alien, 55
performing arts centers in inner cities, 299–300
Performing Arts—The Economic Dilemma (Baumol and Bowen), 190–194
personal contacts, 255, 267
personal power, 134
personnel department, 101
personnel management, 39–41, 290–291
personnel search committees, 120, 143
Peters, Tom, 42, 151
Pew Charitable Trusts, 255
philanthropy, 19, 30, 54; statistics, 254, 265
Philip Morris Company, 271
"Piano Lesson, The" (Wilson), 198
Pierce, Lyman L., 254
planned economy, 178
planning, 10, 11, 69–73, 77, 234, 288–289
plans: short, intermediate, and long range, 70, 288–289; single-use and standing-use, 71; strategic and operational, 70–71
Playwrights Horizons, 23
point-of-purchase systems, 234

political environment, 54–55, 178–179, 288, 292–293
Poor, Henry Varnum, 37
population shifts, 299
position power, 134
possession utility, 232
Potential Rating for Zip Markets, 238–239
power, 132, 133–136
Pracan, Jane J., 271
Pregnancy Discrimination Act of 1978, 117
preindustrial era, 35
prescreening and testing, 234
press, the, 61, 289
price manipulation, 237
prime interest rate, 179
print media, 61, 289
privacy, invasion of, 239–240, 268
Privacy Act of 1974, 117
probationary period, 121–122
problem solving, 78–80
process culture, 103
process theories of motivation, 140–143
production costs, 2
production factors, 183–184
productivity, 193–194
product orientation, 234–235, 289
Professional Performing Arts: Attendance Patterns, Preferences and Motives (Mitchell), 238
professional service organizations, 61
profit, 202, 206
profit-sharing, 198
program development, 289
project budget, 207–208, 210
projection, 149
promote-from-within policy, 118
Protestantism, 35
psychographic profile, 238, 290
psychological satisfaction, 231
publicity, 61, 289
Public Theater (New York City), 25

Quiet Crisis in the Arts, The (McDaniel and Thorn), 52–53, 292

radio, 1, 19, 55–56, 61
railroads, 37
Ramo, Simon, 96
ratio analysis, 223
rational manager, 282–284
Reagan administration, 25, 125
receivable, defined, 217
recession, 52, 53–54, 255, 271
reciprocity concept, 270
recruitment of staff, internal/external, 118–119
Red Cross, 254
Redden, Nigel, 294, 295, 296
reference power, 134
Reformation, 35, 36
Rehabilitation Act of 1973, 117
reinforcement theory, 142–143
Reinhold, Robert, 65–67
Renaissance, 17–18, 36
resource prices, 183–184
restricted fund, 217, 265–266
retirement legislation, 117
retrenchment strategy, 77
revenue, defined, 217
revenue centers, 168
revenue maximization, 187–188, 202
revenue projections, 170
reward power, 134
Rheingold, Howard, 58
Richards, Lloyd, 129
Riley, Joseph P., Jr., 294, 295
risk-taking cultures, 103
rites and rituals, 104–105
Roethlisberger, Fritz, 40
role models, 104
Rome, ancient, 16
Rothstein, Mervyn, 197–199
rumors, 94, 287

Sabater, Juan, 27
Sabinson, Harvey, 197
salaries, 22–23, 116
sales orientation, 230, 233, 235, 289
Salisbury, Wilma, 218
San Diego Symphony Orchestra, 74–75, 259
San Francisco Ballet, 20
San Francisco Museum of Modern Art, 29, 31
satellite communications, 234
Scalar Principle, 99
Scardino, Don, 23
Schaef, Wilson, 105
scheduling, 71, 134
Schein, Edgar H., 146
Schermerhorn, John, 77, 78
Schewe, Charles D., 231, 233, 237
scholarship funds, 266
schools, 58–5
science museums, 20
scientific management, 34, 36, 37–39, 41, 285–287
search committees, 120, 143
Seattle Symphony Orchestra, 11
selective perception, 149
self-censorship, 62, 284–285
Shepard, Chuck, 126
short-term budget, 170
short-term investment, 224
short-term plans, 70, 288–289

Shubert theaters, 19; and Broadway Alliance, 197–199
signals and symbols, 148
Sims, Patterson, 30
Singer, Barbara, 206
single-use plans, 71
situational leadership, 136
Skinner, B. F., 142
Sloan, Charlotte F., 295
Smith, Adam, 36
Smith, Craig, 271
Smoot, Myrna, 31
social system school, 41
societally conscious, 238
Solomon, Muriel, 151
Sonyland, 71
span of control, 100
special events, 109–110, 270
Spoleto Festival U.S.A (Charleston, S. C.), 188, 294–296
stability strategy, 76, 256
Stacy, J. Adams, 140
staffing process, 112–127. *See also* employees; constraints, 117–118; matrix of jobs, 116–117; planning, 114–115
staff manager, 7
standing committees, 144
standing-use plans, 71
Stanford University, 30
state arts agencies, 26
state arts councils, 26, 274
statement of activity, 219–223
Stehle, Vince, 108–110
stereotypes, 149
Stern, Theodore S., 294, 296
Stille, Alexander, 29–31
Stockman, David, 25
Stone, Peter, 197, 199
"storytellers, spies, and priests," 106
strategic fit, 270
Strategic Marketing for Nonprofit Organizations (Kotler and Andreasen), 230
strategic planning, 7, 70–71, 73–77, 94–95; and fund raising, 256–258
Studio Arena Theatre (Buffalo, N. Y.), 277–278
subscribers: complaints, 110; season seats, 232
substitute goods, 180
summary budget, 207, 208
"Summerfolk" (Gorky), 237
supply curve, 183–186
symphony orchestras, 19, 20, 21
Syndermann, B., 140
Syndicate, The, 19
systems approach, 42
Szell, George, 104
target marketing, 237, 238–239, 240, 289
task activities, 146
task groups, 144
tastes, 182–183
tax benefits, 254, 255
tax exemptions, 205–206
Taylor, Frederick W., 37–38
technological environment, 57–58, 95–96, 288
telephone solicitations, 267–268
television, 1, 55–56, 61, 232, 234
theater companies, 19, 21, 51, 119, 237; Broadway Alliance, 197–199; nontraditional casting, 127–130
Theatre Management and Production in America (Langley), 202
theme parks, 71, 234
Theory X and Theory Y approaches (McGregor), 41, 133, 159, 287
Thorn, George, 52–53, 292
ticket pricing, 2, 184–188, 194, 237; Broadway Alliance, 197–199
time and motion studies, 38
time and place utilities, 232
Toledo Repertoire Theatre, 21, 22
top-down planning, 72
total cost, 189
total fixed costs, 189
total variable costs, 189
tough-guy culture, 103
touring groups, 18–19
tracking system and responses (donors), 263–264, 269–270
training: arts manager, 21, 34–35; employees, 122
trait approaches to leadership, 138
transaction, defined, 215
Trinity Repertory Theater, 237
TRW, Inc., 96
Turk, Frederick J., 218, 223, 224
Turner, Rebecca, 236
Tuttle, Edwin E., 217
two-factor theory of motivation (Herzberg and Syndermann), 140

unemployment, technological, 57
unions, 55, 118, 123–125, 141; and arts manager, 290–291
United Scenic Artists, 123
United Way, 271
University of California and Truman Los Angeles, 26, 30
University of Wisconsin, 26
unrelated business income tax, 205
unrestricted account, 217
unrestricted gift, 265–266
urban arts center's problems, 65, 66, 299–300

US West Inc., 271
utilities (form, time, place, and possession), 232

Values and Lifestyles Segment, 238, 239
value system, 103–104, 159
van den Haag, Ernest, 4
variable costs, 189, 212–213, 215
vertical coordination, 98–101
video tape recorder (VCR), 57, 232
virtual reality (VR), 58
Volunteer Lawyers for the Arts, 203, 206
Vroom, Victor, 141

wage discrimination, 117
Walter Kerr Theater (New York City), 198
wants, defined, 231
Ward, Charles S., 254
Ward, Douglas Turner, 127, 128, 130
Wealth of Nations (Smith), 36
Weiss, Michael J., 238, 239
Weissler, Fran and Barry, 127
Wharton School, 38
Whitney Museum (New York City), 123
Who's Who in America, 266
Why Leaders Can't Lead (Bennis), 138
Wildman, Donald, 52
Wilker, Lawrence J., 109
William Penn Foundation, 255
Williams, Brenda, 188
Wilson, August, 128, 130, 198
Wilson, James O., 89–90
WKYC (Cleveland), 124
Wolf, Thomas, 201, 259, 262, 266
Wolfensohn, James D., 108
Wolf Organization, 242, 259
women, 21, 117
Woodside, William S., 123
work groups, permanent/temporary, 143
work hard/play hard culture, 103
Working with Difficult People (Solomon), 151
worth, 206
Wren, Daniel, 36–37, 41–42
wrongful-discharge suits, 123

Yale Repertory Theater, 129, 130
Yale University, 26, 30
YMCA, 254

zero-based budget, 170, 171
zip code analysis, 238–239
zone of indifference, 135